# Extremal Properties of Polynomials and Splines

## Computational Mathematics and Analysis Series

**Series Editors: Niko Vakhania and Henryk Wozniakowski**

**Optimal Recovery** by B. Bojanov and H. Wozniakowski,
ISBN 1-56072-016-6

**Optimization of Methods for Approximate Solution of Operator Equations** by Sergei V. Pereverzev.
ISBN 1-56072-140-5.

**Approximation of Periodic Functions** by V. N. Temlyakov
ISBN 1-56072-131-6.

**Extremal Properties of Polynomials and Splines** by N.P. Korneichuk. ISBN 1-56072-361-0

# Extremal Properties of Polynomials and Splines

**N.P. Korneichuk**
**A.A. Ligun**
**V.F. Babenko**

Nova Science Publishers, Inc.
Commack

Art Director: Maria Ester Hawrys
Assistant Director: Elenor Kallberg
Graphics: Denise Dieterich, Kerri Pfister,
Erika Cassatti and Barbara Minerd
Manuscript Coordinator: Sharyn Schweidel
Book Production: Tammy Sauter, Benjamin Fung
Circulation: Irene Kwartiroff and Annette Hellinger

*Library of Congress Cataloging–in–Publication Data*
*available upon request*

ISBN 1-56072-361-0


*6080 Jericho Turnpike, Suite 207*
*Commack, New York 11725*
*Tele. 516-499-3103 Fax 516-499-3146*
*E Mail Novasci1@aol.com*



*Printed in the United States of America*

# Contents

# Preface

No special class of functions have attracted such attention and have been investigated so thoroughly as polynomials (algebraic and trigonometric) and latter-day - as splines. No wonder. Algebraic polynomial is the simplest by its structure analityc function calculation of value of which requires only three arithmetic operations. At the same time owing to the well-known Weierstrass theorem each continuous on a finite interval function is limit of uniformly convergent sequence of polynomials. And speed of convergence increases, with the rise of smoothness of approximated function. Trigonometric polynomials have analogous properties related periodical functions.

Hence, investigation of continuous functional dependence may be reduced with any desirable degree of accuracy to the investigation of the certain effectively realizable polynomial.

Algebraic polynomials have been studied by mathematicians for a long time. Investigation of their extremal properties was begun by the remarkable papers of P.L.Chebyshev on polynomials of least deviation from zero and was continued by his disciples. As for trigonometric polynomials, investiganion of their properties was to a considerable extent stimulated by the needs of Fourier series theory.

Special interest to polynomials and their properties have been displayed by the mathematicians dealing with the approximation theory and its applications. Choosing the method for approximate representation of functions as a rule they have taken the polynomials (algebraic and trigonometric) thanks to the simplicity of their structure and wide approximative abilities. Profound knownledge of the properties of polynomials has acquired special significance. Begining with P.L.Chebyshev, great mathematicians, who were dealing with the approximation theory, made also a substantial contribution to the investigation of extremal properties of polynomials. In this connection we should specially mention fundamental results obtained by S.N.Bernstein and S.M.Nikolsky and their papers on the approximation of functions.

Comparatively recently, in the fifties, one could have hardly assume that there existed the method of approximation that was able to compete with polynomials in the wide range their application. Nevertheless such method appeared – it were polynomial splines, i.e. functions glued (with the certain degree of smoothness) from the pieces of algebraic polynomials. Though piecewise polynomial functions were known long ago (in particular, they appeared while solving specific problems of variational character on the function theory), but as a method of approximation only broken lines have been sometimes used.

Interest to splines have been increasing rapidly since sixsies. In the avalanche of papers on splines and their applications results characterizing their extremal properties were taken a considerable place.

Properties that are revealed during the solution of variational problems on maximum and minimum and are described by strict inequalities containing this or that characteristics of functions of the class A are said to be extremal properties of one or another special class A of functions. If variational problem is stated on the collection of functions of the considered class A and inequality on this collection is strict, then we may say that this inequality characterizes inner extremal properties of the class A.

Another category of the extremal properties of the class A shows itself when variational problem is stated on the class B of functions, but the extremal belongs to the class A (more often - to the intersection of the classes A and B). In this case inequality which is strict on the class B characterizes class A from outside. Typical example of such a variant is given by the approximational problems for the case when strict estimations of the best approximatoin of classes of functions with bounded $m$-th derivative are realized on perfect splines. Results of solution of such problems are elucidated enough completely in the papers on the approximation theory some of which are quoted in the last chapters of the book as an illustration of applications of inequalities for norms of perfect splines and monosplines. However principle attention in this monograph is paid to inner extremal properties of polynomials and splines. And though many inequalities describing such properies, in particular inequalities of Bernstein's type, may be also found in the literature, while writing this book we were influenced by the following reasons.

First, rather considerable amount of the factual material of these problems at the present is dispersed in magazine articles and as a subsidiary material in books dedicated to the different aspects of mathematics. It seems to us that there is a need for monograph containing coherent elucidation if not of all then at least of the significant part of the gained facts on extremal properties of polynomials and splines.

Secondly, and it seems to be the main thing, recently we have developed principally new methods of solving extremal problems of the functions theory that allowed us to find general approach to investigation of properties of polynomials as well as of splines. Foundation of such approach are theorems of comparison for rearrangements of functions that helped not only to state the general idealogical position extremal properties of polynomials and splines that were established by different methods earlier, but also to prove a series of new strict inequalities for them.

Material elucidated in the book has no pretensions to embrace completely such vast suhject as extremal properties of polynomials and splines.

A series of results is adduced without proofs. Some results that are directly related to this subject as well as some close directions ( for instance, extremal properties of integer and rational functions ) are in the certain degree reflected in commentaries to the chapters, supplementary results, and bibliography. Limited volume of the monograph didn't allow to pay enough attention to applications of extremal properties of polynomials and splines.

For convenience in this book is used a through numbering of theorems, lemmas, statements, and corollaries.

Authors.

# Chapter 1

# Some information from function theory

In this chapter gathered facts of the function theory which define fundamentals of methods of investigation of polynomials and splines extremal problems. These fundamentals compose the main principles of the theory of the best approximatiom and theorems of comparison for rearrangements of functions.

The majority of the general facts about the best approximation of the fixed element in the normed space given in the sections 1.1 and 1.2 is familiar to the specialists, though, for the sake of indetermince of the exposition, it is expedient to provide them with the concise proofs. Material connected with the comparison of rearrangements which has become part of the functions theory is stated a little more detailed in the sections 1.3-1.7. Here is also adduced information about strict inequalities similar to the well-known Kolmogorov's inequality for the derivatives norms which is closely connected with the main subject of the monograph.

## 1.1 Best approximation

Let $X$ be the linear normed space with the norm $||\cdot||_X = ||\cdot||$. Define *the best approximation* of an element $x \in X$ by set $\mathfrak{R} \subset X$ as following quantity:

$$E(x, \mathfrak{R})_X = \inf\{||x - u||_X : u \in \mathfrak{R}\}. \tag{1.1}$$

An element $u_x \in \mathfrak{R}$ realising the exact lower bound in (1.1) i.e. such that

$$\|x - u\|_X = E(x, \Re)_X, \tag{1.2}$$

is said to be *the best approximation element* (b.a.e.) for $x$ in $\Re$. For fixed set $\Re$ the quantity $E(x, \Re)_X$ is a functional defined on $X$. Let us formulate same elementary properties of this functional which proofs one can find, for example, in the monographs of Korneichuk [11,24].

**Proposition 1.1.1.** *For any set $\Re \subset X$ and any elements $x, y \in X$*

$$|E(x, \Re)_X - E(y, \Re)_X| \le \|x - y\| \tag{1.3}$$

*hence, functional $E(x, \Re)_X$ is continuous on $X$.*

**Proposition 1.1.2.** *If $\Re \subset X$ is linear manifold, then the functional $E(x, \Re)_X$ is semiadditive and positively homogeneous, i.a. for any $x, y \in X$*

$$E(x + y, \Re)_X \le E(x, \Re)_X + E(y, \Re)_X, \tag{1.4}$$

*and for any $x \in X$, $\lambda \in \mathbf{R}$*

$$E(\lambda x, \Re)_X = |\lambda| E(x, \Re)_X. \tag{1.5}$$

*If there exists a unique b.a.e. $u_x \in \Re$ for arbitrary element $x \in X$, then operator $P(x) = u_x$, which is called the metric projection of an element $x$ on set $\Re$, is defined.*

**Proposition 1.1.3.** *Let $\Re$ be the subspace of space $X$ and for arbitrary $x \in X$ there exists a unique b.a.e. $P(x) = u_x$, $u_x \in \Re$. Hence, the metric projection operator $P$ is homogeneous, i.e.*

$$P(\lambda x) = \lambda P(x), \tag{1.6}$$

*and if $\Re$ is finite-dimensional subspace, then operator $P$ is continuous on $X$.*

**Proposition 1.1.4.** *If $\Re$ is convex set in $X$, then set $\Re_x$ of all b.a.e. for $x$ in $\Re$ is convex too.*

If for arbitrary $x \in X$ there exists b.a.e. from $\Re \subset X$, then $\Re$ is said to be *the set of existence.*

**Proposition 1.1.5.** *Any finite-dimensional subspace $\mathfrak{R} \subset X$ is a set of existence.*

It must be pointed out that the most general proposition is true too.

**Proposition 1.1.6.** *Any closed locally compact set $\mathfrak{R}$ of linear normed space is a set of existence.*

The space $X$ is said to be *strictly normed* if for any its elements $x \neq 0$, $y \neq 0$, such that $x \neq \lambda y$, $\lambda > 0$ the following inequality holds true

$$||x + y|| < ||x|| + ||y||.$$

**Proposition 1.1.7.** *If space $X$ is strictly normed and for same element $x \in X$ there exists b.a.e. from the subspace $\mathfrak{R} \subset X$, then this b.a.e. is unique.*

As the example of strictly normed space one can consider space $L_p[a, b]$ $(1 < p < \infty)$. As usually we denote by $L_p[a, b]$ the space of functions $x(t)$ with finite norm

$$||x||_p = ||x||_{L_p[a,b]} = \left( \int_a^b |x(t)|^p dt \right)^{1/p}.$$

If for approximation of an element $x \in X$ we are allowed to use only a part of the set $\mathfrak{R}$, changing, may be, with the variable $x$, then there appears the problem of approximation with the restrictions.

The best one-sided approximation problem is its typical example.

Let $L_p = L_p[a, b]$, $\mathfrak{R} \subset L_p$ and for $x \in L_p$

$$\mathfrak{R}_x^{\pm} = \{u(t) : \ u \in \mathfrak{R}, \ \pm u(t) \leq \pm x(t) \ \ alm.everywhere\}.$$

The quantities

$$E^+(x, \mathfrak{R})_p = E(x, \mathfrak{R}_x^+)_p = \inf_{u \in \mathfrak{R}_x^+} ||x - u||_p$$

and

$$E^-(x, \mathfrak{R})_p = E(x, \mathfrak{R}_x^-)_p = \inf_{u \in \mathfrak{R}_x^-} ||x - u||_p$$

we shall consider respectively as the best approximations from below and from above of function $x(t)$ by set $\mathfrak{R}$ in space $L_p$.

As usually we consider that lower bound by empty set is equal to $+\infty$.

Notice, that analogs of propositions 1.1.1-1.1.7 are valid for the best one-sided approximations.

The best and the best one-sided approximations problems in spaces $L_p = L_p[a,b]$ $(1 \le p \le \infty)$ may be decided and from the natural point of view.

Let $\alpha > 0$, $\beta > 0$ are fixed numbers. For $x \in L_p$ we consider, that

$$||x||_{p;\alpha,\beta} = ||\alpha x_+ + \beta x_-||_p, \tag{1.7}$$

where

$$x_\pm(t) = \max\{\pm x(t), 0\}$$

and define the best $(\alpha,\beta)$-approximation in the space $L_p$ of $x \in L_p$ by set $\Re \subset L_p$ in the following way:

$$E(x,\Re)_{p;\alpha,\beta} = \inf_{u\in\Re} ||x-u||_{p;\alpha,\beta}. \tag{1.8}$$

It is clear that $E(x,\Re)_{p;1,1} = E(x,\Re)_p$.

Moreover the following statement is valid.

**Theorem 1.1.8.** *Let $p \in [1,\infty)$, $\Re$ is locally compact subset of space $L_p$. Then for arbitrary function $x \in L_p$ the following equations are executed monotonously on $\alpha$ and $\beta$*

$$\lim_{\beta\to\infty} E(x,\Re)_{p;1,\beta} = E_+(x,\Re)_p, \tag{1.9}$$

$$\lim_{\alpha\to\infty} E(x,\Re)_{p;\alpha,1} = E_-(x,\Re)_p. \tag{1.10}$$

**Proof.** By the definitions (1.7) and (1.8) it follows that if $1 < \beta < \beta'$, then

$$E(x,\Re)_p \le E(x,\Re)_{p;1,\beta} \le E(x,\Re)_{p;1,\beta'}. \tag{1.11}$$

At the same time for all $\beta > 0$

$$E(x,\Re)_{p;1,\beta} \le \inf_{u\in\Re_x^+} ||x-u||_{p;1,\beta} = E^+(x,\Re)_p. \tag{1.12}$$

In consequence of (1.11) and (1.12) the limit in the left part of (1.9) exists and is no more $E^+(x,\Re)_p$.

For every $\beta \ge 1$ we denote by $u_\beta$ such element from $\Re$, that

$$E(x.\Re)_{p;1,\beta} = ||x-u_\beta||_{p;1,\beta}$$

(as the analog of statement 1.1.6 is true for the $(\alpha, \beta)$-approximations, so $u_\beta$ exists). To prove the correlation (1.9) it is enough to show that

$$E^+(x, \Re)_p \leq B := \lim_{\beta \to \infty} E(x, \Re)_{p;1,\beta}. \tag{1.13}$$

If $B = \infty$, then (1.13) is obvious. If $B < \infty$, then for the sufficiently large $\beta$

$$||u_\beta|| \leq ||x - u_\beta||_p + ||x||_p = E(x, \Re)_{p;1,\beta} + ||x||_p \leq B + ||x||_p,$$

and in view of local compactness of $\Re$, there exists the sequence $\{\beta_k\}$ ($\beta_k \to \infty$) and element $u_\infty \in \Re$, such that $u_{\beta_k} \to u_\infty$ when $k \to \infty$.

Next, show that $u_\infty \in \Re_x^+$ or

$$||(x - u)_-||_p = 0. \tag{1.14}$$

Indeed, since

$$E(x, \Re)_{p;1,\beta_k} = ||(x - u_{\beta_k})_+ + \beta_k(x - u_{\beta_k})_-||_p,$$

then using (1.11), we obtaind

$$||(x - u_{\beta_k})_-||_p \leq \beta_k^{-1} E(x, \Re)_{p;1,\beta_k} \leq B \cdot \beta_k^{-1}.$$

In the limit when $k \to \infty$ we obtain (1.14).

Now for $\beta_k \geq 1$ we are having

$$E^+(x, \Re)_p \leq ||x - u_\infty||_p = \lim_{k \to \infty} ||x - u_{\beta_k}||_p \leq$$

$$\lim_{k \to \infty} ||x - u_{\beta_k}||_{p;1,\beta_k} = \lim_{k \to \infty} E(x, \Re)_{p;1,\beta_k},$$

hence, correlations (1.13) and (1.9) are proved.

The equality (1.10) is proved by analogy.

Consequently, the best $(\alpha, \beta)$-approximation problems interpolate the problem of usual best approximation and best one- sided approximation and allow to decide the latest problems from the united point of view.

Let us dwell once more on the characterisation of the best approximation elements. This item is essentially connects with the metric. Let us consider this one in the some concrete functional spaces.

**Theorem 1.1.9.** *Let $\Re$ be the finite-dimensional subspace of space $L_1 = L_1[a,b]$ ; $\alpha, \beta$ are positive numbers. In order to function $u_0(t) \in \Re$ could get the best $(\alpha,\beta)$-approximation in space $L_1$ for the function $x \in L_1$ it is sufficiently, and if*

$$mes\{t : \ t \in [a,b], \ x(t) = u_0(t)\} = 0 \tag{1.15}$$

*then it's also necessary that following equality holds true*

$$\int_a^b u(t) sgn_{\alpha,\beta}\,(x(t) - u_0(t))\,dt = 0, \tag{1.16}$$

*where*

$$sgn_{\alpha,\beta}\,(x(t) - u_0(t)) = \alpha\ sgn\,(x(t) - u_0(t))_+ - \\ \beta\ sgn\,(x(t) - u_0(t))_- .$$

*In particular, if $\alpha = \beta = 1$ (i.e. it is spoken about the usual best $L_1$-approximation), then correlation (1.16) takes form*

$$\int_a^b u(t)\ sgn\,(x(t) - u_0(t))\,dt = 0.$$

**Proof.** If condition (1.16) is held then for any $u \in \Re$ we have

$$||x - u_0||_{1;\alpha,\beta} = \int_a^b (x(t) - u_0(t))\, sgn_{\alpha,\beta}\,(x(t) - u_0(t))\,dt =$$

$$\int_a^b (x(t) - u(t))\, sgn_{\alpha,\beta}\,(x(t) - u_0(t))\,dt \leq$$

$$\int_a^b (x(t) - u(t))\, sgn_{\alpha,\beta}\,(x(t) - u(t))\,dt = ||x - u||_{1;\alpha,\beta}.$$

Let now $u_0(t)$ is being the best $(\alpha,\beta)$-approximation function for $x(t)$ in space $L_1$, moreover $x(t) \neq u_0(t)$ almost everywhere. Let us consider that there exists function $u \in \Re$, such that

$$\int_a^b u(t)\, sgn_{\alpha,\beta}\,(x(t) - u_0(t))\,dt \neq 0. \tag{1.17}$$

Then for any suitable $\epsilon$, which absolute value is sufficiently small, the following inequality is valid

$$\varepsilon \int_a^b u(t)\, sgn_{\alpha,\beta}\,(x(t) - u_0(t) - \varepsilon u(t))\,dt \geq 0. \tag{1.18}$$

Really, there will be $\varepsilon_0 > 0$, such that for all $\varepsilon$, $|\varepsilon| < \varepsilon_0$

$$\int_a^b u(t)\ sgn_{\alpha,\beta}\,(x(t) - u_0(t) - \varepsilon u(t))\,dt \neq 0,$$

Otherwise, from the Lebeg's theorem it follows

$$\int_a^b u(t)\,sgn_{\alpha,\beta}\,(x(t) - u_0(t))\,dt = 0,$$

that conditions to (1.17). Now we should choose the suitable sign for $\varepsilon$.

Further

$$||x - u_0||_{1;\alpha,\beta} = \int_a^b (x(t) - u_0(t))\,sgn_{\alpha,\beta}\,(x(t) - u_0(t))\,dt \geq$$

$$\int_a^b (x(t) - u_0(t))\,sgn_{\alpha,\beta}\,(x(t) - u_0(t) - \varepsilon u(t))\,dt =$$

$$\int_a^b (x(t) - u_0(t) - \varepsilon u(t))\,sgn_{\alpha,\beta}\,(x(t) - u_0(t) - \varepsilon u(t))\,dt +$$

$$\varepsilon \int_a^b u(t) sgn_{\alpha,\beta}\,(x(t) - u_0(t) - \varepsilon u(t))\,dt,$$

and, applying (1.18), we obtain

$$||x - u_0||_{1;\alpha,\beta} \geq ||x - u_0 - \varepsilon u||_{1;\alpha,\beta}.$$

But this contradicts to fact that $u_0(t)$ is the best approximation element for $x(t)$.

*Function* $\Psi(t)$ *is said to be* $N$-function *if* $\Psi(t)$ *is continuous nondecresiang and convex from above* $[0, \infty)$ *function, moverover* $\Psi(0) = 0$.

Let $\omega(t) \in L_1[a, b]$ and $\omega(t) > 0$ almost everywhere on $[a, b]$. Denote by $L_{\Psi,\omega}[a, b]$ the space of all measurable on $[a, b]$ functions $x(t)$, for which

$$||x||_{L_{\Psi,\omega}} := \int_a^b \Psi(|x(t)|)\omega(t)dt < \infty.$$

**Theorem 1.1.10.** *Let* $\Psi(t)$ *be differentiable and strictly convax from above on* $[0, \infty)$ $N$*-function and* $\Re$ *is* $n$*-dimensional subspace of space* $L_{\Psi,\omega}$. *In order to function* $u_0 \in \Re$ *could get the best* $L_{\Psi,\omega}$*-approximatin of function* $x(t)$ *it's necessery and enough that for arbitrary function* $u \in \Re$ *the following condition will be true*

$$\int_a^b u(t)\Psi'(|x(t) - u_0(t)|)sgn(x(t) - u_0(t))\omega(t)dt = 0. \tag{1.19}$$

**Proof.** Let $u_1, \ldots, u_n$ be the basis of subspace $\Re$. Then if

$$\psi(c_1, \ldots, c_n) = \int_a^b \Psi(|x(t) - \sum_{k=1}^n c_k u_k(t)|)\omega(t)dt,$$

then

$$\min\{\int_a^b \Psi(|x(t) - u(t)|)\omega(t)dt : \ u \in \Re\} =$$
$$\min\{\psi(c_1, \ldots, c_n) : \ c_k \in R, \ k = 1, \ldots, n\}.$$

Under the restrictions, which are forced upon the function $\Psi(t)$ and weight $\omega(t)$, function $\psi(c_1, \ldots, c_n)$ is continuously-differentiable and strictly concave at every variable. That is why the following equalities are the necessary and sufficient conditions of its minimum existence

$$\frac{\partial \psi}{\partial c_k} = 0, \ k = 1, \ldots, n,$$

and this is equivalent to execution (1.19) for any function $u \in \Re$.

If $\Psi(t) = t^p$, $1 < p < \infty$, $\omega(t) \equiv 1$ (the case of space $L_p[a, b]$), then condition (1.19) takes form

$$\int_a^b u(t)|x(t) - u_0(t)|^{p-1} sgn(x(t) - u_0(t))dt = 0, \ u \in \Re$$

and if $\Psi(t) = t^2$ (we are having now the space $L_2[a, b]$ with weigh $\omega(t)$), then

$$\int_a^b u(t)(x(t) - u_0(t))\omega(t)dt = 0, \ u \in \Re.$$

As to metric $C$, the characterization of the best uniform approximation polynomial will be given in section 2.1.

## 1.2 Duality for best approximation

In the present paragraph we'll reduce on important theorem of convex analysis and the row of consequences from it for concrete situations. Let $X$ be the real linear space, $p(x)$ is finite real nonpositive function defined everywhere on $X$ and such that

$$p(0) = 0, \ p(x + y) \leq p(x) + p(y), \ p(\alpha) = \alpha p(x), \ \alpha \geq 0,$$

and which will be named as nonsymmetric seminorm.

Put

$$p(x, \Re) = \inf\{p(x-u);\ u \in \Re\}$$

for $\Re \subset X$. Denote by $X'$ the set of all linear functionals defined on $X$, and let for $f \in X'$

$$p^*(f) = \sup\{f(x):\ x \in X,\ p(x) \leq 1\}. \tag{1.20}$$

¿From this definition it immediately follows that for any $f \in X'$ and $x \in X$

$$f(x) \leq p^*(f)p(x). \tag{1.21}$$

If $x$ is normed space and $p(x) = ||x||_X$ then

$$p(x, \Re) = E(x, \Re)_X$$

and for $f \in X^*$ the following inequality

$$p^*(f) = ||f||$$

executed.

On the whole the quantity $p^*(f)$ must not be finite, however (and later it is important) the condition $P(x) \leq 1$ in definition (1.20) may be changed on the condition $p(x) < 1$. Just so for $f \in X'$ and $\eta > 0$

$$sup\{f(x):\ x \in X,\ p(x) < \eta\} = \eta p^*(f). \tag{1.22}$$

Really if we denote the left part of (1.22) as $A$, then due to (1.20) $0 \leq A \leq \eta p^*(f)$, and in the case $p^*(f) = 0$ (1.22) is obvious.

Let $p^*(f) > 0$ and $p^*(f) < \infty$. If in spite of (1.22) $\eta p^*(f) > A$, then from (1.20) it follows that $p(x_0) = \eta$, $f(x_0) > A$, and for all $\alpha \in (0,1)$ sufficiently closed to one, the following inequalities $p(\alpha x_0) < \eta$ and $f(\alpha x_0) > A$ will be executed, but that contradicts to definition of $A$. If the $p^*(f) = \infty$, but $A < \infty$, then for some $x_0$ we obtain that $p(x_0) = \eta$ and $f(x_0) > 2A$ and for $\alpha \in (0,1)$ and sufficiently closed to one we can write $p(\alpha x_0) < \eta$ and $f(\alpha x_0) > 2A$ in spite of definition number $A$.

**Theorem 1.2.1.** *Let $\Re$ be the convex set in linear subspase $X$, $p(x)$ is defined on $X$ nonsymmetric seminorm. Then for any $x \in X$*

$$p(x, \Re) = \sup\{f(x) - \sup_{u \in \Re} f(u):\ f \in X',\ p^*(f) \leq 1\}. \tag{1.23}$$

*Moreover there exists functional $f_0 \in X'$, $p^*(f_0) = 1$ realising the upper bound in (1.23).*

**Proof.** Denote for brevity the right part of (1.23) by $N(x)$. If $u \in \Re$ and $p^*(f) \leq 1$ then owing to (1.21)

$$f(x) - sup\{f(y): \ y \in \Re\} \leq f(x) - f(u) = f(x-u) \leq$$

$$p^*(f)p(x-u) \leq p(x-u),$$

and passing in the left part to the upper bound by $f \in X'$, $p^*)f) \leq 1$ and in the right part to the lower bound by $u \in \Re$ we obtain the inequality

$$N(x) \leq p(x, \Re). \tag{1.24}$$

Let us show that $p(x, \Re) \leq N(x)$, moreover we can consider that $p(x, \Re) :$ $= \mu > 0$.

For fixed $x \in X$ examine the set

$$S(x, \mu) = \{y: \ y \in X, \ p(x-y) < \mu\}.$$

It's not difficult to check that $S(x, \mu)$ is convex body in $X$. $S(x, \mu)$ doesn,t intersect with $\Re$ so far as if $u \in \Re$, then $p(x-u) \geq \mu$. According to the separability theorem (look, for example, A.N.Kolmogorov and S.V.Fomin [1.p.130]) there exists unzero functional $f \in X'$ and such that

$$f(u) \leq f(y), \ u \in \Re, \ y \in S(x, \mu). \tag{1.25}$$

Now we show that for given functional $p^*(f) < \infty$. It's easy to check the coincidence of the sets

$$\{z: \ p(z) < \mu\} = \{z: \ z \in x - S(x, \mu)\}. \tag{1.26}$$

Fix an element $u' \in \Re$ and let $p(v) < 1$. Then in consequence of (1.26) $\mu v = x - y$, where $y \in S(x, \mu)$ and in account of (1.25) we obtain

$$f(x) - \mu f(v) = f(x - \mu v) = f(y) \geq f(u').$$

Therefore, for any $v \in X$ such that $p(v) < 1$

$$f(v) \leq \frac{1}{\mu}(f(x) - f(u')) = c < \infty,$$

and this, owing to (1.22), means that $p^*(f) < \infty$.

Functional $f_0 = f/p^*(f)$ belongs to $X'$, moreover $p^*(f_0) = 1$ and $f_0(u) \leq f_0(y)$, $u \in \Re$, $y \in S(x, \mu)$. According to (1.26) and (1.22) we can write now

$$\sup\{f_0(u): \ u \in \Re\} \leq \inf\{f_0(y): \ y \in S(x, \mu)\}$$

$$\inf\{f_0(x-z):\ p(z)<\mu\} = f_0(x) - \sup\{f_0(z):\ p(z)<\mu\} = \\ f_0(x) - \mu = f_0(x) - p(x,\Re),$$

so that

$$p(x,\Re) \le f_0(x) - \sup\{f_0(u):\ u\in\Re\} \le N(x).$$

Together with (1.24) it signifies that

$$p(x,\Re) = N(x) = f_0(x) - \sup\{f_0(u):\ u\in\Re\}.$$

Note the particular case of the theorem 1.2.1, when $\Re$ is the subspace.

**Theorem 1.2.2.** *Let $\Re$ be the subspace of linear space $X$, $p(x)$ be the setting on $X$ nonsymmetric seminorm. Then for any $x\in X$*

$$p(x,\Re) = \sup\{f(x):\ f\in\Re^{\perp}, p^*(f)\le 1\}, \tag{1.27}$$

*where*

$$\Re^{\perp} = \{f:\ f\in X',\ f(u)=0\ \ \forall u\in\Re\}.$$

*If $p(x,\Re) > 0$, then there exists functional $f_0 \in \Re^{\perp}$,realising the upper bound in (1.27), moreover $p^*(f_0) = 1$.*

**Proof.** If $\Re$ is subspace in $X$ and functional $f\in X'$, $p^*(f)\le 1$, such that, for some $u'\in\Re$, $f(u')=\alpha\ne 0$, then for every $x\in X$ the followihg inequality is executed

$$f(x) - \sup\{f(u):\ u\in\Re\} \le f(x) - \sup\{f(\gamma u'):\ \gamma\in R\} = -\infty.$$

This means that under the calculation of upper bound in (1.23) we can consider only functionals $f\in\Re^{\perp}$. But for these functionals

$$\sup\{f(u):\ u\in\Re\} = 0,$$

and (1.23) will be written in the form of (1.27).

The accessibility of upper bound follows from corresponding fact in the theorem 1.2.1.

If $X$ is normed space and there exists $c>0$ such that for any $x\in X$, $p(x) < c||x||$, then in the formulations of theorems 1.2.1 and 1.2.2 $X'$ can be substituted for conjugate space $X^*$, because in this case under the condition $p^*(f)\le 1$ we'll have

$$||f|| \le \sup\{f(x):\ p(x)<c\} = c\ \sup\{f(x):\ p(x)<1\} = cp^*(f) \le c.$$

¿From the theorems 1.2.1 and 1.2.2 the next statement immediately follows .

**Corollary 1.2.3.** *If $X$ is normed space and $\mathfrak{R}$ is convex set in $X$, then for any $x \in X$*

$$E(x, \mathfrak{R})_X = \sup_{f \in X^*, \|f\| \le 1} \{f(x) - \sup_{u \in \mathfrak{R}} f(u)\}. \tag{1.28}$$

*If, in particular, $\mathfrak{R}$ is subspace, then*

$$E(x, \mathfrak{R})_X = \sup\{f(x) : \ f \in X^*, \ f \in \mathfrak{R}^\perp, \ \|f\| \le 1\}. \tag{1.29}$$

*At the same time there exist functionals from $X^*$ realising upper bounds in (1.28) and (1.29).*

In view of common form for linear continuous functionals in the space $L_p[a, b]$, $1 \le p \le \infty$, the duality correlations for the best $L_p$-approximations by convex set and subspace can be written as follows:

**Theorem 1.2.4.** *If $\mathfrak{R}$ is convex set in space $L_p = L_p[a, b]$, $1 \le p \le \infty$, then for any function $x \in L_p$*

$$E(x, \mathfrak{R})_{L_p} = \sup_{y \in L_{p'}, \|y\|_{p'} \le 1} \left( \int_a^b x(t)y(t)dt - \sup_{u \in \mathfrak{R}} \int_a^b u(t)y(t)dt \right), \tag{1.30}$$

*where $p' = p/(p-1)$. In particular, if $\mathfrak{R}$ subspace, then*

$$E(x, \mathfrak{R})_{L_p} = \sup \left( \int_a^b x(t)y(t)dt : \ y \perp \mathfrak{R}, \ \|y\|_{p'} \le 1 \right), \tag{1.31}$$

*where $y \perp \mathfrak{R}$ means, that*

$$\int_a^b u(t)y(t)dt = 0, \quad \forall\, u \in \mathfrak{R}.$$

**Corollary 1.2.5.** *For best approximation of function $x \in L_p[a, b]$, $1 \le p < \infty$ by constant the following equality*

$$E_0(f)_p := \inf\left(\|x - \lambda\|_p : \ \lambda \in R\right) =$$

$$\sup \left( \int_a^b x(t)y(t)dt : \ \|y\|_{p'} \le 1, \ \int_a^b y(t)dt = 0 \right), \ \frac{1}{p} + \frac{1}{p'} = 1.$$

*is true.*

If numbers $\alpha > 0$ and $\beta > 0$ are fixed and $p \in [1, \infty)$, then quantity

$$p(x) = ||x||_{p;\alpha,\beta} = ||\alpha x_+ + \beta x_-||_p$$

is nonsymmetric norm in $L_p[a,b]$, moreover $p(x) \le \max\{\alpha, \beta\}||x||_p$. Due to theorem 1.2.1 for convex set $\Re \in L_p[a,b]$

$$E(x,\Re)_{p;\alpha,\beta} = \sup\{\left(f(x) - \sup_{u\in\Re} f(u)\right) : \ f \in L_p^*, \quad \sup_{||v||_{p;\alpha,\beta}\le 1} f(v) \le 1\},$$

Taking into account that for any function $z(t) \in L_p[a,b]$

$$\sup_{||v||_{p;\alpha,\beta}\le 1} \int_a^b v(t)z(t)dt = ||z||_{p';\alpha^{-1},\beta^{-1}}.$$

(see N.P.Korneichuk [24,p.393]) we obtain the following statement.

**Theorem 1.2.6.** *Let $\alpha > 0$, $\beta > 0$, $p \in [1,\infty)$ and $\Re$ is convex set in $L_p = L_p[a,b]$. Then for any function $x \in L_p$*

$$E(x,\Re)_{p;\alpha,\beta} = \sup_{||y||_{p';\alpha^{-1},\beta^{-1}}\le 1} \left(\int_a^b x(t)y(t)dt - \sup_{u\in\Re}\int_a^b u(t)y(t)dt\right). \tag{1.32}$$

*In particular, if $\Re$ is subset, then*

$$E(x,\Re)_{p;\alpha,\beta} = \sup\left(\int_a^b x(t)y(t)dt : \ ||y||_{p';\alpha^{-1},\beta^{-1}} \le 1, \ y \perp \Re\right). \tag{1.33}$$

The upper bounds in (1.32) and (1.33) are got on functions $y \subset L'_p$ with norm $||y||_{p';\alpha^{-1},\beta^{-1}} = 1$.

¿From the theorems 1.1.8 and 1.2.6 the following statement can be easy obtained.

**Theorem 1.2.7.** *Let $1 \le p < \infty$ and $\Re$ be the convex locally compact set in $L_p = L_{p[a,b]}$. Then for any $x \in L_p$*

$$E^{\pm}(x,\Re)_{L_p} = \sup_{y\in L_{p'}, ||y_\pm||_{p'}\le 1} \left(\int_a^b x(t)y(t)dt - \sup_{u\in\Re}\int_a^b u(t)y(t)dt\right). \tag{1.34}$$

*In particular, if $\Re$ is subset, then*

$$E^{\pm}(x,\Re)_{L_p} = \sup\left(\int_a^b x(t)y(t)dt : \ y \in L_{p'}, ||y_\pm||_{p'} \le 1, y\perp\Re\right). \tag{1.35}$$

## 1.3 Rearrangements of functions

The device of rearrangements will be very significant under the investigation of extremal properties for functions from different classes, in particular, the polynomials and splines. That is why one can find here detailed account both well-known properties of equally measurable rearrangements and new facts which will be used in future.

Let $f(t)$ be nonnegative summable on interval $(a,b)$ (finite or infinite) and, hence, measurable and almost everywhere finite on $(a,b)$ function.

Comparing every $y \geq 0$ with number

$$m(f,y) = mes\{t: \ t \in (a,b), f(t) > y\}, \tag{1.36}$$

we obtain nonincreasing on the interval $[0,\infty)$ continuous from the right function $m(f,y)$, which is named as a function of distribution for $f(t)$.

In common case the rearrangement $r(f,t)$ of function $f(t)$ is defined by equality

$$r(f,t) = \inf\{y: \ m(f,y) \leq t\}, \tag{1.37}$$

and as a function of distribution $m(f,t)$ it is continuous from right.

¿From the definition it follouws that for all $y \geq 0$

$$mes\{t: \ t \in (0,b-a), r(f,t) > y\} = m(f,y), \tag{1.38}$$

that is functions $f(t)$ and $r(f,t)$ have one and the same function of distribution, that is why they are named equally measurable.

Assume that

$$\pi(f,t) = r(f_+,t) - r(f_-,b-a-t), \ 0 \leq t \leq b-a, \tag{1.39}$$

for any summable on finite segment $[a,b]$ function $f(t)$. Here as above $f_\pm = \max\{\pm f(t), 0\}$ are positive and negative parts of function $f(t)$.

¿From the definition and from the equality (1.38) it follows, that

$$\pi(f+c,t) = \pi(f,t) + c, \ \int_a^b f(t)dt = \int_a^b \pi(f,t)dt,$$

and for essentially restricted on $[a,b]$ function $f(t)$

$$\pi(f,0) = r(f_+,0) = vrai \sup_{a \leq t \leq b} f(t),$$

and

$$\pi(f,b-a) = -r(f_-,0) = vrai \inf_{a \leq t \leq b} f(t).$$

Directly from the definitions it follows also, that if $|f(t)| \leq |g(t)|$, $a \leq t \leq b$ then

$$r(|f|,t) \leq r(|g|,t), \; 0 \leq t \leq b-a$$

and for $f, g \in C[a,b]$

$$||\pi(f) - \pi(g)||_{C(0,b-a)} \leq ||f-g||_{C[a,b]}.$$

Now we note, that if $\Phi(t)$ is continuous and nonnegative on $[0,\infty)$ function and $\Phi(0) = 0$, then

$$\int_a^b \Phi(|f(t)|)dt = \int_0^{b-a} \Phi(r(|f|,t))dt. \tag{1.40}$$

Really, if, for instance, $\Phi(t)$ strictly increases, then for any $y > 0$

$$mes\{t: \; t \in (0,b-a), \; r(\Phi(|f|),t) > y\} =$$

$$mes\{t: \; t \in (a,b), \; \Phi(|f(t)|) > y\} =$$

$$mes\{t: \; t \in (a,b), \; |f(t)| > \Phi^{-1}(y)\} =$$

$$mes\{t: \; t \in (0,b-a), \; \Phi(r(|f|,t)) > y\}.$$

Now it remains to take into consideration the definition of Lebeg's integral.

¿From (1.40), in particular, it follows that

$$\int_a^b |f(t)|^p dt = \int_0^{b-a} r^p(|f|,t)dt, \; p > 0. \tag{1.41}$$

**Proposition 1.3.1.** *For any summable on $[a,b]$ function $f(t)$ and by all $t \in [0,b-a]$ the following correlations*

$$\sup\{\int_E |f(u)|du: \; E \subset [a,b], mesE = t\} = \int_0^t r(|f|,u)du; \tag{1.42}$$

*and*

$$\inf\{\int_E |f(u)|du: \; E \subset [a,b], mesE = t\} = \int_{b-a-t}^{b-a} r(|f|,u)du; \tag{1.43}$$

*hold true.*

**Proof.** Really, let $E \in [a,b]$, $mes\ E = t$ and

$$f_E(t) = \begin{cases} |f(t)|, & t \in E, \\ 0, & t \in [a,b] \backslash E. \end{cases}$$

Then

$$\int_E |f(t)|dt = \int_a^b f_E(t)dt = \int_0^{b-a} r(f_E, t)dt =$$

$$\int_0^{mesE} r((f_E, u)du = \int_0^t r(f_E, u)du.$$

Moreover, $f_E(t) \leq |f(t)|$, $a \leq t \leq b$ and therefore

$$r(f_E, u) \leq r(|f|, u), \ 0 \leq u \leq b-a.$$

Consequently,

$$\int_E |f(u)|du \leq \int_0^t r(|f|, u)du. \tag{1.44}$$

In the same time, if $y_t = r(|f|, t)$ then

$$E_1 = \{u: \ u \in [a,b], \ |f(u)| > y_t\},$$

and

$$E_2 = \{u: \ u \in [a,b], \ |f(u)| \geq y_t\},$$

then $mesE_1 \leq t \leq mesE_2$. That is why there exists the set $E_t$, such that

$$E_1 \subset E_t \subset E_2, \ mesE_t = t$$

and

$$\int_{E_t} |f(u)|du = \int_0^t r(|f|, u)du,$$

and this together with (1.44) proves (1.42).

The correlation (1.43) is setted analogously.

**Proposition 1.3.2.** *Let $f(t)$ be the continuously differentiable on segment $[a,b]$ function and point $\theta \in (0, b-a)$ such, that the set*

$$E = \{t: \ t \in [a,b], |f(t)| = r(|f|, \theta)\}$$

*consists of finite number points $t_1, t_2, \ldots, t_m$ at which $f'(t_k) \neq 0$. Then*

$$\frac{1}{|r'(|f|, \theta)|} = \sum_{k=1}^m \frac{1}{|f'(t_k)|}.$$

**Proof.** Really, for sufficiently small $\Delta\theta > 0$ a set

$$\{t: \ r(|f|, \theta) \geq |f(t)| \geq r(|f|, \theta + \Delta\theta)\}$$

consists from $m$ intervals $(\alpha_k, \beta_k)$, $k = 1, 2, \ldots, m$. At the same time for any $k$ either $\alpha_k = t_k$ or $\beta_k = t_k$. It's clear, that

$$\sum_{k=1}^{m} (\beta_k - \alpha_k) = \Delta\theta.$$

That is why

$$\frac{\Delta\theta}{r(|f|, \theta) - r(|f|, \theta + \Delta\theta)} = \sum_{k=1}^{m} \frac{\beta_k - \alpha_k}{|f(\beta_k) - f(\alpha_k)|}.$$

Passing to the limit when $\Delta\theta \to 0$ and taking into account, that under this $\beta_k - \alpha_k \to 0$, $k = 1, 2, \ldots, m$, and $\alpha_k = t_k$ or $\beta_k = t_k$ we obtain

$$|r'(|f|, \theta + 0)|^{-1} = \sum_{k=1}^{m} |f'(t_k)|^{-1}.$$

Analogously we set, that

$$|r'(|f|, \theta - 0)|^{-1} = \sum_{k=1}^{m} |f'(t_k)|^{-1}.$$

**Proposition 1.3.3.** *Let measurable on segment $[a, b]$ functions $f(t)$ and $g(t)$ are such that product $\pi(f, t)\pi(g, t)$ is summable on $[0, b - a]$. Then*

$$\int_a^b f(t)g(t)dt \leq \int_0^{b-a} \pi(f, t)\pi(g, t)dt. \tag{1.45}$$

**Proof.** Let at first $\{a_k\}_1^n$ and $\{b_k\}_1^n$ are arbitrary nonnegative numbers and $\{\pi(a_k)\}_1^n$ and $\{\pi(b_k)\}_1^n$ are the rearrangements of these numbers in nonincreasing order, then

$$\sum_{k=1}^{n} a_k b_k \leq \sum_{k=1}^{n} \pi(a_k)\pi(b_k) \tag{1.46}$$

Let now $f_n(t)$ and $g_n(t)$ are arbitrary nonnegative on segment $[a, b]$ piece-constant functions with breaks at points $t_k = a + kh$, $h = (b - a)/n$, $k = 1, \ldots, n$ and

$$a_k = f_n(t), \ b_k = g_n(t), \ t \in (t_{k-1}, t_k), \ t_0 = a, t_n = b.$$

Then

$$\int_a^b f_n(t)g_n(t)dt = h\sum_{k=1}^n a_k b_k$$

and

$$\int_0^{b-a} \pi(f_n,t)\pi(g_n,t)dt = h\sum_{k=1}^n \pi(a_k)\pi(b_k)$$

and owing to (1.46)

$$\int_a^b f_n(t)g_n(t)dt \leq \int_0^{b-a} \pi(f_n,t)\pi(g_n,t)dt.$$

With the help of limited transition we satisfy ourselves that inequality (1.45) is true for any nonnegative measurable on $[a,b]$ functions.

Let now $f(t)$ and $g(t)$ are arbitrary measurable on $[a,b]$ and restricted from below functions. Then puting that $\lambda = \inf\{f(t) : t\}$ and $\mu = \inf\{g(t) : t\}$, in consequesce from proved above facts we obtain

$$\int_a^b f(t)g(t)dt = \int_a^b (f(t)-\lambda)(g(t)-\mu)dt + \lambda\int_a^b g(t)dt +$$

$$\mu\int_a^b f(t)dt - \lambda\mu(b-a) \leq \int_0^{b-a} \pi(f-\lambda,t)\pi(g-\mu,t)dt +$$

$$\lambda\int_a^b g(t)dt + \mu\int_a^b f(t)dt - \lambda\mu(b-a) =$$

$$\int_0^{b-a} \pi(f-\lambda,t)\pi(g-\mu,t)dt + \lambda\int_0^{b-a} \pi(g,t)dt +$$

$$\mu\int_0^{b-a} \pi(f,t)dt - \lambda\mu(b-a) = \int_0^{b-a} \pi(f,t)\pi(g,t)dt.$$

**Proposition 1.3.4.** *For all functions $f, F \in L_1[a,b]$ such that production $\pi(f,t)\pi(F,t)$ is summable on $[0,b-a]$ the next inequality is valid*

$$\sup\{\int_a^b g(t)F(t)dt : \ g \in L_1[a,b], \ \pi(g,t) \equiv \pi(f,t)\} =$$

$$\int_0^{b-a} \pi(f,t)\pi(F,t)dt. \tag{1.47}$$

Really from the proposition 1.3.3 it follows that left part of the equality (1.47) is not more than right one. At the same time for the piece-constant functions considered in the proof of proposition 1.3.3 the equality (1.47) is obvious. With the help of limited transition this equality is easy established in common case.

Enter the next important meaning. For summable on segment $[a,b]$ functions $f(t)$ and $g(t)$ we shall write

$$(f \prec g)_{[a,b]}, \tag{1.48}$$

if for all $t \in [0, b-a]$ the following inequality is valid

$$\int_0^t \pi(f,t)dt \leq \int_0^t \pi(g,t)dt.$$

This inequality in the case when $f(t) \geq 0$, $g(t) \geq 0$ can be represented in such form

$$\int_0^t r(f,t)dt \leq \int_0^t r(g,t)dt.$$

The relations as (1.48) we are settling to name by the inequalities. In the case when the segment is known we'll write $f \prec g$ in particular for $2\pi$-periodical functions instead of $(f \prec g)_{[0,2\pi]}$ we'll write $f \prec g$. Further it's clear that from $f \prec g$ and $g \prec h$ it follows that $f \prec h$.

**Proposition 1.3.5.** *For summable on $[a,b]$ functions $f(t)$ and $g(t)$ the following inequality*

$$r(|f+g|) \prec r(|f|) + r(|g|)$$

*holds true.*

**Proof.** Really, if $E_t$ is arbitrary set from $[a,b]$ with measure that is equal to $t$, then due to 1.3.1

$$\int_0^t r(|f+g|), u)du = \sup_{E_t} \int_{E_t} |f(u)+g(u)|du \leq$$

$$\sup_{E_t} \int_{E_t} |f(u)|du + \sup_{E_t} \int_{E_t} |g(u)|du = \int_0^t r(|f|), u)du + \int_0^t r(|g|), u)du.$$

**Proposition 1.3.6.** *If functoins $f(t)$ and $g(t)$ are summable on $[a,b]$ and $f_+ \prec g_+$, $f_- \prec g_-$, then $|f| \prec |g|$.*

**Proof.** Really, considering the graphs of monotonous functions $r(|f|, t)$, $r(|f_+|, t)$ and $r(|f_-|, t)$ one can easy understand, that if $t \in (0, b-a)$ and $t = t_1 + t_2$, $t_1, t_2 \geq 0$, then

$$\int_0^t r(|f|), u)du \geq \int_0^{t_1} r(|f_+|), u)du + \int_0^{t_2} r(|f_-|), u)du, \qquad (1.49)$$

Moreover, there exists such points $t_1^*$ and $t_2^*$ that (1.49) turns in to equality. Then

$$\int_0^t r(|f|), u)du = \int_0^{t_1^*} r(|f_+|), u)du + \int_0^{t_2^*} r(|f_-|), u)du \leq$$

$$\int_0^{t_1^*} r(|g_+|), u)du + \int_0^{t_2^*} r(|g_-|), u)du \leq \int_0^t r(|g|), u)du.$$

**Proposition 1.3.7.** *Let $f(t)$ and $g(t)$ are summable on $[a, b]$ functions and*

$$\int_a^b f(t)dt = \int_a^b g(t)dt = 0. \qquad (1.50)$$

*Then from the inequalities $f_+ \prec g_+$ and $f_- \prec g_-$ it follows that $f \prec g$.*

**Proof.** Let at first $f(t)$ and $g(t)$ are continuous on $[a, b]$. Now we must state that for difference $\delta(t) = \pi(g, t) - \pi(f, t)$ for all $t \in [0, b-a]$ the inequality

$$\int_0^t \delta(u)du \geq 0 \qquad (1.51)$$

holds true.

It's enough to prove this fact only for points $t \in (0, b-a)$ such that $\delta(t) = 0$. Let $t$ is such point. If $\pi(g, t) = \pi(f, t) \geq 0$ then (1.51) follows from the inequality $f_+ \prec g_+$, if $\pi(g, t) = \pi(f, t) < 0$ then owing to (1.50),

$$\int_0^t \delta(u)du = -\int_t^{b-a} \delta(u)du = \int_0^{b-a-t} \left(r(g_-, u) - r(f_-, u)\right) du \geq 0$$

as $f_- \prec g_-$.

To prove proposition 1.3.7 in common case, we should account that set of continuous on $[a, b]$ functions is dense everywhere in $L_{1[a,b]}$.

**Proposition 1.3.8.** *Let $f, g \in L_{1[a,b]}$, the equalities (1.50) are executed and $f \prec g$. Then for any $\lambda \in R$*

$$(f - \lambda)_\pm \prec (g - \lambda)_\pm. \qquad (1.52)$$

**Proof.** Let us fix $\lambda \in R$. Assuming that

$$\xi(f) = mes\{t: \ t \in [a,b], f(t) > \lambda\},$$

we can write

$$\pi(f,t) - \lambda = r((f-\lambda)_+, t), \ 0 \le t \le \xi(f),$$

and analogous inequality for $g(t)$. If $\xi(f) \le \xi(g)$, then due to inequality $f \prec g$ for $0 \le t \le \xi(f)$

$$\int_0^t r((f-\lambda, u)_+)du \le \int_0^t r((g-\lambda, u)_+)du \tag{1.53}$$

and so far as for $t \in (\xi(f), b-a)$ $r((f-\lambda, t)_+) = 0$ but $r((g-\lambda)_+, t) \ge 0$ then inequality (1.53) is true for all $t \in [0, b-a]$. If $\xi(g) < \xi(f)$ then for $t \in [0, \xi(f)]$

$$\pi(g,t) - \lambda = \pi(g-\lambda, t) \le r((g-\lambda, t)_+, t)$$

and, taking into account, that $f \prec g$ we obtain for $t \in [0, \xi(f)]$

$$\int_0^t r((f-\lambda)_+, u)du = \int_0^t \pi(f,u)du - \lambda t \le \int_0^t \pi(g,u)du - \lambda t =$$

$$\int_0^t (\pi(g,u) - \lambda)\,du = \int_0^t r((g-\lambda, u)_+)du.$$

It remains to note, that for $t \in [\xi(f), b-a]$

$$r((f-\lambda)_+, t) = r((g-\lambda)_+, t) = 0.$$

To put the second inequality in (1.52) note that $f \prec g$ signifies also that

$$\int_0^{b-a-t} \pi(f,u)du \le \int_0^{b-a-t} \pi(g,u)du, \ 0 \le t \le b-a,$$

and besides that $\pi(-f,t) = -\pi(f, b-a-t)$. Out of (1.50) we find

$$\int_0^t \pi(-f,u)du = \int_0^{b-a-t} \pi(f,u)du \le \int_0^{b-a-t} \pi(g,u)du =$$

$$\int_0^t \pi(-g,u)du$$

that is $-f \prec -g$ and this fact together with first inequality in (1.52) gives us second one.

Further we will need in following lemma.

**Lemma 1.3.9.** *Let $\psi(t)$ and $\mu(t)$ are summable on $[a,b]$ functions, moreover $\mu(t) \geq 0$, $a \leq t \leq b$ and $\mu(t)$ doesn't increase. If*

$$\int_a^t \psi(u)du \geq 0, \ a \leq t \leq b,$$

*and product $\psi(t)\mu(t)$ is summable on $[a,b]$, then*

$$\int_a^b \psi(u)\mu(t)du \geq 0.$$

Really if $\mu(t)$ is restricted, then all follows out of theorem about mean value

$$\int_a^b \psi(u)\mu(t)du = \mu(a+0)\int_a^\xi \psi(u)du \geq 0, \ a < \xi < b.$$

In common case we construct the consequence $\mu_N(t) = \min\{\mu(t), N\}$ and use theorem of limited transition under the sign of Lebeg's integral.

**Proposition 1.3.10.** *Let $f, g \in L_{q[a,b]}$, $1 \leq q \leq \infty$ and $|f| \prec |g|$. Then*

$$\int_a^b |f(t)|^q dt \leq \int_a^b |g(t)|^q dt, \ 1 \leq q < \infty. \tag{1.54}$$

*If, besides that, $f(t)$ and $g(t)$ are essentielly restricted on $[a,b]$, then*

$$vrai \sup_{a \leq t \leq b} |f(t)| \leq vrai \sup_{a \leq t \leq b} |g(t)|. \tag{1.55}$$

**Proof.** In view of (1.41) $r(|f|), r(|g|) \in L_{q[a,b]}$. Puting in lemma 1.3.9

$$\psi(t) = r(|g|,t) - r(|f|+,t)$$

and

$$\mu(t) = r^{q-1}(|f|,t)$$

and taking into account the inequality $|f| \prec |g|$ we obtain

$$\int_0^{b-a} r^q(|f|,t)dt = \int_0^{b-a} r^{q-1}(|f|,t)r(|f|,t)dt$$

$$\leq \int_0^{b-a} r^{q-1}(|f|,t)r(|g|,t)dt,$$

and owing to Gelder's inequality

$$\int_0^{b-a} r^q(|f|,t)dt \leq \left(\int_0^{b-a} r^q(|f|,t)dt\right)^{1-1/q} \left(\int_0^{b-a} r^q(|g|,t)dt\right)^{1/q}.$$

Hence it immediately follows that

$$\int_0^{b-a} r^q(|f|,t)dt \leq \int_0^{b-a} r^q(|g|,t)dt,$$

and to obtain (1.54) we ought to account (1.41). Correlation (1.55) is obvious.

*Remined, that $\Phi(t)$ is $N$-function, if $\Phi(t)$ is continuous, convex from above, nonnegative on semiaxis $[0,\infty)$ function and $\Phi(0) = 0$.*

*Also settle that $\Phi(t)$ is $N^*$-function if $\Phi(t)$ is continuously differetiable and strictly convex from above $N$-function.*

Function $\Phi(t) = t^q$ $(q > 1)$ is typical example of $N$-function, the function

$$\Phi_\alpha(t) = \begin{cases} 0, & t \in [0,\alpha], \\ t-\alpha, & t > \alpha. \end{cases} \tag{1.56}$$

is typical one too.

Now consider following statement that will play further the essential role.

**Theorem 1.3.11.** *If for summable on $[a,b]$ functions $f(t)$ and $g(t)$ the inequality*

$$|f| \prec |g|, \tag{1.57}$$

*is valid, then for any $N$-function $\Phi(t)$ the inequality*

$$\int_a^b \Phi(|f(t)|)dt \leq \int_a^b \Phi(|g(t)|)dt, \tag{1.58}$$

*holds true, and vice versa, if for any n-function $\Phi(t)$ inequality (1.58) holds true, then inequality (1.57) is valid too.*

Note, that out of theorem 1.3.11 and proposition 1.3.6 it follows, that if for summable on $[a,b]$ functions $f(t)$ and $g(t)$ the inequalities

$$f_+ \prec g_+, \quad f_- \prec g_-,$$

hold true, then for any $N$-function $\Phi(t)$ inequality (1.58) is true, and consequently, for any $q \in [1,\infty)$ the inequalities (1.54) and (1.55) are valid.

**Proof.** Let first $f(t)$ and $g(t)$ are continuous on $[a, b]$ functions vanished at least into ont point out of $[a, b]$, and let condition (1.57) is executed.

Then

$$r(|f|, 0) \leq r(|g|, 0) \tag{1.59}$$

and

$$r(|f|, b-a) = r(|g|, b-a). \tag{1.60}$$

For avery $x \in [0, b-a]$ define $y = y(x)$ by equality

$$y = mes\{t : \ t \in [0, b-a], \ r(|g|, t) \geq r(|f|, x)\}$$

and prove, that

$$\int_0^x (r(|f|, t) - r(|f|, x))\, dt \leq \int_0^x (r(|g|, t) - r(|g|, x))\, dt. \tag{1.61}$$

Note, befor all, that due to (1.59)

$$r(|f|, x) = r(|g|, y)$$

and if $x \leq y$, then

$$\int_0^x (r(|f|, t) - r(|f|, x))\, dt \leq \int_0^x (r(|g|, t) - r(|f|, x))\, dt =$$

$$\int_0^x (r(|g|, t) - r(|g|, y))\, dt \leq \int_0^y (r(|g|, t) - r(|g|, y))\, dt.$$

Besides that, from (1.60) it follows, that $y(b-a) = b-a$ and, consequently, inequality (1.61) holds true for $x = b-a$. Thus, if inequality (1.61) is not valid for some $x_0$, then

$$x_0 > y_0 = y(x_0), \tag{1.62}$$

and

$$\int_0^{x_0} (r(|f|, t) - r(|f|, x_0))\, dt > \int_0^{y_0} (r(|g|, t) - r(|g|, y_0))\, dt \tag{1.63}$$

and there exissts apoint $x_* \in (x_0, b-a)$ such that

$$r(|f|, x_*) = r(|g|, x_*), \ r(|f|, t) > r(|g|, t), \ t \in (y, x_*). \tag{1.64}$$

Thruough the condition (1.57) and equality (1.64) it follows, thet

$$\int_0^{x_*} (r(|f|, t) - r(|f|, x_*))\, dt \leq \int_0^{x_*} (r(|g|, t) - r(|g|, x_*))\, dt.$$

But due to (1.62)-(1.64)

$$\int_0^{x_*} (r(|g|,t) - r(|g|,x_*))\,dt = \int_0^{y_0} (r(|g|,t) - r(|g|,y_0))\,dt +$$

$$\int_{y_0}^{x_0} (r(|g|,t) - r(|g|,x_*))\,dt + y_0\,(r(|g|,y_0) - r(|g|,x_*)) +$$

$$\int_{x_0}^{x_*} (r(|g|,t) - r(|g|,x_*))\,dt < \int_0^{x_0} (r(|g|,t) - r(|g|,x_*))\,dt +$$

$$y_0\,(r(|g|,y_0) - r(|g|,x_*)) + (x_0 - y_0)\,(r(|g|,y_0) - r(|g|,x_*)) +$$

$$\int_{x_0}^{x_*} (r(|f|,t) - r(|g|,x_*))\,dt = \int_0^{x_0} (r(|f|,t) - r(|f|,x_0))\,dt +$$

$$y_0\,(r(|f|,x_0) - r(|f|,x_*)) + (x_0 - y_0)\,(r(|f|,x_0) - r(|f|,x_*)) +$$

$$\int_{x_0}^{x_*} (r(|f|,t) - r(|f|,x_*))\,dt = \int_0^{x_*} (r(|f|,t) - r(|f|,x_*))\,dt,$$

that contradicts to orevious equality.

Thus, if condition (1.57) is executed, then inequality (1.61) is valid for any $y \in [0, b-a]$.

Believing that $\alpha = r(|f|, x)$, define function $\Phi_\alpha(t)$ with the help of equality (1.56). Then in view of (1.42) and (1.61)

$$\int_a^b \Phi_\alpha(|f(t)|)dt = \int_a^{b-a} \Phi_\alpha(r(|f|,t)dt = \int_0^x (r(|f|,t) - r(|f|,x))\,dt \leq$$

$$\int_0^y (r(|g|,t) - r(|g|,y))\,dt = \int_a^{b-a} \Phi_\alpha(r(|g|,t)dt =$$

$$\int_a^b \Phi_\alpha(|g(t)|)dt.$$

Thus, if functions $f(t)$ and $g(t)$ are continuous, have zeros on $[a, b]$, $r(|f|, t)$ strictly decreases on $[0, b-a]$ and condition (1.61) is axecuted, then for any $\alpha \in [0, b-a]$ the following inequality

$$\int_a^b \Phi_\alpha(|f(t)|)dt \leq \int_a^b \Phi_\alpha(|g(t)|)dt. \tag{1.65}$$

holds true.

For any summable on $[a,b]$ functions $f(t), g(t)$ and out of arbitrariness $\varepsilon > 0$ we can select continuous and having zeros at $[a,b]$ functions $f_\varepsilon$ and $g_\varepsilon$ so, that inequalitiaes

$$\int_a^b |f(t) - f_\varepsilon(t)|dt \leq \varepsilon, \int_a^b |g(t) - g_\varepsilon(t)|dt \leq \varepsilon,$$

will be true and rearrangement $r(|f_\varepsilon|, t)$ would be strictly monotonous. But then

$$\int_a^b \Phi_\alpha(|f(t)|)dt \leq \varepsilon + \int_a^b \Phi_\alpha(|f_\varepsilon(t)|)dt \leq$$

$$\varepsilon + \int_a^b \Phi_\alpha(|g_\varepsilon(t)|)dt \leq 2\varepsilon + \int_a^b \Phi_\alpha(|g(t)|)dt.$$

¿From this fact and out of arbitriness $\varepsilon > 0$ it follows, that for any summable on $[a,b]$ functions $f(t)$ and $g(t)$ for which the condition (1.57) is executed, the inequality (1.65) holds true for all $\alpha \in [0, b-a]$.

To complete the proof for the first part of theorem it remains to show that for any $\varepsilon > 0, M > 0$ and $N$-functuon $\Phi(t)$ here exist points $\alpha_i$, $i = 0, 1, \ldots, n$; $0 = \alpha_0 < \alpha_1 < \ldots < \alpha_n = M$ and numbers $\beta_i \geq 0$, $i = 1, 2, \ldots, n$ such that

$$||\Phi - \sum_{i=1}^{n} \beta_i \Phi_{\alpha_i}||_{C[0,M]} \leq \varepsilon. \tag{1.66}$$

Denote by $l_n(\Phi, t)$ the broken line with the knots at points $\alpha_i = iM/n$, $i = 0, 1, \ldots, n$ interpolating function $\Phi(t)$ in this knots. If $\Phi(t)$ is $N$-function, then $l_n(\Phi, t)$ is also $N$-function on $[0, M]$. This means, that if $l_n(\Phi, t) = A_i t + B_i$ for $t \in [(i-1)M/n, iM/n)$, then

$$A_0 = 0 \leq A_1 \leq \ldots \leq A_n.$$

and believing $\beta_i = A_i - A_{i-1}$, $i = 1, 2, \ldots, n$, we obtain

$$l_n(\Psi, t) = \sum_{i=1}^{n} \beta_i \Phi_{\alpha_i}(t)$$

and

$$||\Phi - l_n(\Phi)||_{C[0,M]} \leq \omega(\Phi, M/n),$$

here $\omega(\Phi, \delta)$ is modulus of continuity for function $\Phi$. Choosing $n$ such that $\omega(\Phi, M/n) \leq \varepsilon$ we'll obtain the inequality (1.66), and first part of theorem will be valid.

Let now the inequality (1.58) holds true for any $N$-function $\Phi(t)$. Then, thinking that $\Phi(t) = t$, we obtain

$$\int_0^{b-a} r(|f|,t)dt = \int_a^b |f(t)|dt \leq \int_a^b |g(t)|dt = \int_0^{b-a} r(|g|,t)dt$$

Therefore, the assumption that inequality (1.57) is not true, means that if

$$\delta(x) = \int_0^x (r(|g|,t) - r(|f|,t))\, dt$$

then

$$\min\{\delta(x) : \ x \in [0, b-a]\} = \delta(x_0) < 0.$$

But so far as $\delta(0) = 0$ and $\delta(b-a) \geq 0$ then $\delta'(x_0) = 0$.

Thus

$$\int_0^{x_0} r(|f|,t)dt > \int_0^{x_0} r(|g|,t)dt$$

and

$$r(|g|, x_0) = r(|f|, x_0).$$

Besides that, if $\alpha = r(|f|, x_0)$ and $N$-function $\Phi_\alpha(t)$ is defined by equality (1.56) then

$$\int_0^{x_0} r(|f|,t)dt = \int_0^{x_0} (r(|f|,t) - r(|f|,x_0))\, dt + x_0 r(|f|,x_0) =$$

$$\int_0^{b-a} \Phi_\alpha(r(|f|),t)dt + x_0 r(|g|,x_0) \leq \int_0^{b-a} \Phi_\alpha(|f(t)|)dt +$$

$$x_0 r(|g|,x_0) \leq \int_0^{b-a} \Phi_\alpha(|g(t)|)dt + x_0 r(|g|,x_0) = \int_0^{x_0} r(|g|,t)dt$$

that contradicts to previous facts.

¿From the theorem 1.3.11 and equality (1.41) (when $p = 1$) one can easy obtain the following statement.

**Corrolary 1.3.12.** *If $f, g \in L_{1[a,b]}$, $a < c < b$ then from inequalities $(|f| \prec |g|)_{[a,c]}$ and $(|f| \prec |g|)_{[c,b]}$ it follows that $(|f| \prec |g|)_{[a,b]}$.*

¿From the comparing of propositions 1.3.7 and 1.3.8 and theorem 1.3.11 it's easy to formulate the following statement.

**Theorem 1.3.13.** *Let for summable on $[a,b]$ functions $f(t)$ and $g(t)$ such that their mean values are equal to zero on $[a,b]$, the inequalities $f_+ \prec g_+$ and $f_- \prec g_-$ hold true. Then for any $\lambda \in R$*

$$(f-\lambda)_\pm \prec (g-\lambda)_\pm, \ (f-\lambda) \prec (g-\lambda).$$

*and if $f, g \in Lp[a,b]$ then*

$$\|f-\lambda\|_p \leq \|d-\lambda\|_p.$$

Call the rearrangement for narrowing of $2\pi$-periodic function $f(t)$ on period $[0,2\pi]$ as its decreasing rearrangement. It's clear that rearrangement of $2\pi$-periodic function is invariant concerning to the shift of argument, that is for any $\alpha \in R$ and $t \in [0.2\pi)$ $\pi(f(\cdot+\alpha),t) \equiv \pi(f(\cdot),t)$.

Denote by $L_p$, $1 \leq p \leq \infty$ the space of all $2\pi$-periodic functions $f(t)$ with finite norm $\|f\| = \|f\|_{L_p[0,2\pi]}$.

**Proposition 1.3.14.** *Let be continuous $2\pi$-periodic functions $f(t)$ and $F(t)$ have mean values that are equal to zero on period and such that for all $\lambda \in R$*

$$(f-\lambda)_\pm \prec (F-\lambda)_\pm \tag{1.67}$$

*Then for any function $g \in L_1$*

$$\int_0^{2\pi} \pi(f,t)\pi(g,t)dt \leq \int_0^{2\pi} \pi(F,t)\pi(g,t)dt$$

**Proof.** It's obvious, that we can consider $g \perp 1$. Choose now $\lambda$ from the conditions

$$mes\{t: \ t \in [0,2\pi], f(t) > \lambda\} \leq a := mes|\{t: \ t \in [0,2\pi], g(t) \geq 0\};$$

$$mes\{t: \ t \in [0,2\pi], \ F(t) < \lambda\} \leq 2\pi - a,$$

and put

$$b = mes\{t: \ t \in [0,2\pi], f(t) > \lambda\}.$$

Let $a \geq b$ (the case, when $a < b$, is considered by analogy). Then

$$\int_0^{2\pi} \pi(f,t)\pi(g,t)dt = \left(\int_0^b + \int_b^a + \int_a^{2\pi}\right) \pi(f-\lambda,t)\pi(g,t)dt$$

and second item at latter sum $\leq 0$.

That is why

$$\int_0^{2\pi} \pi(f,t)\pi(g,t)dt \leq \left(\int_0^b + \int_a^{2\pi}\right) \pi(f-\lambda,t)\pi(g,t)dt \leq$$

$$\int_0^b r(g_+,t)r((f-\lambda)_+,t)dt + \int_0^{2\pi-a} r(g_-,t)r((f-\lambda)_-,t)dt.$$

Owing to (1.67) and lemma 1.3.9 we obtain

$$\int_0^b r(g_+,t)r((f-\lambda)_+,t)dt \leq \int_0^b r(g_+,t)r((F-\lambda)_+,t)dt =$$

$$\int_0^b \pi(g,t)\pi((F-\lambda),t)dt;$$

$$\int_0^{2\pi-a} r(g_-,t)r((f-\lambda)_-,t)dt \leq \int_0^{2\pi-a} r(g_-,t)r((F-\lambda)_-,t)dt =$$

$$\int_a^{2\pi} \pi(g,t)\pi((F-\lambda),t)dt.$$

Therefore

$$\int_0^{2\pi} \pi(f,t)\pi(g,t)dt \leq \left(\int_0^b + \int_a^{2\pi}\right) \pi(g,t)\pi((F-\lambda),t)dt.$$

Here the right part in consequesce of choice $\lambda$ is not more

$$\int_0^{2\pi} \pi(F-\lambda,t)\pi(g,t)dt = \int_0^{2\pi} \pi(F,t)\pi(g,t)dt.$$

## 1.4 Properties of rearrangements

Main theorems of this paragraph are directed to the clarification of the extremal properties of the broad functional classes mainly in the terms of rearrangements. As a rule, the properties are described by the strict inequalities between the characteristics of the arbitrary function and certain standard function of comparison.

Function $\phi(t)$ of the space $C(-\infty,\infty)$ will be called regular if it has a certain period $2l$ and at the interval $(a, a+2l)$ where $a$ is the point of absolute extremum of the function $\phi(t)$ there exists point $c$ such that $\phi(t)$ is strictly monotone at each of the intervals $(a,c)$ and $(c, a+2l)$. If it is necessary to emphasize the length $2l$ of the period we will call $\phi(t)$ a $2l$-regular function.

*We will say that function $f(t)$ of $C(-\infty,\infty)$ possesses the $\mu$-property with respect to the regular function $\phi(t)$ if for any $\alpha \in R$ difference $\phi(t) - f(t+\alpha)$ at each interval of monotony of the function $\phi(t)$ either doesn't change the sing or changes it only once, and what is more, the sing changes from plus to minus when $\phi(t)$ decreases, and from minus to plus when $\phi(t)$ increases.*

It is clear that if function $f(t)$ possesses the $\mu$-property with respect to $\phi(t)$, then $f(t)$ possesses the $\mu$-property with respect to the function $\phi(t+\beta)$ for any $\beta$.

Directly from the definitions may be drawn the following lemma on which one way or another will be based the main results of this and some other paragraphs.

**Lemma 1.4.1.** *Let function $f(t)$ from $C(-\infty,\infty)$ possesses the $\mu$-property respectively regular function $\phi(t)$. Moreover let for all $t$*

$$\min_t \phi(t) := m \le f(t) \le M := \max_t \phi(t). \tag{1.68}$$

*If on segment $[\alpha,\beta]$ the function $\phi(t)$ is monotonous and $f(\xi) = \phi(\eta)$, where $\alpha \le \eta \le \beta$, then in the case of increasing $\phi(t)$ on $[\alpha,\beta]$*

$$f(\xi+u) \le \phi(\eta+u), \ \ 0 \le u \le \beta - \eta \tag{1.69}$$

*and*

$$f(\xi-u) \ge \phi(\eta-u), \ \ 0 \le u \le \eta - \alpha, \tag{1.70}$$

*and in the case of decreasing of $\phi(t)$ on $[\alpha,\beta]$ the signs in (1.69) and (1.70) are opposite.*

**Proof.** Taking into account for clearness that $\phi(t)$ is increasing on $[\alpha,\beta]$ and $\eta \in [\alpha,\beta]$ suppose that inequality (1.69) is not true. Then for some $u' \in [0,\beta-\eta]$ it will be $f(\xi+u') > \phi(\eta+u')$. As so $f(t) < M$ then on composing $[\alpha,\beta]$ monotony interval of $\phi(t)$ there exists the point $\eta_1 = \eta + u' + \delta$, $\delta > 0$ such that $\phi(\eta_1) = f(\xi+u')$.

Put $\xi_1 = \eta + \delta/2$ and $\xi_2 = \eta_1 - \delta/2$. For the fuction $f_1(t) = f(t+\xi-\xi_1)$ the next equations

$$f_1(\xi_1) = f(\xi) = \phi(\eta), \ \ f_1(\xi_2) = f(\xi+u') = \phi(\eta_1)$$

are valid and as $\phi(t)$ strictly increases on $[\eta,\eta_1]$ and $\eta < \xi_1 < \xi_2 < \eta_1$, then

$$f_1(\xi_1) < \phi(\xi_1), \ \ f_1(\xi_2) > \phi(\xi_2),$$

i.e. on increase segment of $\phi(t)$ the difference $\phi(t) - f(t + \xi - \xi_1)$ changes sign from plus to minus. But this is impossible owing to $\mu$-property.The inequality (1.69) is proved, the other variants of lemma are established analogously.

¿From lemma 1.4.1 we can infer the row of other results about comparative conduct either function $f(t)$ itself or its derivative correspondingly with $\phi(t)$ and $\phi'(t)$.

**Proposition 1.4.2.** *Let $f(t)$ and $\phi(t)$ are locally absolute continuous on $(-\infty, \infty)$, $\phi(t)$ is regular function and $f(t)$ possesses the $\mu$-property with respect to $\phi(t)$,moreover the inequality (1.68) is true. Then,if $f(\xi) = \phi(\eta)$ (in assumption that $f'(\xi)$ and $\phi'(\eta)$ exist) and $f'(\xi)\phi'(\eta) \geq 0$, then*

$$|f'(\xi)| \leq |\phi'(\eta)|, \tag{1.71}$$

*and, hence, almost everywhere on $(-\infty, \infty)$ the inequalities*

$$vrai \inf_t \phi'(t) \leq f'(t) \leq vrai \sup_t \phi'(t) \tag{1.72}$$

*hold true.*

Really, taking into account for clearness that $f'(\xi) \geq 0$ and $\phi'(\eta) \geq 0$ due to (1.69) and (1.70) for sufficiently small by absolute value $h$ we obtain

$$\frac{f(\xi + h) - f(\xi)}{h} \leq \frac{\phi(\eta + h) - \phi(\eta)}{h},$$

thus $f'(\xi) \leq \phi'(\eta)$. Analogously one can consider the case when $f'(\xi) \leq 0,\ \phi'(\eta) \leq 0$.

**Proposition 1.4.3.** *Let conditions of lemma 1.4.1. are valid. Then if $f(\xi) = \phi(\eta), f(\xi_1) = \phi(\eta_1)$, while $(\xi_1 - \xi)(\eta_1 - \eta) > 0$ and points $\eta$ and $\eta_1$ are on the same segment of monotony for $\phi(t)$ then*

$$|\xi_1 - \xi| \geq |\eta_1 - \eta|.$$

Taking into consideration for clearness that $\xi_1 > \xi,\ \ \eta_1 > \eta$ and $f(\xi_1) > f(\xi)$(the other variants studied by analogy) we admit that $\xi_1 - \xi < \eta_1 - \eta$. Then for $u = \xi_1 - \xi$ we get

$$\phi(\eta + u) < \phi(\eta_1) = f(\xi + u),$$

that contradicts to (1.69).

**Proposition 1.4.4.** *Let conditions of lemma 1.4.1 are valid, $\phi(t)$ is $2\pi/n$-regular function and*

$$\phi(a) = \min\{\phi(t) : t\}, \ \ \phi(c) = \max\{\phi(t) : t\}, \ \ b = a + 2\pi/n.$$

*Then for each $h \in (0, \min\{b-c, c-a\})$*

$$||f(\cdot + h) - f(\cdot)||_C \leq ||\phi(\cdot + h) - \phi(\cdot)||_C. \tag{1.73}$$

**Proof.** Choose the point $t_*$ from the condition

$$||f(\cdot + h) - f(\cdot)||_C = |f(t_* + h) - f(t_*)|,$$

and let for clearness $f(t_* + h) - f(t_*) > 0$. From the condition (1.68) it follows that there exists the point $u_* \in [a, c]$ such that $f(t_*+h) = \phi(u_*)$.But then owing to lemma 1.4.1 for all $\xi \in [0, u_*-a]$ the inequality $f(t_*+h-\xi) > \phi(u_* - \xi)$ holds true and if $h < u_* - a$ then for $\xi = h$ we immediately obtain

$$f(t_* + h) - f(t_*) \leq \phi(u_*) - \phi(u_* - h) \leq ||\phi(\cdot + h) - \phi(\cdot)||_C.$$

If $h > u_* - a$ then

$$f(t_* + h) - f(t_*) \leq f(t_* + h) - \phi(a) = \phi(u_*) - \phi(a) \leq$$

$$||\phi(\cdot + h) - \phi(\cdot)||_C.$$

**Proposinion 1.4.5.** *Let $f, \phi \in C^1(-\infty, \infty)$ and conditions of lemma 1.4.1 have been valid for this functions. Then for each $\lambda \in R$*

$$\min_t (\phi'(t) + \lambda\phi(t)) \leq f'(t) + \lambda f(t) \leq \max_t (\phi'(t) + \lambda\phi(t))$$

*and*

$$||f' + \lambda f|| \leq ||\phi' + \lambda\phi||.$$

**Proof.** For any $t \in [0, 2\pi]$ there exists the point $u \in [0, 2\pi]$ such that $f(t) = \phi(u)$ and $f'(t)\phi'(u) \geq 0$. Let for clearness $\phi'(u) \geq 0$ and $f'(t) \geq 0$.In view of proposition 1.4.2 $f'(t) \leq \phi'(u)$ and therefore

$$f'(t) + \lambda f(t) \leq \phi'(u) + \lambda\phi(u) \leq \max_t (\phi'(t) + \lambda\phi(t))$$

Besides that, if

$$\phi(t_*) = \min\{\phi(t) : t \in [0, 2\pi]\}, \phi(t_{**}) = \max\{\phi(t) : t \in [0, 2\pi]\},$$

then for $\lambda > 0$

$$f'(t) + \lambda f(t) \geq \lambda f(t_*) = \phi'(t_*) + \lambda\phi(t_*) \geq$$

$$\min\{\phi'(t) + \lambda\phi(t) : \ t\},$$

and for $\lambda < 0$

$$f'(t) + \lambda f(t) \geq \lambda f(t_{**}) = \phi'(t_{**}) + \lambda\phi(t_{**}) \geq$$

$$\min\{\phi'(t) + \lambda\phi(t) : \ t\}.$$

In the case when $f'(t) \leq 0$, $\phi'(u) \leq 0$ the reasoning is analogous.

Further we'll denote by $C$ and $C^m$, $m = 1, 2, \ldots$ the set of continuous and defined on all real axis, accordingly $m$ times continuously differentiable $2\pi$-periodic functions.

Formulate now the main comparison theorem for rearrangements.

**Theorem 1.4.6.** *Let $f, g \in C^1$, $g(t)$ and $g'(t)$ are $2\pi/n$-regular functions such that the functions $f(t)$ and $f'(t)$ possess the $\mu$-property with respect to $g(t)$ and $g'(t)$ correspondingly and*

$$\min\{g(t) : t\} \leq f(t) \leq \max\{g(t) : t\}. \tag{1.74}$$

*Then for any $\lambda \in R$ the next inequalities*

$$(f' - \lambda f)_\pm \prec (g' - \lambda g)_\pm \tag{1.75}$$

*hold true.*

**Proof.** As so the derivative of periodic function has zeros on period and the rearrangements of its positive and negative parts are translate invariant, then we can transit to changes of functions $f(t)$ and $g(t)$ by argument and consider that $f'(0) = g'(0) = 0$.

Let $[0, \eta_+)$ and $[0, \eta_-)$ are most semiintervals for which

$$r(f'_+, t) > 0, \quad r(f'_-, t) > 0;$$

$[0, \eta_+^*)$, $[0, \eta_-^*)$ are semiintervals correspondingly for $g'_+(t)$ and $g'_-(t)$. As so $\eta_+ + \eta_- \leq 2\pi = \eta_+^* + \eta_-^*$, then at least one from the inequalities $\eta_+ \leq \eta_+^*$ or $\eta_- \leq \eta_-^*$ holds true.

Taking into account for clearness that $\eta_+ \leq \eta_+^*$ we prove that

$$f'_+ \prec g'_+, \quad (f'_+ := (f')_+). \tag{1.76}$$

Put

$$\delta(t) = r(g'_+, t) - r(f'_-, t)$$

and note that justice of inequality

$$\int_0^t \delta(u)du \geq 0$$

it is sufficiently to prove only for $t \in (0, 2\pi)$ for which $\delta(t) = 0, \;\; \delta(u) < 0$ for $u \in (t - \varepsilon, t)$ for some $\varepsilon > 0$. Let $t = \xi$ is namely such point and

$$z = r(f'_+, \xi) = r(g'_+, \xi). \tag{1.77}$$

If for $y \geq 0$

$$E_y = \{t : \;\; t \in (0, 2\pi], \;\; f'_+(t) > 0\},$$

then $mes\, E_z = \xi$ and

$$\int_0^\xi r(f'_+, t)dt = \int_{E_z} f'_+(t)dt = \int_{E_z} f'(t)dt.$$

For composing intervals $(a_k, b_k) \subset (0, 2\pi)$ of the set $E_z$ the next correlations

$$f'(a_k) = f'(b_k) = z; \tag{1.78}$$

$$f'(t) > z, \;\; a_k < t < b_k \tag{1.79}$$

hold true. Prove that number of such intervals is no more than $n$.

Suppose that in the set $E_z$ there exist $n+1$ composing intervals $(a_k, b_k)$, $k = 1, 2, \ldots, n$. Select the number $h > 0$ so small that:

1) set $E_{z+h}$ has nonempty intersection with each of $n + 1$ intervals $(a_k, b_k)$;

2) if

$$r(f'_+, \xi - \gamma) = z + h$$

then $\delta(t) < 0$ for $\xi - \gamma \leq t \leq \xi$.

¿From condition 2) and strictly monotony of rearrangement $r(g'_+, t)$ it follows that if

$$r(g'_+, \xi - \gamma_0) = z + h, \tag{1.80}$$

then $\gamma > \gamma_0$.

Moreover, let $a'_k$ and $b'_k$ are the points from intervals $(a_k, b_k)$ defined by correlations

$$f'(a'_k) = f'(b'_k) = z + h; \tag{1.81}$$

$$z < f'(t) < z + h, \;\; t \in (a_k, a'_k) \cup (b'_k, b_k).$$

It is clear that

$$\gamma = mes \; E_z - mes \; E_{z-h} = \sum_{k=1}^{n+1} (a'_k - a_k + b_k - b'_k). \tag{1.82}$$

The graph of function $g'_+(t)$ (as well as the graph of function $g'_-(t)$) on $[0, 2\pi]$ consists from $n$ identical "caps". Take one from their with the bearer $[\alpha, \beta]$, i.e. $g'(\alpha) = g'(\beta) = 0$, $g'(t)$ strictly increases on $(\alpha, c)$, $\alpha < c < b$ and strictly decreases on $(c, \beta)$. There exist the points $\eta_1, \eta_2, \eta', \eta''$ such that

$$\alpha < \eta_1 < \eta' < c < \eta'' < \eta_2 < \beta; \tag{1.83}$$

$$g'(\eta_1) = g'(\eta_2) = z; \;\; g'(\eta') = g'(\eta'') = z + h. \tag{1.84}$$

Passing to rearrangement $r(g'_+, t)$ and taking into account (1.78), (1.80) and (1.84) it is easy to understand that

$$\gamma_0 = n(\eta' - \eta_1 + \eta_2 - \eta''). \tag{1.85}$$

Compare the right parts (1.82) and (1.85). Owing to (1.78), (1.81) and (1.84) the functions $f'(t)$ and $g'(t)$ are in conditions of proposition 1.4.3, in consequence of which $a'_k - a_k \geq \eta' - \eta_1$ and $b_k - b'_k \geq \eta_2 - \eta'_1$ and therefore

$$\gamma \geq (n+1)(\eta' - \eta_1 + \eta_2 - \eta'') > \gamma_0,$$

but that contradicts to previous.

Thus, we established that on $(0, 2\pi)$ there would be only finite number $l$, $l \leq n$ of intervals $(a_k, b_k)$ which satisfy the conditions (1.78) and (1.79). If $(c_i, di)$, $i = 1.2, \ldots, m$ are so composing intervals of the set

$$\{t : \; t \in (0, 2\pi), \;\; f'(t) > 0\}$$

which contain at least one interval $(a_k, b_k)$ then $m \leq l \leq n$. Fix the interval $(c_i, d_i)$ and let $a_\nu$ and $b_\mu$ are correspondingly first from left and last from right of the ends of intervals $(a_k, b_k)$ in $(c_i, d_i)$. Then considering again the "cap" of function $g'_+(t)$ on $(\alpha, \beta)$ and points (1.83), we find that

$$f'(c_i) = g'(\alpha) = 0, \;\; f'(a_\nu) = g'(\eta_1) = z,$$

$$f'(d_i) = g'(\beta) = 0, \;\; f'(b_\mu) = g'(\eta_2) = z,$$

moreover the function $g(t)$ increases on $(\eta_1)$ and decreases on $(t_2, \beta)$.

In view of lemma 1.4.1

$$f'(a_\nu - u) \geq g'(\eta_1 - u), \;\; 0 \leq u \leq \eta_1 - \alpha,$$

and

$$f'(b_\mu + u) \geq g'(t_2 + u), \;\; 0 \leq u \leq \beta - \eta_2,$$

therefore

$$\int_{(c_i,d_i)\setminus E_z} f'(t)dt \geq \int_{c_i}^{a_\nu} f'(t)dt + \int_{\beta_\mu}^{d_i} f'(t)dt \geq$$

$$\int_a^{\eta_1} g'(t)dt + \int_{\eta_2}^{\beta} g'(t)dt = g(\eta_1) - g(\eta_2) + g(\beta) - g(\alpha).$$

Note now that owing to condition of theorem

$$\int_{c_i}^{d_i} f'(t)dt = f(d_i) - f(c_i) \leq \max_t f(t) - \min_t f(t) \leq$$

$$\max_t g(t) - \min_t g(t) = g(\beta) - g(\alpha)$$

we obtain

$$\int_0^{\xi} r(f'_+, t)dt = \int_{E_z} f'(t)dt = \sum_{k=1}^{l} \int_{a_k}^{b_k} f'(t)dt =$$

$$\sum_{i=1}^{m} \left( \int_{c_i}^{d_i} f'(t)dt - \int_{(c_i,d_i)\setminus E_z} f'(t)dt \right) \leq$$

$$m[g(\eta_2) - g(\eta_1)] = m \int_{\eta_1}^{\eta_2} g'(t)dt =$$

$$(m/n) \int_0^{\xi} r(g'_+, t)dt \leq \int_0^{\xi} r(g'_+, t)dt. \tag{1.86}$$

Inequality (1.76) is proved. The justice of inequality

$$\int_0^t r(f'_-, u)du \leq \int_0^t r(g'_-, u)du$$

for $t \in (0, \eta_1)$ is established quite analogously. If $t \in (\eta_1, \eta_2)$ then taking into account that $f' \perp 1$ for such $t$ we obtain

$$\int_0^t r(f'_-, u)du \le \int_0^{2\pi} f'_-(u)du = \int_0^{2\pi} f'_+(u)du = \int_0^{2\pi} r(f'_+, u)du \le$$

$$\int_0^{2\pi} r(g'_+, u)du = \int_0^{2\pi} g'_+(u)du = \int_0^{2\pi} g'_-(u)du =$$

$$\int_0^{\eta_1} r(g', u)du = \int_0^t r(g'_-, u)du.$$

Thus we proved that $f'_\pm \prec g'_\pm$. From here due to theorem 1.3.5 it follows that (1.75) is true.

**Remark.** Observing the proof of theorem 1.4.6 (see (1.86)) it is easy to note that if functions $f(t)$ and $g(t)$ satisfy the conditions of this theorem and $f'(t)$ has at most $2m$ changes of sign on period then the most exact than (1.75) inequalities

$$(f' - \lambda)_\pm \prec \min\{m/n, 1\}(g' - \lambda)_\pm$$

hold true. Besides that in the case when $m \le n$ their proof get simplified, as it's not necessary to prove that the number of intervals $(a_k, b_k)$ is no more $n$.

Pass now to consideration of concrete classes of functions extremal properties.

Let $L_p^m$ is the set of all $2\pi$-periodic functions $f(t)$, for wich $(m-1)$-th derivative $f^{(m-1)}(t)$ is locally absolutely continuous on all axis and $f^{(m)} \in L_p$, and $W_p^m$ is the class of functions $f(t) \in L_p^m$, for which $\|f^{(m)}\|_p \le 1$. If we deal with $2\pi$-periodic functions then everywhere instead of $\|\cdot\|_{L_p[0.2\pi]}$ we'll write $\|\cdot\|_p$.

Introduce into consideration now functions $\varphi_{n,m}(t)$ which will play the role of standard functions in the comparison theorems. Put now for $n = 1, 2, \ldots$ and $m = 0, 1, \ldots$

$$\varphi_{n,m}(t) = \frac{4}{\pi n^m} \sum_{\nu=1}^{\infty} \frac{\sin[n(2\nu+1)t - \pi m/2]}{(2\nu+1)^{m+1}}. \tag{1.87}$$

Function $\varphi_{n,m}(t)$ is $m$-th periodic integral with zero mean value on period of function $\varphi_{n,0}(t) = sgn \sin nt$.

It is clear that $\varphi_{n,m} \in W_\infty^m$ has period equal to $2\pi/n$. Moreover

$$\varphi_{n,m}^\nu(t) = \varphi_{n,m-\nu}(t).$$

For even $m$ function $\varphi_{n,m}(t)$ is similar to $(-1)^{m/2}\sin nt$: has simple zeros at points $k\pi/n$, $k = 0, \pm 1. \pm 2, \ldots$, it graph is symmetric as to points $(k\pi/n, 0)$ and straight lines $t = (2k+1)\pi/(2n)$. For odd $m$ function $\varphi_{n,m}(t)$ is similar to $(-1)^{(m+1)/2}\cos nt$: simple zeros are at points $((2k+1)\pi/(2n))$, the symmetry as to points $(2k+1)\pi/2n, 0$ and direct lines $t = k\pi/n$. The function $\varphi_{n,m}(t)$ hasn't other zeros, when $m \geq 1$ it is strictly monotonous between the neighbouring points of extremum and keeps the convexity on each intervals between zeros.Thus function $\varphi_{n,m}(t)$ is $2\pi/n$-regular. Note that

$$\varphi_{n,m}(t) = n^{-m}\varphi_{1,m}(nt). \tag{1.88}$$

Furthermore under the decision of extremal problems there will often appear the constant values (Favard's constants)

$$K_m = \|\varphi_{n,m}\|_C = \frac{4}{\pi}\sum_{\nu=0}^{\infty}\frac{(-1)^{\nu(m+1)}}{(\nu+1)^{m+1}}, \quad m = 0, 1, \ldots. \tag{1.89}$$

Moreover

$$K_{2j} = |\varphi_{1,2j}(\pi/2)|, \quad K_{2j-1} = |\varphi_{1,2j-1}(0)|$$

One can simply check that

$$K_0 = 1, \quad K_1 = \frac{\pi}{2}, \quad K_2 = \frac{\pi^2}{8}, \quad K_3 = \frac{\pi^3}{24}, \quad K_4 = \frac{5\pi^4}{384}, \quad K_5 = \frac{\pi^5}{240}, \ldots$$

and

$$1 = K_0 < K_2 < K_4 < \ldots < 4/\pi < \ldots < K_3 < K_1 = \pi/2.$$

In view of (1.88)

$$\|\varphi_{n,m}\|_\infty = K_m n^{-m}, \quad m = 0, 1, \ldots. \tag{1.90}$$

Note that

$$\|\varphi_{n,m}\|_1 = \bigvee_0^{2\pi}(\varphi_{n,m+1}) = 4n\|\varphi_{n,m+1}\|_\infty = \frac{4K_{m+1}}{n^m}.$$

It is easy to calculate also that

$$\|\varphi_{n,m}\|_2 = \frac{\sqrt{4K_{2m+1}}}{n^m}.$$

**Theorem 1.4.7.** *For any function $f \in W_\infty^m$, $m = 1, 2, \dots$ and $\lambda \in R$ the next inequality*

$$|f' - \lambda| \prec \left( \frac{||f||_\infty}{||\varphi_{1,m}||_\infty} \right)^{(m-1)/m} |\varphi_{1,m-1} - \lambda|, \tag{1.91}$$

*is valid. And for all $p \in [1, \infty]$*

$$\left( \frac{||f' - \lambda||_\infty}{||\varphi_{1,m-1} - \lambda||_\infty} \right)^{1/(m-1)} \leq \left( \frac{||f||_\infty}{||\varphi_{1,m}||_\infty} \right)^{1/m}. \tag{1.92}$$

*All inequalities are exact and turn into equality for $f(t) = \varphi_{n,m}(t + \alpha)$.*

Now and further we will use the next lemma.

**Lemma 1.4.8.** *For arbitrary function $f \in W_\infty^m$, $m = 1, 2, \dots$ and number $\alpha$ the difference $f(t) - \varphi_{n,m}(t + \alpha)$ has on period $[0, 2\pi)$ at most $2n$ changes of sign and $f(t)$ possesses the $\mu$-property with respect to $\varphi_{n,m}(t)$.*

Really on each from intervals $((k-1)\pi/n, k\pi/n)$, $k = 1, 2, \dots, 2n$

$$\varphi_{n,m}^{(m)}(t) = (-1)^{k-1}, \quad |f^{(m)}(t - \alpha)| \leq 1$$

almost everywhere. Therefore on each from intervals $(2k\pi/n, (2k+1)\pi/n)$ the difference

$$\varphi_{n,m}^{(m-1)}(t) - f^{(m-1)}(t - \alpha)$$

doesn't decrease and on each from intervals $((2k-1)\pi/n, 2k\pi/n)$ it doesn't increase. But then on period $[0, 2\pi)$ this difference has at most $2n$ changes of sign. From here and from Roll's theorem (if we'll reason from contrary) it's easy to infer that the difference $\varphi_{n,m}(t) - f(t - \alpha)$ has on period at most $2n$ changes of sign and $f(t)$ possesses the $\mu$-property with respect to $\varphi_{n,m}(t)$.

**Proof of theorem 1.4.7.** From the theorem 1.4.6 an lemma 1.4.8 it follows that if $f \in W_\infty^m$ and for some $n = 1, 2, \dots$, $||f||_\infty \leq ||\varphi_{n,m}||_\infty$, then

$$f'_\pm \prec n^{1-m}(\varphi_{1,m-1})_\pm, \tag{1.93}$$

as so it is evident that owing to definition of rearrangement we have

$$r(g_\pm(n\cdot), t) \equiv r(g_\pm, t). \tag{1.94}$$

Let $f \in W_\infty^m$,

$$b = (\|f\|_\infty / \|\varphi_{1,m}\|_\infty)^{1/m},$$

$\varepsilon > 0$ and integral numbers $p$ and $n$ are chosen so that

$$b^{m-1} < (p/n)^{m-1} < b^{m-1} + \varepsilon. \tag{1.95}$$

Put

$$\psi(t) = p^{-m} f(pt). \tag{1.96}$$

It is clear, that $\psi \in W_\infty^m$ and

$$\|\psi\|_\infty = p^{-m}\|f\|_\infty = (b/p)^m \|\varphi_{1,m}\|_\infty < n^{-m}\|\varphi_{1,m}\|_\infty = \|\varphi_{n.m}\|_\infty.$$

¿From (1.93)-(1.96) it follows that

$$f'_\pm \prec (\varepsilon + b^{m-1})(\varphi_{1,m})_\pm$$

and due to arbitrariness of $\varepsilon > 0$ we obtain

$$f'_\pm \prec b^{m-1}(\varphi_{1,m})_\pm.$$

To complete the proof it remains to use the statement of the theorem 1.3.12.

Introduce the comparison function which services the nonsymmetric case. It will be conveniently to us begin from an analog of $\varphi_{1,m}(t)$.

For positive numders $\alpha$ and $\beta$ by $\varphi_{1,0}(\alpha, \beta; t)$ we denote $2\pi$-periodic function which is equal to $\alpha$ for $t \in [0, 2\pi\beta/(\alpha+\beta))$ and which is equal to $\beta$ for $t \in [2\pi\beta/(\alpha+\beta), 2\pi)$. For $m = 1, 2, \dots$ define the function $\varphi_{1,m}(\alpha, \beta; t)$ as the $m$-th periodic integral with zero mean value on period of function $\varphi_{1,0}(\alpha, \beta; t)$. It is clear that $\varphi_{1,m}(1, 1; t) = \varphi_{1,m}(t)$.

Give a few properties of the functions $\varphi_{1,m}(\alpha, \beta, t)$ which are analogous to the properties of function $\varphi_{1,m}(t)$.

1) for each $m \in N$

$$\varphi'_{1,m}(\alpha, \beta; t) = \varphi_{1,m-1}(\alpha, \beta; t);$$

2) for odd $m$ the graph of function $\varphi_{1,m}(\alpha, \beta; t)$ is symmetric as to points $(t_k, 0)$, $t_k = \pi\beta/(\alpha+\beta) + k\pi$, $k \in Z$, and for even $m$ it graph is symmetric as to straight lines $t = t_k$;

3) for odd $m$ the function $\varphi_{1,m}(\alpha, \beta; t)$ vanishes at points $t_k = \pi\beta/(\alpha + \beta) + k\pi$, $k \in Z$. If $m$ is even number then on interval $(t_0, t_0 + 2\pi)$ there exist exactly two points $y'$ and $y''$ which are symmetric with respect to $t_1$ and such that $\varphi_{1,m}(\alpha, \beta; t)$ vanishes at points $y'_k = y' + 2\pi k$, $y''_k = y'' + 2\pi k$ and only at these points;

4) the function $\varphi_{1,m}(\alpha,\beta;t)$ is strictly monotonous between the neighbouring extremum points.

Put now for $n = 2, 3, \dots$

$$\varphi_{n,m}(\alpha,\beta;t) = n^{-m}\varphi_{1,m}(\alpha,\beta;t), \quad m = 1, 2, \dots. \tag{1.97}$$

It is clear that the function $\varphi_{n,m}(\alpha,\beta;t)$ for each $m = 1, 2, \dots$ is $2\pi/n$-regular. Note that for the best approximation by constant for $p \geq 1$ the next correlations

$$E_0(\varphi_{n;2\nu-1}(\alpha,\beta))_p = \|\varphi_{n,2\nu-1}(\alpha,\beta)\|_p, \quad \nu = 1, 2, \dots;$$

$$E_0(\varphi_{n,2\nu}(\alpha,\beta))_p < \|\varphi_{n,2\nu}(\alpha,\beta)\|_p, \quad \nu = 1, 2, \dots, \quad \alpha \neq \beta;$$

hold true.

Furthermore everywhere $f_{\pm}^{(m)}$ denotes $(f^{(m)})_{\pm}$.

**Lemma 1.4.9.** *If $f \in W_{\infty}^m$, $m \in N$ then for any $\tau \in R$,*

$$\alpha \geq \|f_+^{(m)}\|_\infty$$

*and*

$$\beta \geq \|f_-^{(m)}\|_\infty$$

*the difference*

$$f(t-\tau) - \varphi_{n,m}(\alpha,\beta;t)$$

*has on period $[0, 2\pi)$ at most $2n$ changes of sign and the function $f(t)$ possesses the $\mu$-property with respect to $\varphi_{n,m}(\alpha,\beta;t)$.*

To verify in justice of lemma it is necessary to repeat with evident modifications the reasoning, conducted in the proof of theorem 1.4.8.

**Theorem 1.4.10.** *For any function $f \in W_{\infty}^m$, $m \in N$, arbitrary numbers*

$$\alpha \geq \|f_+^{(m)}\|_\infty, \quad \beta \geq \|f_-^{(m)}\|_\infty$$

*and $\lambda \in R$ the next inequality*

$$(f' - \lambda)_\pm \prec \left(\frac{E_0(f)_\infty}{E_0(\varphi_{1,m}(\alpha,\beta))_\infty}\right)^{(m-1)/m} (\varphi_{1,m-1}(\alpha,\beta) - \lambda)_\pm. \tag{1.98}$$

*holds true. This inequality turns into equality for any function of the form*

$$f(t) = a\varphi_{1,m}(\alpha,\beta;nt), \quad a \geq 0, \quad n \in N.$$

**Proof.** From the lemma 1.4.9 and theorem 1.4.6 it follows that if $g \in L_\infty^m$

$$E_0(g)_\infty \leq E_0(\varphi_{n,m}(\alpha,\beta;\cdot))_\infty, \tag{1.99}$$

then

$$(g' - \lambda)_\pm \prec n^{1-m}(\varphi_{1,m-1}(\alpha,\beta;n\cdot) - \lambda)_\pm, \tag{1.100}$$

and as

$$r(g_\pm(n\cdot),t) = r(g_\pm,t) \tag{1.101}$$

we obtain

$$(g' - \lambda)_\pm \prec n^{1-m}(\varphi_{1,m-1}(\alpha,\beta;\cdot) - \lambda)_\pm. \tag{1.102}$$

If $f$ is arbitrary function from $L_\infty^m$ , then

$$b = (E_0(f)_\infty / E_0(\varphi_{1,m}(\alpha,\beta))_\infty)^{1/m},$$

$\epsilon \geq 0$ and the fraction $k/n$ such that

$$b^{m-1} < (k/n)^{m-1} < b^{m-1} + \epsilon, \tag{1.103}$$

then puting

$$g(t) = k^{-m} f(kt), \tag{1.104}$$

we'll have

$$E_0(g)_\infty = k^{-m} E_0(f(k\cdot))_\infty = k^m E_0(f)_\infty =$$
$$(b/k)^m E_0(\varphi_{1,m}(\alpha,\beta))_\infty < n^{-m} E_0(\varphi_{1,m}(\alpha,\beta))_\infty = E_0(\varphi_{n,m}(\alpha,\beta))_\infty$$

Therefore the inequality (1.99) holds true. From (1.101)-(1.104) we obtain

$$(f' - \lambda)_\pm \prec (b^{m-1} + \epsilon)(\varphi_{1,m-1}(\alpha,\beta) - \lambda)_\pm$$

and owing to arbitrariness of $\varepsilon > 0$

$$(f' - \lambda)_\pm \prec b^{m-1}(\varphi_{1,m-1}(\alpha,\beta) - \lambda)_\pm$$

that is equivalent to (1.98).

## 1.5 A comparison of rearrangements

Remind that

$$E(f,\Re)_{p;\alpha,\beta} = \inf\{||f - \varphi||_{p;\alpha,\beta}\} : \varphi \in \Re\}$$

is the best $(\alpha,\beta)$-approximation of function $f$ by the set $\Re$ in the space $L_p$ and

$$E_0(f)_{p;\alpha,\beta} = \inf\{||f - \lambda||_{p;\alpha,\beta} : \lambda \in R\}$$

is the best $(\alpha,\beta)$-approximation of $f$ by constant.

**Theorem 1.5.1.** *Let the functions $f, g \in C$ have equal mean values on period. Inequalitits*

$$E_0(f)_{1;\alpha,\beta} \leq E_0(g)_{1;\alpha,\beta} \tag{1.105}$$

*hold true if and only if for all $\alpha, \beta > 0$ when for all $\lambda \in R$ the inequalities*

$$(f - \lambda)_{\pm} \prec (g - \lambda)_{\pm} \tag{1.106}$$

*are valid.*

**Proof.** Let first for all $\alpha, \beta > 0$ the inequalities (1.105) hold true. Proving the correlation (1.106) at first we may consider without the loss of community that $f(t)$ and $g(t)$ have zero mean values on period. In view of theorem 1.3.13 the inequalities (1.106) it's sufficiently to prove for $\lambda = 0$, i.e. it's sufficiently to prove that for all $t \in [0, 2\pi]$

$$\int_0^t r(f_+, u)du \leq \int_0^t r(g_+, u)du \tag{1.107}$$

and

$$\int_0^t r(f_-, u)du \leq \int_0^t r(g_-, u)du. \tag{1.108}$$

Let

$$a_f^{\pm} = mes\{t : \ t \in [0, 2\pi], f_{\pm} > 0\},$$

$\alpha = (a_f^+)^{-1}$ and $\beta = (a_f^-)^{-1}$. Then

$$\int_0^{2\pi} (\alpha \ sgn\, f_+(t) - \beta \ sgn\ \ f_-(t))dt = 0$$

and due to the criterion of the best $(\alpha, \beta)$-approximation in $L_1$

$$E_0(f)_{1;\alpha,\beta} = ||f||_{1;\alpha,\beta}.$$

Therefore

$$(\alpha + \beta)||f_{\pm}||_1 = ||f||_{1;\alpha,\beta} \leq E_0(f)_{1;\alpha,\beta} \leq E_0(g)_{1;\alpha,\beta} \leq$$
$$||g||_{1;\alpha,\beta} = (\alpha + \beta)||g_{\pm}||_1 \tag{1.109}$$

i.e.

$$||f_{\pm}||_1 \leq ||g_{\pm}||_1.$$

Let

$$A^+ = [0, \min\{a_f^+, a_g^+\}], \quad t \in A^+$$

and $\chi_{[0,t]}(u)$ is a characteristic function of segment $[0,t]$. Taking into account the proposition 1.3.1 we obtain

$$\int_0^t r(g_+,u)du = \sup\left(\int_0^{2\pi} g_+(u)h(u)du;\ r(h,u)\equiv\chi_{[0,t]}(u)\right) =$$

$$\sup\left(\int_0^{2\pi} g(u)h(u)du;\ r(h,u)\equiv\chi_{[0,t]}(u)\right) =$$

$$\sup\left(\int_0^{2\pi} g(u)[h(u)-h_0]du;\ r(h,u)\equiv\chi_{[0,t]}(u)\right) =$$

$$\sup\left(\int_0^{2\pi} g(u)h(u)du;\ r(h,u)\equiv r(\varphi_{1,0}(\alpha,\beta),u)\right),$$

where

$$h_0 = \frac{1}{2\pi}\int_0^{2\pi} h(t)dt,\quad \alpha = 1-t/(2\pi),\ \beta = t/(2\pi).$$

¿From here with the help of duality theorem (theorem 1.2.12) for the best $(\alpha,\beta)$-approximations in the space $L_1$ we find

$$\int_0^t r(g_+,u)du = E_0(g)_{1;\alpha,\beta}.$$

Analogously for $t\in A_+$ we have

$$\int_0^t r(f_+,u)du = E_0(f)_{1;\alpha,\beta}.$$

Therefore for $t\in A_+$

$$\int_0^t r(f_+,u)du = E_0(f)_{1;\alpha,\beta} \le E_0(g)_{1;\alpha,\beta} = \int_0^t r(g_+,u)du$$

If $a_f^+ \le a_f^+$, then for $t\ge a_f^+$ it will be $r(f_+,t)\equiv 0$ and $r(g_+,t)\ge 0$ and consequently

$$\int_0^t f_+(u)du = \int_0^{a_f^+} r(f_+,u)du \le \int_0^{a_f^+} r(g_+,u)du \le \int_0^t r(g_+,u)du$$

If $a_g^+ < a_g^+$, then for $t\ge a_g^+$ owing to (1.109)

$$\int_0^t r(f_+,u)du \le \int_0^{a_f^+} r(f_+,u)du = \|f_+\|_1 = \|g_+\|_1 =$$

$$\int_0^{a_g^+} r(g_+,u)du = \int_0^t r(g_+,u)du$$

so that (1.107) is true for all $t \in [0, 2\pi]$. Inequality (1.108) is proved by analogy.

Let now for all $\lambda \in R$ the inequalities (1.106) are valid. Then for all $\lambda \in R$

$$||(f-\lambda)_\pm||_1 \le ||(f-\lambda)_\pm||_1$$

and consequently, for all $\alpha, \beta > 0$

$$||f-\lambda||_{1;\alpha,\beta} = \alpha||(f-\lambda)_+||_1 + \beta||(f-\lambda)_-||_1 \le$$

$$\alpha||(g-\lambda)_+||_1 + \beta||(g-\lambda)_-||_1 = ||g-\lambda||_{1;\alpha,\beta},$$

i.e.

$$||f-\lambda||_{1;\alpha,\beta} \le ||g-\lambda||_{1;\alpha,\beta}$$

and, if $\lambda_0$ such that

$$||g-\lambda_0||_{1;\alpha,\beta} = E_0(g)_{1;\alpha,\beta}$$

then

$$E_0(f)_{1;\alpha,\beta} \le ||f-\lambda_0||_{1;\alpha,\beta} \le ||g-\lambda_0||_{1;\alpha,\beta} = E_0(g)_{1;\alpha,\beta}$$

Further it will be proved one comparison theorem for rearrangements useful for studying the extremal properties of polynomials and splines. Give necessary for its proving informations about $\Sigma$-rearrangements of functions. In detail with the theory of $\Sigma$-rearrangements and its application to the investigations of extremal problems one can become acquainted by monographs of N.P. Korneichuk [11, 22, 24] and N.P.Korneichuk, A.A.Ligun, V.G.Doronin [1]. Continuous on all axis function $\varphi(t)$ will w'll call simple function if $\varphi(t) = 0$ out of some interval $(\alpha, \beta)$, $|\varphi(t)| > 0$ $(\in (\alpha, \beta))$ and for each $y$ such that

$$0 < y < \max\{|\varphi(t)| : t \in (\alpha, \beta)\},$$

the equation $|\varphi(t)| = y$ has exactly two roots.

Any function $f \in C^1$ may be represented in the form of finite or calculated sum (see lemma 1.5.2)

$$f(t) = \sum_k \varphi_k(t) + d, t_0 \le t < 2\pi, \tag{1.110}$$

where $d = f(t_0)$ and $t_0$ such that $|f(t_0)| = \min\{|f(t()| : t\}$; $\varphi_k(t)$ are simple functions (with the bearers $[\alpha_k, \beta_k]$) and such that $f'(t) = \varphi_k'(t)$ everywhere, where the function $\varphi_k'(t)$ exists and is different from zero.

Proceeding from (1.106) define the $\Sigma$-rearrangement $R(f,t)$ of function $f \in C_1$ with the help of equality

$$R(f,t) = \sum_k r(|\varphi_k|,t) + |d|. \tag{1.111}$$

Moreover consider that for $t \geq 2\pi$ $R(f,t) = R(f,2\pi) = |d|$.

It is clear that $R(f,t)$ doesn't increase on $[0,2\pi]$ and $R(f,t) \equiv |d|$ for $t > \max\{\beta_k - \alpha_k : k\}$. Besides that $\Sigma$-rearrangement $R(f,t)$ is translate invariant, i.e.

$$R(f(\cdot + \alpha),t) \equiv R(f(\cdot),t).$$

It is clear also that

$$R(f,0) = \frac{1}{2}V_0^{2\pi}(f) + |d| = \frac{1}{2}||f'|| + |d| \tag{1.112}$$

and

$$||r(f)||_1 = ||f||_1. \tag{1.113}$$

Let $\ell(f)$ is the number of local maximums of function $f \in C$ on period. In this time a segment $[\alpha,\beta]$ on which the function reaches a local maximum we calculate as one local maximum.

**Lemma 1.5.2.** *For any function $f \in C$ for which $\ell(f)$ is finite there exists the decomposition (1.110) of function $|f(t)|$ in sum at most $\ell(f)$ simple functions.*

**Proof.** Without the restriction of reasoning community we may assume that $f(0) = 0$ and $f(t) \geq 0$, $0 \leq t \leq 2\pi$. In the case when $\ell(f) = 1$ it is sufficiently to put $\varphi_1(t) = f(t)$, $t \in (0,2\pi)$.

Suppose that statement of lemma is true for $\ell(f) \leq n-1$ and function $f(t)$ is such that $\ell(f) = n$. Denote by $\alpha$ one from local minimums of function $f(t)$ (if it isn't one then we'll take any from its) in which the function $f(t)$ has the most value and by $(\beta,\gamma)$ any from composing intervals of the set

$$E = \{t : \ t \in (0,2\pi), \ f(t) > f(\alpha)\},$$

and put $\varphi_1(t) = f(t) - f(\alpha)$ if $t \in (\beta,\gamma)$, $\varphi_1(t) = 0$ if $t \notin (\beta,\gamma)$ and $f_1(t) = f(t) - \varphi_1(t)$. Then $\ell(f_1) \leq n-1$ and consequently by induction supposing there exist the simple functions $\{\varphi_k(t)|\}_{k=2}^{n_1}$, $(n_1 \leq n)$ such that $\varphi_k'(t) = f_1'(t) = f(t)$ everywhere, where $\varphi_k(t)$ exists and is different from zero and

$$f_1(t) = \sum_{k=2}^{n_1} \varphi_k(t).$$

But then

$$f(t) = \sum_{k=1}^{n_1} \varphi_k(t).$$

**Lemma 1.5.3.** *Let the decomposition of function $f(t)$ at the sum of simple functions consists of $m$ simple functions. Then*

$$R(f,t) \prec m\, r(|f|, mt) \tag{1.114}$$

*or that is the same*

$$\int_0^t R(f,u)du \leq \int_0^{mt} r(|f|,u)du, \quad t \in [0,\infty). \tag{1.115}$$

**Proof.** Let

$$|f(t)| = \sum_{k=1}^{n} \varphi_k(t) + d, \quad d = \min\{|f(t)| : t\},$$

is the decomposition of function $|f(t)|$ at the sum of simple functions. Then for all $t > 0$ there exist intervals $(\eta_k, \theta_k)$, $k = 1, 2, \ldots$ such that $\theta_k - \eta_k = t$ and

$$\int_0^t r(\varphi_k, u)du = \int_{\eta_k}^{\theta_k} \varphi_(u)du.$$

That is why

$$\int_0^t R(f,u)du = td + \sum_{k=1}^{m} \int_0^t r(\varphi_k, u)du = td + \sum_{k=1}^{m} \int_{\eta_k}^{\theta_k} \varphi_k(u)du =$$

$$\int_{E_t} \{d + \sum_{k=1}^{m} \varphi_k(u)\}du == \int_{E-t} f(u)du$$

where

$$E_t = \bigcup_{k=1}^{m} (\eta_k, \theta_k).$$

¿From here and from proposition 1.3.1 finally we obtain

$$\int_0^t R(f,u)du \leq \int_0^{mes\, E_t} r(|f|,u)du \leq \int_0^{mt} r(|f(u)|du$$

Note that inequality (1.115) is exact for each $t$ because it becomes in equality, for example, for $f(t) = \sin nt$, $m = 2n$.

Comparing the lemmas 1.5.2 and 1.5.3 we obtain the next result.

**Lemma 1.5.4.** *For any function $f \in C$ the inequality*

$$R(f, \cdot) \prec \ell(|f|)\, r(|f|, \ell(|f|), \cdot).$$

*holds true.*

Furthermore we will be needed in wellknown comparison theorem for $\Sigma$-rearrangements of functions (see, for example, Korneichuk [11]).

**Theorem 1.5.5.** *Let $m \in N$, $f \in W_1^m$ and number $b$ is chosen from the next condition*

$$\|f\|_1 = b^{m-1}\|\varphi_{1,m-1}\|_\infty.$$

*Then for all $t \in [0, 2\pi]$ the inequality*

$$R(f,t) \geq \frac{1}{4} b^{m-1} R\left(\varphi_{1,m-1}, \frac{t}{b}\right).$$

*holds true.*

**Theorem 1.5.6.** *Let $m \in N$, $k = 1, 2, \ldots, m-1$, $f \in W_1^m$, $\ell = \ell(|f|)$ and*

$$a_k = (\|f^{(m-k)}\|_1 / \|\varphi_{1,k}\|_\infty)^{1/k}, \quad k = 1, 2, \ldots, m. \tag{1.116}$$

*Then*

$$r\left(\varphi_{1,m-1}, \frac{2t}{a_k\, \ell}\right) \prec \frac{2\ell}{a_k^{m-1}} r(f,t),$$

*i.e. for all $t > 0$*

$$\int_0^t r(\varphi_{1,m-1}, \frac{2u}{a_k\, \ell}) du \leq \frac{2\ell}{a_k^{m-1}} \int_0^t r(f,u) du. \tag{1.117}$$

**Proof.** From the lemma 1.5.4 and theorem 1.5.5 we obtain that for all $t > 0$

$$\int_0^{\ell t} r(f,u) du \geq \int_0^t R(f,u) du \geq \frac{1}{4} a_{m-1}^{m-1} \int_0^t R\left(\varphi_{1,m-1}, \frac{u}{a_{m-1}}\right) du,$$

$$R(\varphi_{1,m-1}, t) = 2\; r(\varphi_{1,m-1}, 2t).$$

Consequently

$$\int_0^{t\ell} r(f,u) du \geq 2 a_{m-1}^{m-1} \int_0^t r\left(\varphi_{1,m-1}, \frac{2u}{a_{m-1}}\right) du =$$

$$\frac{a_{m-1}^{m-1}}{2\ell} \int_0^t r\left(\varphi_{1,m-1}, \frac{2u}{\ell\ a_{m-1}}\right) du.$$

¿From here in particular it follows that

$$||f||_1 \geq \int_0^\infty r(f,u)du \geq \frac{a_{m-1}^{m-1}}{2\ell} \int_0^\infty r\left(\varphi_{1,m-1}, \frac{2u}{\ell\ a_{m-1}}\right) du =$$

$$\frac{a_{m-1}^m}{4} \int_0^\infty r(\varphi_{1,m-1},u)du \frac{a_{m-1}^m}{4} ||\varphi_{1,m-1}|\text{-}1 = a_{m-1}^m ||\varphi_{1,m}||_\infty$$

i.e.

$$a_{m-1} = \left(\frac{||f'||_1}{||\varphi_{1,m-1}||_\infty}\right)^{1/(m-1)} \leq \left(\frac{||f||_1}{||\varphi_{1,m}||_\infty}\right)^{1/m}. \tag{1.118}$$

But then

$$a_1 \leq a_2 \leq \ldots \leq a_{m-1} \leq a_m,$$

and to complete the proof of (1.117) it remains to note that function

$$\Psi(x) = x^{m-1} \int_0^x r(\varphi_{1,m-1}, \theta u/x)du$$

inrceases on semiaxis $[0,\infty)$.

Let $\nu(f)$ is a number of sign changes for $2\pi$-periodic function $f$ on peroid. ¿From theorems 1.5.6 and 1.3.11 the next statement follows.

**Corollary 1.5.7.** *Let $m \in N$, $k = 1,2,\ldots,m-1$, $f \in W_1^m$, $\ell = \ell(|f|)$, $\nu = \nu(f')$ and numbers $a_k$ are defined by equalities (1.116). Then for arbitrary $N$-function $\Phi$ the following inequalities*

$$\int_0^{2\pi} \Phi(|f(t)|)dt \geq \frac{\ell\, a_k}{2} \int_0^{2\pi} \Phi\left(\frac{a_k^{m-1}}{2\ell}|\varphi_{1,m-1}(t)|\right) dt \tag{1.119}$$

*hold true. And consequently (as $\ell \leq \nu$)*

$$\int_0^{2\pi} \Phi(|f(t)|)dt \geq \frac{\nu\, a_k}{2} \int_0^{2\pi} \Phi\left(\frac{a_k^{m-1}}{2\nu}|\varphi_{1,m-1}(t)|\right) dt. \tag{1.120}$$

*If,besides that, $f(t) \geq 0$, $(t \in R)$, then $\ell \leq \nu/2$ and consequently*

$$\int_0^{2\pi} \Phi(|f(t)|)dt \geq \frac{\nu a_k}{4} \int_0^{2\pi} \Phi\left(\frac{a_k^{m-1}}{\nu}|\varphi_{1,m-1}(t)|\right) dt.$$

*The inequalities (1.119) and (1.120) are exact on the class $W_1^m$.*

## 1.6 Rearrangement of conjugate functions

In this section we shall tell only about $2\pi$-periodic functions and for the abbreviation here we write $C$, $L_p$ and $||f||_p$ instead of $C[0,2\pi]$, $L_{p[0,2\pi]}$ and $||f||_{L_{p[0,2\pi]}}$ correspondingly.

Together with trigonometrical series

$$\frac{a_0}{2}+\sum_{k=1}^{\infty}(a_k\cos kt+b_k\sin kt)=\sum_{k=-\infty}^{\infty}c_k e^{ikt} \qquad (1.121)$$

we consider the conjugate series

$$\sum_{k=1}^{\infty}(a_k\sin kt+b_k\cos kt)=-i\sum_{k=-\infty}^{\infty}c_k e^{-ikt}. \qquad (1.122)$$

Its name connects with the fact that real and imaginary parts of power series

$$\frac{a_0}{2}+2\sum_{k=1}^{\infty}c_k z^k$$

in the disk $|z|\leq 1$ give us the pair of conjugate garmonic functions.

Let series (1.122) is Fourier's series of some function $f(t)$. If series (1.122) is also a Fourier's series of some function, then this function is called as trigonometrical conjugate (or simply conjugate) with function $f(t)$ and denote by $\tilde{f}(t)$.

There exists another, more general definition of conjugate function as an integral:

$$\tilde{f}(t)=-\frac{1}{\pi}[f(t+u)-f(t-u)]\cot\frac{u}{2}du, \qquad (1.123)$$

which exists for any summable function $f(t)$ almost everywhere and is finite. However, defined like this function $\tilde{f}(t)$ is not always summable. In that case, when this function is summable, its Fourier's series will be (1.122) and hence in this case the definition (1.123) coincides with previous one. ¿From here it follows that if $f(t)$ and $\tilde{f}(t)$ are summable functions and mean value of $f(t)$ is equal to zero on period, then an equality $\tilde{\tilde{f}}(t)=f(t)$ holds true.

The definition of conjugate function $\tilde{f}(t)$ by equality (1.123), as it is easy to verify with the help of identity

$$\frac{t}{2}\cot\frac{t}{2}=1+t\sum_{k=1}^{\infty}\left(\frac{1}{t-2k\pi}+\frac{1}{t+2k\pi}\right),$$

is equivalent to the definition

$$\tilde{f}(t) = -\frac{1}{\pi}\int_0^\infty \frac{f(t+u)-f(t-u)}{u}du \qquad (1.124)$$

This equality we may consider as the definition of cojugate function, defined on all real axis (not only periodic).

Consider some examples of conjugate functions. Let $\varphi_{1,m}$ is Euler's spline which has been given in section 1.4. From (1.87) it follows that

$$\tilde{\varphi}_{1,m}(t) = \frac{4}{\pi}\sum_{k=1}^{\infty}\frac{\sin[(2k+1)t-\pi(m+1)/2]}{(2k+1)^{m+1}}. \qquad (1.125)$$

It is clear that $\tilde{\varphi_{1,m}}(t)$ is odd function for odd $m$ and even function for even $m$. It is easy to verify that

$$\tilde{\varphi}_{1,0}(t) = \frac{1}{\pi}\ln|\tan(t/2)|,\ 0<t<\pi, \qquad (1.126)$$

and when $m = 1,2,\ldots$, $\varphi_{1,m}(t)$ is $m$-th periodic integral of $\varphi_{1,0}(t)$, and its mean value is equal to zero on period.

For odd $m > 1$ the function $\varphi_{1,m}(t)$ vanishes at points $\pi n$, $n \in Z$, and for even $m$ at points $\pi/2 + \pi n$, $n \in Z$. ¿From this fact, equality (1.126) and Roll's theorem it follows that under odd $m$ function $\tilde{\varphi}_{1,m}(t)$ changes sign at points $\pi n$, $\in Z$, and only at this points and for even $m$ at points $\pi/2 + \pi n$, $n \in Z$ and only in this points. Therefore for $m = 1,2,\ldots$

$$sgn\tilde{\varphi}_{1,0}(t) = sgn\sin(t-\pi(m+1)/2) \qquad (1.127)$$

Note, moreover, that for any function $f \in L_1$

$$f(\tilde{n(\cdot)}) = \tilde{f}(nt),$$

that is why if

$$\varphi_{n,m}(t) = n^{-m}\varphi_{1,m}(nt),$$

then

$$\tilde{\varphi_{n,m}}(t) = n^{-m}\tilde{\varphi_{1,m}}(nt).$$

Thus

$$sgn\, q\tilde{\varphi}_{n,m}(t) = sgn\sin n(t-\pi(m+1)/2). \qquad (1.128)$$

The next Stain's and Way's theorem we will use for examination the extremal properties of functions which are trigonometrically conjugate with polynomials and splines. However this result has special interest as it has another important applications in functions theory.

**Theorem 1.6.1.** *Let $E$ is a measurable subspace of segment $[0,1]$ and $\chi(t)$ is $2\pi$-perioic continuation of characteristic function for set $E$ on all real axis. Then for any $\tau \in [0, 2\pi]$*

$$r(\tilde{\chi}_E, \tau) = r(\tilde{\chi}_{[0, mes\, E]}, \tau) = \frac{2}{\pi} ar\, sh \frac{\sin(mes E/2)}{\tan(\tau/2)}. \tag{1.129}$$

**Proof.** Let at first for function $f(t) = \chi'_E(t)$ the set $E$ consists of finite numbers of intervals $(x_k, y_k)$, $k = 1, 2, \ldots$,

$$0 \leq x_1 < y_1 < x_2 < y_2 < \ldots < x_n < y_n \leq 3\pi. \tag{1.130}$$

Then with the help of equality (1.123) we obtain

$$\tilde{f}(t) = \frac{1}{2\pi} \sum_{k=1}^{n} \int_{x_k}^{y_k} \cot \frac{t-u}{2} du =$$

$$-\frac{1}{2\pi} \sum_{k=1}^{n} \left( \ln \left| 2 \sin \frac{y_k - t}{2} \right| - \ln \left| 2 \sin \frac{x_k - t}{2} \right| \right) =$$

$$-\frac{1}{2\pi} \ln \left| \prod_{k=1}^{n} \sin \frac{y_k - t}{2} \left( \prod_{k=1}^{n} \sin \frac{y_k - t}{2} \right)^{-1} \right|.$$

¿From here, in particular, it follows that

$$\tilde{f}(x_k \pm 0) = -\infty, \quad \tilde{f}(y_k \pm 0) = +\infty, \tag{1.131}$$

and on every of intervals $(x_k, y_k)$, $(x_{k+1}, y_{k+1})$, $k = 1, 2, \ldots, n$ the function $\tilde{f}(t)$ is continuous. Therefore for any $a$ the equation

$$\tilde{f}(t) = a \tag{1.132}$$

has at least $2n$ real different zeros on period $[0, 2\pi]$. Besides that for any $a$ the equation (1.132) may be represented in such form

$$\left| \prod_{k=1}^{n} \sin \frac{y_k - t}{2} \right| \left| \left( \prod_{k=1}^{n} \sin \frac{y_k - t}{2} \right) \right|^{-1} = e^{a\pi/2}. \tag{1.133}$$

Every root of equation (1.133) also is a root of next trigonometrical equation of order $n$:

$$\prod_{k=1}^{n} (1 - \cos(x_k - t)) = e^{at} \prod_{k=1}^{n} (1 - \cos(y_k - t)) \tag{1.134}$$

which has at most $2n$ roots on period. Thus, for any $a$ the equation (1.132) (or (1.134)) has exactly $2n$ roots on period and all these roots are different. From here and from correlations (1.131) in view of that function $\tilde{f}(t)$ is continuous on each from intervals $(x_k, y_k)$ and $(y_k, x_{k+1})$ it follows that for any $a$ an equation $\tilde{f}(t) = a$ has one and only one root on each from intervals $(x_k, y_k)$ and $(y_k, x_{k+1})$, and hence (here we consider again (1.131)) on each from intervals $(x_k, y_k)$ function $\tilde{f}(t)$ monotonously increases from $-\infty$ to $+\infty$, and on each from intervals $(y_k, x_{k+1})$ it monotonously decreases from $+\infty$ to $-\infty$. But then the function

$$\left|\prod_{k=1}^{n} \sin \frac{y_k - t}{2}\right| \left|\prod_{k=1}^{n} \sin \frac{y_k - t}{2}\right|^{-1}$$

monotonously increases from $o$ to $+\infty$ on each from intervals $(x_k, y_k)$ and monotonously decreases on each of intervals $(y_k, x_{k+1})$ from $+\infty$ to $o$. Therefore function

$$\phi(t) = \left|\left(\prod_{k=1}^{n} \sin \frac{y_k - t}{2}\right) \left(\prod_{k=1}^{n} \sin \frac{y_k - t}{2}\right)\right|^{-1}$$

monotonously decreases from $+\infty$ to $-\infty$ on each from intervals $(y_k, y_{k+1})$. Now taking into account that $\phi(x_k) = 0$, we conclude that for any $a$ the equation

$$\phi(t) = e^{a\pi/2} \tag{1.135}$$

has exactly $n$ roots $\alpha_k \in (y_k, x_{k+1})$ on period, and the equation

$$\phi(t) = -e^{a\pi/2} \tag{1.136}$$

has exactly $n$ roots $\beta_k \in (x_k, y_k)$ on period. Moreover if $a > 0$ then

$$E_a := \{t : t \in [x_1, x_2 + 2\pi), \ \ \tilde{f}(t) > a\} = \bigcup_{k=1}^{n} (\beta_k, \alpha_k).$$

Thus

$$m(\tilde{f}, a) = mesE_a = \sum_{k=1}^{n} (\alpha_k - \beta_k) := \theta.$$

But $r(f, t)$ is a revers function to distribution function $m(f, t)$. Therefore $a = r(\tilde{f}, \theta)$. It remains to prove that for

$$a = \frac{2}{\pi} ar\, sh \frac{\sin(m/2)}{\tan(\tau/2)}, \quad m = mesE, \tag{1.137}$$

the equality

$$\theta = \tau \tag{1.138}$$

holds true.

Put

$$z = e^{it},\ z_k = e^{ix_k},\ \ w_k = e^{iy_k},\ \ k = 1, 2, \ldots, n.$$

Then

$$2i \sin \frac{x_k - t}{2} = (z_k - z)e^{-i(t+x_k)/2}$$

and

$$2i \sin \frac{y_k - t}{2} = (w_k - z)e^{-i(t+y_k)/2},$$

and equation (1.135) takes form

$$\prod_{k=1}^{n} (z_k - z) = A \prod_{k=1}^{n} (w_k - z) \tag{1.139}$$

and equation (1.136) takes such form

$$\prod_{k=1}^{n} (z_k - z) = -A \prod_{k=1}^{n} (w_k - z) \tag{1.140}$$

where

$$A = exp\left(\frac{a\pi}{2} - \frac{i}{2}\sum_{k=1}^{n}(y_k - x_k)\right) = exp\left(\frac{a\pi}{2} - \frac{im}{2}\right). \tag{1.141}$$

Owing to Viet's theorem for roots $e^{i\alpha_k}$, $k = 1, 2, \ldots, n$ of equation (1.139) we obtain

$$\prod_{k=1}^{n} e^{i\alpha_k} = \frac{(-1)^n}{1+A}\left(\prod_{k=1}^{n} z_k - A \prod_{k=1}^{n} w_k\right).$$

Analogously from (1.140) we find

$$\prod_{k=1}^{n} e^{i\beta_k} = \frac{(-1)^n}{1+A}\left(\prod_{k=1}^{n} z_k + A \prod_{k=1}^{n} w_k\right).$$

¿From here

$$\frac{\prod_{k=1}^{n} e^{i\alpha_k}}{\prod_{k=1}^{n} e^{i\beta_k}} = \frac{1+A}{1-A}\,\frac{\prod_{k=1}^{n} z_k - A\prod_{k=1}^{n} w_k}{\prod_{k=1}^{n} z_k + A\prod_{k=1}^{n} w_k}.$$

Thus

$$e^{i\theta} = exp\left(i\sum_{k=1}^{n}(\alpha_k - \beta_k)\right) = \frac{1+A}{1-A}\frac{1-AB}{1+AB},$$

where

$$B = \prod_{k=1}^{n}\frac{w_k}{z_k} = exp\left(i\sum_{k=1}^{n}(y_k - x_k)\right) = e^{im}.$$

Consequently

$$\frac{1+A}{1-A}\frac{1-Ae^{im}}{1+Ae^{im}},$$

or (in view of (1.141))

$$\tan\frac{\theta}{2} = \frac{1}{i}\frac{e^{i\theta}-1}{e^{i\theta}+1} = \frac{1}{i}\frac{e^{-im/2}-e^{im/2}}{e^{-ia/2}e^{ia/2}}\ \frac{\sin(m/2)}{\sin(a\pi/2)}.$$

Selecting $a$ according to (1.137) we obtain equality (1.138), and thus the statement of theorem when $a > 0$. The case $a < 0$ is proved by analogy.

Adduce now still one statement at the type of theorem 1.6.1.

**Theorem 1.6.2.** *Let*

$$f(t) = \sum_{k=1}^{n} d_k \cot\frac{t - x_k}{2}, \tag{1.142}$$

*where*

$$0 \le x_1 < x_2 < \ldots < x_n \le x_1 + 2\pi = x_{n+1}; \tag{1.143}$$

$$d_k > 0,\ k = 1, 2, \ldots, n;\ d_1 + d_2 + \ldots + d_n = d.$$

*Then for any* $t \in [-, 2\pi]$

$$\pi(f, t) = d\cot\frac{t}{2}. \tag{1.144}$$

**Proof.** As $f(x_k + 0) = +\infty$ and $f(x_k - 0) = -\infty$, and

$$f'(t) = -\sum_{k=1}^{n}\frac{d_k}{1-\cos(t-x_k)} < 0$$

almost everywhere, then for any $y$ the equation $f(t) = y$ has exactly $n$ roots $\alpha_k$, $x_k < \alpha_k < x_{k+1}$ on period and

$$E_y := \{t : t \in [x_1, x_1 - 2\pi], f(t) > y\} = \bigcup_{k=1}^{n} (x_k, \alpha_k).$$

Hence $m(f, y) = mesE_y = \theta$, where

$$\theta = \sum_{k=1}^{n} (\alpha_k - x_k). \tag{1.145}$$

Therefore

$$y = \pi(f, \theta). \tag{1.146}$$

Calculate now $\theta$.

Put

$$z = e^{it} \quad z_k = e^{it_k}, \quad k = 1, 2, \ldots, n.$$

Then equation $f(t) = y$ takes form

$$i \sum_{k=1}^{n} d_k \frac{z + z_k}{z - z_k} = y,$$

or

$$i \sum_{k=1}^{n} d_k (z + z_k) \prod_{l \neq k} (z - z_l) = y \prod_{l=1}^{n} (z - z_l).$$

It is clear that numbers $e^{i\alpha_k}$, $k = 1, 2, \ldots, n$ and only they are the roots of this equation. Therefore by Viet's theorem

$$\exp\left(i \sum_{k=1}^{n} \alpha_k\right) = \prod_{k=1}^{n} e^{i\alpha_k} =$$

$$(-1)^n \left(i(-1)^{n-1} \sum_{k=1}^{n} d_k \prod_{l=1}^{n} z_l - (-1)^n \prod_{l=1}^{n} z_l\right) \left(i \sum_{k=1}^{n} d_k\right)^{-1} =$$

$$\frac{y + i\sum_{k=1}^{n} d_k}{y - i\sum_{k=1}^{n} d_k} \prod_{l=1}^{n} z_l = \frac{y + id}{y - id} \prod_{l=1}^{n} z_l = \frac{y + id}{y - id} \exp\left(i \sum_{k=1}^{n} x_k\right)$$

or

$$\exp\left(i \sum_{k=1}^{n} (\alpha_k - x_k)\right) = \frac{y + id}{y - id},$$

and this fact together with the (1.145) gives us

$$e^{i\theta} = (y + id)/(y - id)$$

or

$$\tan(\theta/2) = d/2.$$

Comparing the last equality with (1.146) we get the statement of our theorem.

**Theorem 1.6.3.** *Let functions $f, g \in C$ are such that $\tilde{f}, \tilde{g} \in C$ and the function $f(t)$ almost everywhere is different from zero,*

$$E_{\pm} = \{t : t \in [0, 2\pi], \pm\tilde{g}(t) > 0\}, \ \ mesE_{\pm} = \pi; \tag{1.147}$$

$$||\tilde{g}||_1 = \int_0^{2\pi} r(|g|, t) r(|s\tilde{g}n\, \tilde{g}|, t) dt. \tag{1.148}$$

*Then, if*

$$|f| \prec |g| \tag{1.149}$$

*then*

$$||\tilde{f}||_1 \leq ||\tilde{g}||_1. \tag{1.150}$$

**Proof.** Let at first $\tilde{f}(t)$ is different from zero almost everywhere .As

$$\int_0^{2\pi} \phi(t)\tilde{\psi}(t)dt = -\int_0^{2\pi} \psi(t)\tilde{\phi}(t)dt \tag{1.151}$$

then

$$||\tilde{f}||_1 = \int_0^{2\pi} \tilde{f}(t) sgn\, \tilde{f}(t) dt = -\int_0^{2\pi} f(t)(\widetilde{sgn}\, \tilde{f}(t) dt. \tag{1.152}$$

Using the proposition 1.3.3 we obtain

$$||\tilde{f}||_1 \leq \int_0^{2\pi} r(|f|, t) r(|s\tilde{g}n\, \tilde{f}(t) dt. \tag{1.153}$$

¿From the inequality (1.153) and lemma 1.3.14 we find

$$||\tilde{f}||_1 \leq \int_0^{2\pi} r(|g|, t) r(|\widetilde{sgn}\, \tilde{f}(t) dt. \tag{1.154}$$

Moreover, due to condition (1.147) and theorem 1.6.1 we obtain

$$r(|\widetilde{sgn\, \tilde{g}}|, t) = 2r\left(\tilde{\chi}_{E_+(g)}, \frac{t}{2}\right) = \frac{\pi}{4} ar\, sh\left(\cot\frac{t}{4}\right),$$

$$r(|\widetilde{sgn\, \tilde{f}}|, t) = 2r\left(\tilde{\chi}_{E_+(f)}, \frac{t}{2}\right) = \frac{\pi}{4} ar\, sh\left(\frac{\sin(mesE_+(f)/2)}{tan(t/4)}\right),$$

thus for all $t \in (0, 2\pi)$

$$r(|\widetilde{sgn\, \tilde{f}}|, t) \leq r(|\widetilde{sgn\, \tilde{g}}|, t) \tag{1.155}$$

¿From (1.153) thanks to (1.154) and (1.148), we establish

$$||\tilde{f}||_1 = \int_0^{2\pi} r(|g|, t) r(|\widetilde{sgn \tilde{g}}|, t) dt$$

and inequality (1.150) is proved for any function $f(t)$ for which $\tilde{f} \neq 0$ almost everywhere.

Let now $\tilde{f}(t)$ is'nt equivalent to zero but $\tilde{f}(t) = 0$ on some set with positive measure. Note that transition operator to conjugate function continuously acts from $C$ to $L_1$. Therefore, there exsist some absolute constant $K > 0$ such that if $g, h \in C$ and $||f - h||_C < \varepsilon$ then $||\tilde{f} - \tilde{h}||_C < K\varepsilon$ for all $\varepsilon > 0$. Further, as

$$\xi(t) = \int_0^t r(g, u) du$$

is nondecreasing convex from below and positive on $(0, 2\pi]$ function, then there exists a constant $K_1 > 0$ such that for $0 \leq t \leq 2\pi$ the inequality $t \leq K_1 \xi(t)$ holds true.

Fix arbitrary $\varepsilon > 0$ and find different from constant trigonometric polynomial $\tau(t)$ such that $||f - \tau||_C < \varepsilon$. Then polynomial $\tau(t)$ vanishes only at finite number of points on period. Besides that $||f - \tau||_1 < K_1\varepsilon$ and in view of proposition 1.3.5 and condition (1.147) for $t \in [0, 2\pi]$

$$\int_0^t r(|\tau, u) du \leq \int_0^t r(|f|, u) du \int_0^t r(|f - \tau|, u) du \leq$$

$$\int_0^t r(|g|, u) du + t\varepsilon \leq \int_0^t r(|g|, u) du + K_1\varepsilon \int_0^t r(|g|, u) du =$$

$$(1 + K_1) \int_0^t r(|g|, u) du,$$

i.e.

$$|\tau| \prec (1+K_1)|g|.$$

¿From proved facts it follows that

$$||\tilde{f}||_1 \le ||\tilde{\tau}||_1 + ||\tilde{\tau} - \tilde{f}||_1 \le K\varepsilon + (1+K_1\varepsilon)||\tilde{g}||_1$$

due to arbitrariness of $\varepsilon > 0$ and inequality

$$||\tilde{f}||_1 \le ||\tilde{g}||_1.$$

In conclusion we note that condition of theorem 1.6.3 is satisfied, for example, by functions $g(t) = \varphi_{n,m}(t)$.

## 1.7 Inequality of Kolmogorov's type

As in present section we'll tell only about $2\pi$-periodic functions then we'll write $L_p$, $L_p^m$ and $||f||$ instead of $L_p[0,2\pi]$, $L_p^m[0,2\pi]$ and $||f||_{L_p[0,2\pi]}$ correspondingly.

A great attention in function theory is paid to inequalities between the norms of successive derivatives of functions or inequalities of Kolmogorov's tipe:

$$||f^{(k)}||_p \le M||f^{(r)}||_q^\alpha ||f||_s^\beta, \tag{1.156}$$

$$\alpha+\beta = 1, \ \ \alpha,\beta > 0, \ \ k \le r.$$

Here we adduce a few exact inequalities in the form (1.156), which will be useful under the receiving the exact estimates for norms of derivatives of trigonometric polynomials and splines.

**Theorem 1.7.1.** *(Kolmogorov's inequality) Let $r = 2,3,\ldots$ and $k = 1,2,\ldots,r-1$. For any function $f \in L_\infty^r$, $f(t) \ne const$ the exact on $L_\infty^r$ inequality*

$$\left(\frac{||f^{(k)}||_\infty}{||\varphi_{1,r-k}||_\infty ||f^{(r)}||_\infty}\right)^{1/(r-k)} \le \left(\frac{||f||_\infty}{||\varphi_{1,r}||_\infty ||f^{(r)}||_\infty}\right)^{1/r}. \tag{1.157}$$

*holds true.* here and furthermore $\varphi_{1,m}(t)$ is Euler's spline.

**Proof.** Let $f \in L^r_\infty$,

$$f_1(t) = f(t)/\|f^{(r)}\|_\infty$$

and

$$b = (\|f_1\|_\infty / \|\varphi_{1,r}\|_\infty)^{1/r}.$$

For any $\varepsilon > 0$ the fraction $m/n$ may be selected so that

$$b^{r-1} < (m/n)^{r-1} < b^{r-1} + \varepsilon.$$

Consider the function

$$\psi(t) = m^{-r} f_1(mt),$$

then $\psi \in W^r_\infty$ and

$$\|\psi\|_\infty = m^{-r}\|f_1(m\cdot)\|_\infty = m^{-r}\|f_1\|_\infty =$$

$$m^{-r} b^r \|\varphi_{1,r}\|_\infty < n^{-r}\|\varphi_{n,r}\|_\infty = \|\varphi_{n,r}\|_\infty,$$

and consequently in view of lemma 1.4.8 the function $\psi(t)$ possesses the $\mu$-property with respect to function $\varphi_{n,r}$. ¿From this fact and from proposition 2.4.2 it follows that

$$\|\psi'\|_\infty \leq \|\varphi'_{n,r}\|_\infty = \|\varphi_{n,r-1}\|_\infty = n^{1-r}\|\varphi_{1,r-1}\|_\infty.$$

However

$$\|\psi'\|_\infty = m^{1-r}\|f_1'\|_\infty = m^{1-r}\|f'\|_\infty / \|f^{(r)}\|_\infty.$$

That is why

$$\|f'\|_\infty \leq (m/n)^{r-1}\|\varphi_{1,r-1}\|_\infty \|f^{(r)}\|_\infty < (b^{r-1} + \varepsilon)\|\varphi_{1,r-1}\|_\infty \|f^{(r)}\|_\infty.$$

¿From here and from the arbitrariness of $\varepsilon > 0$ it follows that

$$\frac{\|f'\|_\infty}{\|\varphi_{1,r-1}\|_\infty \|f^{(r)}\|_\infty} \leq b^{r-1} =\leq \left( \frac{\|f\|_\infty}{\|\varphi_{1,r}\|_\infty \|f^{(r)}\|_\infty} \right)^{(r-1)/r}.$$

Thus

$$\frac{\|f^{(r-1)}\|_\infty}{\|\varphi_{1,1}\|_\infty \|f^{(r)}\|_\infty} \leq \ldots \leq \left( \frac{\|f'\|_\infty}{\|\varphi_{1,r-1}\|_\infty \|f^{(r)}\|_\infty} \right)^{1/(r-1)} \leq$$

$$\leq \left( \frac{\|f\|_\infty}{\|\varphi_{1,r}\|_\infty \|f^{(r)}\|_\infty} \right)^{1/r},$$

that is equivalent to inequality (1.157).

The exactness of the inequality (1.157) follows from fact that it turns into equality for

$$f(t) = a\varphi_{n,r}(t + \alpha). \tag{1.158}$$

It's easy to prove that inequality (1.157) turns into equality only for functions in the form (1.158).

**Theorem 1.7.2.** *(Stein's inequality) Let $r = 2, 3, \dots$ and $k = 1, 2, \dots, r-1$. For any function $f \in L_1^r$, $f(t) \neq const$ the exact on $L_1^r$ inequality*

$$\left(\frac{\|f^{(k)}\|_1}{\|\varphi_{1,r-k}\|_\infty \|f^{(r)}\|_1}\right)^{1/(r-k)} \leq \left(\frac{\|f\|_1}{\|\varphi_{1,r}\|_\infty \|f^{(r)}\|_1}\right)^{1/r}. \qquad (1.159)$$

*holds true.*

¿From inequality (1.159) it follows that for any function $f \in V^r$ ($V^m$ is the set of all $2\pi$-periodic functions for which $(m-1)$-th derivative is locally absolutely continuous and $m$-th has bounded variation on period) the exact on $V^r$ inequality

$$\left(\frac{V_0^{2\pi}(f^{(k)})}{\|\varphi_{1,r-k}\|_\infty V_0^{2\pi}(f^{(r)})}\right)^{1/(r-k)} \leq \left(\frac{V_0^{2\pi}(f)}{\|\varphi_{1,r}\|_\infty V_0^{2\pi}(f^{(r)})}\right)^{1/r}. \qquad (1.160)$$

is valid.

**Proof.** Note that under the proving of theorem 1.5.6 inequality (1.159) is established in fact (it immediatelly follows from (?? ) ). But we'll demonstrate the Stein's way which allows us to reduce the inequality (1.159) directly from (1.157). This way will be used and in other problems. Let $f(t) \neq const$ is arbitrary function from $L_1^1$ and

$$g(t) = \int_0^{2\pi} f(t+u) sgn\, f^{(r-k)}(u) du$$

Then

$$g^{(m)}(t) = \int_0^{2\pi} f^{(m)}(t+u) sgn\, f^{(r-k)}(u) du, \quad m = 1, 2, \dots, r,$$

and therefore

$$\|g\|_\infty \leq \|f\|_1, \quad \|g^{(m)}\| \leq \|f^{(m)}\|_1, \quad m = 1, 2, \dots, r,$$

¿From here and from theorem 1.7.1 we obtain

$$\|f^{(r-1)}\|_1 = g^{(r-k)}(0) \leq \|g^{(r-k)}\|_\infty \leq$$

$$\|\varphi_{1,k}\|_\infty \left(\frac{\|g\|_\infty}{\|\varphi_{1,r}\|_\infty}\right)^{k/r} \|g^{(r)}\|_\infty^{(r-k)/r} \leq$$

$$\|\varphi_{1,k}\|_\infty \left(\frac{\|f\|_1}{\|\varphi_{1,r}\|_\infty}\right)^{k/r} \|f^{(r)}\|_1^{(r-k)/r},$$

i.e.

$$\left(\frac{\|f^{(r-k)}\|_1}{\|\varphi_{1,k}\|_\infty \|f^{(r)}\|_1}\right)^{1/k} \le \left(\frac{\|f\|_1}{\|\varphi_{1,r}\|_\infty \|f^{(r)}\|_1}\right)^{1/r}.$$

that is equivalent to the inequality (1.157).

The exactness of the inequality (1.159) follows from the next reasoning. Let

$$f_h(t) = \frac{a}{4hn^r}\int_{t-h/2}^{t+h/2} \varphi_{1,r-1}(nu)du, \quad a > 0, \, h > 0.$$

Then for $h < \pi/(2n)$, $\|f_h^{(r)}\|_1 = a$ and for $h \to 0$

$$\|f_h^{(k)}\|_1 \to \frac{a}{4n^{r-k}}\|\varphi_{1,r-k}\|_1 = an^{r-k}\|\varphi_{1,r-k}\|_\infty \tag{1.161}$$

and

$$\|f_h\|_1 \to \frac{a}{4n^r}\|\varphi_{1,r-k}\|_1 = an^{-r}\|\varphi_{1,r}\|_\infty. \tag{1.162}$$

**Theorem 1.7.3.** *(Hardy's, Littlwood's and Polya's inequality). Let $r > 0$ and $k \in (0, r)$. For any function $f \in L_2^r$, $f \neq const$ (for any $r > 0$ $f^{(r)}(x)$ we consider Weil's derivative (see, for example, Ahieser [6,p.169])) the exact on $L_2^r$ inequality*

$$\left(\frac{\|f^{(k)}\|_2}{\|f^{(r)}\|_2}\right)^{1/(r-k)} \le \left(\frac{\|f\|_2}{\|f^{(r)}\|_2}\right)^{1/r} \tag{1.163}$$

*holds true.*

**Proof.** Let $f(t) \neq const$ is arbitrary function from $L_2^r$ and

$$\sum_{l=0}^{\infty} \rho_l \cos(lt + t_l)$$

is its Fourier's series. Then

$$\|f(k)\|_2^2 = \|\sum_{l=1}^{\infty} l^k \rho_l \cos(l \cdot + t_l + k\pi/2)\|_2^2 = \pi \sum_{l=1}^{\infty} l_{2k}\rho_l^2 =$$

$$\pi \sum_{l=1}^{\infty} (l_{2k}\rho_l^2)^{k/r}(\rho_l^2)^{(r-k)/r}.$$

Using Helder's inequality for the serieses, we obtain

$$\|f(k)\|_2^2 \pi \left(\sum_{l=1}^{\infty} l_{2k}\rho_l^2\right)^{k/r} \left(\sum_{l-1}^{\infty} \rho_l^2\right)^{(r-k)/r} \le$$

$$\left(\pi\sum_{l=1}^{\infty} l_{2k}\rho_l^2\right)^{k/r}\left(\pi\left(\frac{1}{2}\rho_0^2+\sum_{l-1}^{\infty}\rho_l^2\right)\right)^{(r-k)/r} = ||f^{(r)}||_2^{k/r}||f||_2^{(r-k)/r}.$$

The exactness of the inequality (1.163) on the set $L_2^r$ follows from fact that it turns into equality for

$$f(t) = a\cos(nt+\alpha). \tag{1.164}$$

It's easy to verify that inequality turns into equality only for functions in the form of (1.164).

**Theorem 1.7.4.** *Let $r = 2,3,\dots$ and $k = 1,2,\dots,r-1$. Then for any function $f \in L_\infty^r$, $f \neq const$ the exact on $L_\infty^r$ inequality*

$$\left(\frac{||\tilde{f}^{(k)}||_1}{||\tilde{\varphi}_{1,r-k}||_1||f^{(r)}||_\infty}\right)^{1/(r-k)} \le \left(\frac{||f||_\infty}{||\varphi_{1,r}||_\infty||f^{(r)}||_\infty}\right)^{1/r}. \tag{1.165}$$

*holds true.*

**Proof.** . Let $f \in L_\infty^r$, $f/neconst$ and $||f^{(r)}||_\infty = 1$.Put

$$b = \left(\frac{||f^{(k-1)}||_\infty}{||\varphi_{1,r-r+1}||_\infty}\right)^{1/(r-k+1)}.$$

Due to theorems 1.4.7 and 1.6.3

$$||\tilde{f}^{(k)}||_1 \le b^{r-k}||\tilde{\varphi}_{1,r-k}||_1 = \left(\frac{||f^{(k-1)}||_\infty}{||\varphi_{1,r-r+1}||_\infty}\right)^{(r-k)/(r-k+1)} ||\tilde{\varphi}_{1,r-k}||_1.$$

Therefore for any function $f \in L_\infty^r$, $f/neconst$

$$\left(\frac{||\tilde{f}^{(k)}||_1}{||\tilde{\varphi}_{1,r-k}||_1||f^{(r)}||_\infty}\right)^{1/(r-k)} \le \left(\frac{||f^{(k-1)}||_\infty}{||\varphi_{1,r-k+1}||_\infty||f^{(r)}||_\infty}\right)^{1/r}$$

and to complete the proof of (1.165) it remains to use the inequality (1.157).

The inequality (1.165) becomes to equality for

$$f(t) = a\varphi_{1,r}(nt+\alpha). \tag{1.166}$$

**Theotem 1.7.5.** *Let $r = 2, 3, \ldots$ and $k = 1, 2, \ldots, r-1$. Then for any function $f \in L^r_\infty$, $f \neq const$ the exact on $L^r_\infty$ inequality*

$$\frac{|f^{(k)}|}{\|f^{(r)}\|_\infty} \prec \left(\frac{\|f\|_\infty}{\|\varphi_{1,r}\|_\infty \|f^{(r)}\|_\infty}\right)^{(r-k)/r} |\varphi_{1,r-k}| \tag{1.167}$$

*holds true. For any $N$-function $\Phi(t)$ the exact on $L^r_\infty$ inequality*

$$\int_0^{2\pi} \Phi\left(\frac{|f^{(k)}(t)|}{\|f^{(r)}\|_\infty}\right) dt \leq$$

$$\int_0^{2\pi} \Phi\left(\left(\frac{\|f\|_\infty}{\|\varphi_{1,r}\|_\infty \|f^{(r)}\|_\infty}\right)^{(r-k)/r} |\varphi_{1,r-k}(t)|\right) dt \tag{1.168}$$

*holds true and therefore for any $p \in [1, \infty]$*

$$\left(\frac{\|f^{(k)}\|_p}{\|\varphi_{1,r-k}\|_p \|f^{(r)}\|_\infty}\right)^{1/(r-k)} \leq \left(\frac{\|f\|_\infty}{\|\varphi_{1,r}\|_\infty \|f^{(r)}\|_\infty}\right)^{1/r}. \tag{1.169}$$

**Proof.** From theorem 1.4.7 it follows that if $\psi(t) = f^{(k-1)}(t)$, then

$$\frac{|\psi'|}{\|\psi^{(r-k+1)}\|_\infty} \prec \left(\frac{\|\psi\|_\infty}{\|\varphi_{1,r-k+1}\|_\infty \|\psi^{(r)}\|_\infty}\right)^{(r-k)/(r-k+1)} |\varphi_{1,r-k}|$$

or

$$\frac{|f^{(k)}|}{\|f^{(r)}\|_\infty} \prec \left(\frac{\|f^{(k-1)}\|_\infty}{\|\varphi_{1,r-k+1}\|_\infty \|f^{(r)}\|_\infty}\right)^{(r-k)/(r-k+1)} |\varphi_{1,r-k}|$$

¿From here and fron inequality (1.157) we obtain the inequality (1.167). Ther inequalities (1.168) and (1.169) follow from (1.167) and from theorem 1.3.11. The exactness of these inequalities follows from the fact that they turn into equalities for functions in the form of (1.166). By analogy from theorem 1.4.10 we can infer the next statements.

**Theorem 1.7.6.** *If $r \in N$, $k = 1, 2, \ldots, r-1$ and $f \in L^r_\infty$, then for all $\lambda \in R$ and $1 \leq p \leq \infty$ the following inequalities*

$$\left(\frac{\|(f^{(k)} - \lambda)_\pm\|_p}{\|(\varphi_{1,r-k}(\alpha, \beta) - \lambda)_\pm\|_p}\right)^{1/(r-k)} \leq \left(\frac{E_0(f)_\infty}{E_0(\varphi_{1,r}(\alpha, \beta))_\infty}\right)^{1/r}$$

*hold true. In these inequalities $\alpha = \|f_+^{(r)}\|_\infty$ and $\beta = \|f_-^{(r)}\|_\infty$.*

**Theorem 1.7.7.** *If $r \in N$, $k = 1, 2, \ldots, r-1$ and $f \in L^r_\infty$, then for all $\lambda \in R$ and $1 \le p \le \infty$ the following inequalities*

$$\left(\frac{E_0(f^{(k)})_p}{2\pi E_0(B_{r-k})_p \|f_\pm^{(r)}\|_\infty}\right)^{1/(r-k)} \le \left(\frac{E_0(f)_\infty}{2\pi E_0(B_r)_\infty \|f_\pm^{(r)}\|_\infty}\right)^{1/r}$$

*($B_r(t)x$ is Bernully's function) hold true. For any $\alpha, \beta > 0$ the next inequalities*

$$\left(\frac{E_0(f^{(k)})_p}{E_0(\varphi_{1,r-k}(\alpha,\beta))_p \|f_+^{(r)}/\alpha + f_-^{(r)}/\beta\|_\infty}\right)^{1/(r-k)} \le$$

$$\left(\frac{E_0(f)_\infty}{2\pi E_0(\varphi_{1,r}(\alpha,\beta))_\infty \|f_+^{(r)}/\alpha + f_-^{(r)}/\beta\|_\infty}\right)^{1/r}$$

*are valid.*

Note still that from the corollary 1.5.7 it immediately follows the justice of the next statement.

**Theorem 1.7.8.** *Let $r = 2, 3, \ldots$, $k = 1, 2, \ldots, r-1$, $\Phi(t)$ is $N$-function, $m(f)$ is the number of local maximums for $f$ on period and $\mu(f)$ is the number of sign changes for $f$ on period. Then for any function $f \in V^r$ the exact on $V^r$ inequalities*

$$\int_0^{2\pi} \Phi\left(\frac{|f(t)|}{V_0^{2\pi}(f^{(r)})}\right) dt \ge$$

$$\frac{m(f)}{2}\left(\frac{G_{r,k}(f)}{\|\varphi_{1,r-k}\|_\infty}\right)^{1/(r-k)} \int_0^{2\pi} \Phi\left(\left(\frac{G_{r,k}(f)}{\|\varphi_{1,r-k}\|_\infty}\right)^{r/(r-k)} \frac{2|\varphi_{1,r}(t)|}{m(f)}\right) dt \ge$$

$$\frac{\mu(f')}{2}\left(\frac{G_{r,k}(f)}{\|\varphi_{1,r-k}\|_\infty}\right)^{1/(r-k)} \int_0^{2\pi} \Phi\left(\left(\frac{G_{r,k}(f)}{\|\varphi_{1,r-k}\|_\infty}\right)^{r/(r-k)} \frac{2|\varphi_{1,r}(t)|}{\mu(f')}\right) dt$$

*where*

$$G_{r,k}(f) = V_0^{2\pi}(f^{(k)})/V_0^{2\pi}(f^{(r)}).$$

*And in particular for all $1 \le p \le \infty$ and $\mu(f') = 2m$*

$$\left(\frac{4mV_0^{2\pi}(f^{(r-k)})}{V_0^{2\pi}(f^{(r)})V_0^{2\pi}(\varphi_{m,r})}\right)^{1/k} \le \left(\frac{4m\|f\|_p}{\|\varphi_{m,r}\|_p V_0^{2\pi}(f^{(r)})}\right)^{1/(1+1/p)}.$$

## 1.8 Stain's method

Under the proving of theorem 1.7.2 we used the method allowing us to obtain from the inequality between the norms of derivatives in the space $L_\infty$ the analoguos inequality in another space. The development of this method is very useful and in another situations. In this section we'll adduce the general proposition of the same type. Let $X \subset L_\infty$ be a real linear normed space of periodic functions with the translate invariant norm, i.e. for any $f \in X$ and $\alpha \in R$

$$||f(\cdot + \alpha)||_X = ||f(\cdot)||_X,$$

$H$ and $\aleph$ are linear manifolds in $L_\infty$ and $X$ correspondingly, translate invariant and such that for any $F \in X^\rho$ and $f \in H$ function $F(f(\cdot + \alpha))$ of variable $t$ belongs to $H$.

Let moreover, the mappings

$$A_0, A_1, \dots, A_n : H \to X$$

are translate unvariant, i.e. for all $t \in R$

$$A_i(f(\cdot + t)) = (A_i f)(\cdots + t)$$

and such that for all $F \in X^\rho$ and $f \in \aleph$ the equalities

$$F[A_i(f(\cdot + t))] = A_i[F(f(\cdot + t))] \tag{1.170}$$

hold true. Then the next statement takes place.

**Theorem 1.8.1.** *Let for each function $f \in H$ the inequalities*

$$||A_0 f||_\infty \le \varphi(||A_1 f||_\infty, \dots, ||A_n f||_\infty) \tag{1.171}$$

*hold true, where $\varphi(x_1, x_2, \dots, x_n)$ is real function monotonously nondecreasing by every variable. Then for any function $f \in \aleph$*

$$||A_0 f||_X \le \varphi(||A_1 f||_X, \dots, ||A_n f||_X). \tag{1.172}$$

**Proof.** Let $f \in \aleph$. Then under the conditions of theorem for each functional $f \in X^\rho$ we have that $F(f(\cdot + t)) \in H$ and, consequently in view of (1.171) we obtain

$$||A_0 F(f(\cdot + t))||_\infty \le \varphi(||A_1 F(f(\cdot + t))||_\infty, \dots, ||A_n F(f(\cdot + t))||_\infty). \tag{1.173}$$

Taking into account the equality (1.170), we may write (1.173) in the form

$$||F[A_0(f(\cdot + t))]||_\infty \le \varphi(||F[A_1(f(\cdot + t))]||_\infty, \ldots, ||F[A_n(f(\cdot + t))]||_\infty). \tag{1.174}$$

Passing to upper bounds by $F \in X^\rho$ for which $||F|| \le 1$ in both parts of inequality (1.174) and taking into account the monotony of function $\phi$, we obtain

$$\sup_{||F|| \le 1} ||F[A_0(f(\cdot + t))]||_\infty \le$$

$$\varphi( \sup_{||F|| \le 1} ||F[A_1(f(\cdot + t))]||_\infty, \ldots, \sup_{||F|| \le 1} ||F[A_n(f(\cdot + t))]||_\infty). \tag{1.175}$$

But for any $i = 0, 1. \ldots, n$

$$\sup_{||F|| \le 1} ||F[A_i(f(\cdot + t))]||_\infty =$$

$$vrai \sup_t ||A_i(f(\cdot + t))||_X = vrai \sup_t ||A_i(f)||_X = ||A_i(f)||_X, \tag{1.176}$$

that together with previous inequality completes the proof of theorem 1.8.1.

Furthermore we'll adduce a few applications of theorem 1.8.1. Here we note only that to prove the theorem 1.7.2 we are to put in the theorem 1.8.1 the following: $n = 2, X = L_1$, $H = L^r_\infty$, $\aleph = L^r_1$ ,

$$A_1(f) = f, \quad A_2(f) = \frac{d^r}{dt^r} f, \quad A_0(f) = \frac{d^k}{dt^k} f$$

and

$$\varphi(x_1, x_2) = c_{r,k} x_1^{(r-k)/r} x_2^{k/r},$$

where

$$c_{r,k} = ||\varphi_{1,r-k}||_\infty ||\varphi_{1,r}||_\infty^{(r-k)/r}$$

and employ the Kolmogorov's inequality.

## 1.9 Commentaries

Sec. 1.1. The general properties of the best one-sided approximations are stated in detail in the monograph of Korneichuk N.P., Ligun A.A., Doronin V.G. [1]. The best $\alpha, \beta$-approximations in the spaces of functions and its tie with the best one-sided approximations were studied in the papers of V.F.Babenko [1, 3, 8].

Sec.1.2. The duality correlations for the best approximations were effectively used for the decision of extremal problems of approximation theory by S.M.Nikolski [1].

The proof of the theorem 1.2.1 (basing only on separability theorem) is conducted according to scheme proposing by V.M.Tihomirov. The wording of this theorem was given by V.F.Babenko [3]. The duality correlations for the best one-sided approximations were considered in the monograph of N.P.Korneichuk, A.A.Ligun, V.G.Doronin [1]. As to duality for the best $\alpha, \beta$-approximations one can familiarize oneself with this results with the help of V.F.Babenko works [1, 3, 8].

The monographs of A.D.Joffe, V.M.Tihomirov [1] and E.G.Golstain are devoted to the duality extremal problems and its applications in the approximation theory.

Sec.1.3. A great number of useful properties for rearrangement are contained in the monographs of G.H.Hardy, D.I.Littelwood and D.Polia [2], A.Zigmund [2], N.P.Korneichuk, A.A.Ligun, V.G.Doronin [1]. The idea of $\Sigma$-rearrangement was proposed by N.P.Korneichuk [5]. The theorem 1.3.11 belongs to G.H.Hardy, D.I.Littelwood, D. Polia [1].

Sec.1.4. Lemma 1.4.1 and proposition 1.4.2 rise to the comparison theorem of A.N.Kolmogorov [2], its different variants were established by many authors.

Theorem 1.4.6 in the important case when $b = n^-1$ was proved by N.P.Korneichuk. Furthermore its different variants were established by V.G. Doronin and A.A.Ligun [2], V.F.Babenko [12, 18] and etc.

Theorems 1.4.7 and 1.4.10 rise to the paper of N.P. Korneichuk [5] in which the analogous results was proved for the functions orthogonal to the trigonometric polynomials. Given here theorems 1.1.7 and 1.4.10 were proved by A.A.Ligun [7] and V.F.Babenko [5, 12, 18] correspondingly.

Sec.1.5. Theorem 1.5.1 was proved by V.F.Babenko [19]. The main results of $\Sigma$-rearrangements theory are stated in detail in the monographs of of N.P.Korneichuk [11, 22, 24], N.P.Korneichuk, A.A.Ligun and V.G.Doronin [1], where the effectiveness of $\Sigma$-rearrangements device under the investigations of different extremal problems of approximation theory was also shown. Theorem 1.5.6 is proved by A.A.Ligun.

Sec.1.6. Therem 1.6.1 was proved by E.Stain and G.Weiss [1]. Theorem 1.6.2 was proved by O.D.Zeretely [1]. Theorem 1.6.3 is proved in the paper of V.F.Babenko [13]. In the same place one can find a great number of the applications of these theorems to the decision of the extremal properties of approximation theory on the classes of conjugate functions.

Sec.1.7. Theorem 1.7.1 belongs A.N.Kolmogorov [2]. A great number of articles are devoted to establishing the inequalities analogous to the Kolmogorov'v inequality in the spaces $L_p$ in mixed metrics and also on semiaxis and on finite segment. Side by side with the given in the text result of E.Stain [1] (theorem 1.7.2) one can find the row of the results that are established in the works of S.B.Stechkin [5-7], V.N.Gabushin [1, 2],

V.V.Arestov [1, 2], Y.N.Subbotin [3],Y.N.Subbotin, L.V.Taikov [1]. Theorems 1.7.5 and 1.7.8 are proved by A.A.Ligun [7, 10]. Theorem 1.7.4 is proved by V.F.Babenko [13].

Nonsymmetric situations in the Kolmogorov's type inequalities at first was investigated by L.Hermander [1]. As to further results then one can familirize oneself with it in the monograph of N.P.Korneichuk, A.A.Ligun, V.G.Doronin [1] and in the works of V.F.Babenko [3, 12].

Sec.1.8. Formulated in this sestion theorem is obtained by V.F.Babenko and S.A.Pichugov. Its proof is the development of Stain's method which was used him under the proving of theorem 1.7.2. Another applications of this method are considered in the work of V.F.Babenko and S.A.Pichugov [1].

# Chapter 2

# Polynomials deviating least from zero

The classical extremal problems on the set of polynomials are problems about the polynomials of the least deviations from zero either in the same metric or another one. The organization of such problems and first results rise to the works of P.L. Chebyshev whose investigations were continued by V.A.Markov and A.A.Markov, A.I.Korkin E.I.Zolotarev and other authors. A great number of results on these problems have been obtained at present. Though we don't claim for completeness of account, then we present in this chapter the typical result.

## 2.1 General properties of polynomials

Integral function $p(x)$ of the form

$$p(x) = a_0 x^n + a_1 x^{n-1} + \ldots + a_{n-1} x + a_n, \; a_0 \neq 0, \tag{2.1}$$

is called by algebraic polynomial of power $n$.

The set of all algebraic polynomials $p(x)$ of power no less then $n$, with $n+1$ real coefficients $a_k$, $k = 0, 1, \ldots, n$ denots by $P_n$. The set $P_n$ is $(n+1)$-th subspace on the space $C[a,b]$ for all $a$ and $b$.

Furthermore we shall study extremal problems on such subsets of the set $P_n$:

$$P_n^+[a,b] = \{p : p \in P_n, \, p(x) \geq 0 \, (x \in [a,b])\};$$

$$P_n^-[a,b] = \{p : p \in P_n, \, p(x) \leq 0 \, (x \in [a,b])\};$$

$$P_n^{\pm} = P_n^{\pm}[-1,1];$$

$P_n(1)$ is a set of all polynomials in the form (2.1) for $a_0 = 1$; $P_n(1, \sigma)$ is a set of all polynomials in the form (2.1) for $a_0 = 1$ and $a_1 = \sigma$;

$$P_n^0 = \{p : p \in P_n, \, p(-1) = 0\};$$

$$P_n^{00} = \{p : p \in P_n, \, p(-1) = p(1) = 0\};$$

$$P_n^0(1) = P_n^0 \bigcap P_n(1);$$

$$P_n^{00}(1) = P_n^{00} \bigcap P_n(1);$$

$$P_n^{0,\pm} = P_n^0 \bigcap P_n^\pm;$$

$$P_n^{00,\pm} = P_n^{00} \bigcap P_n^\pm.$$

We denote $2\pi$-periodic function

$$\tau(x) = a_0 + \sum_{k=1}^{n} (a_k \cos kx + b_k \sin kx), \; |a_n|^2 + |b_n|^2 \neq 0, \tag{2.2}$$

by trigonometric polynomial of order $n$.

The set of all trigonometric polynomials of order no less then $n$ with real coefficients $a_k$, $k = 0, 1, \ldots, n$ and $b_k$, $k = 1, 2, \ldots, n$ denote by $T_{2n+1}$. It is clear that $T_{2n+1}$ is $(2n+1)$-th dimensional subspace of space $C$.

¿From Euler's formula

$$e^{ix} = \cos x + i \sin x$$

it follows that polynomial (2.14) we may represent in the form

$$t(x) = \sum_{k=-n}^{n} c_k e^{ikx}, \tag{2.3}$$

where

$$c_0 = a_0; \; c_k = \frac{a_k - ib_k}{2}; \; c_{-k} = \frac{a_k + ib_k}{2}; \; k = 1, 2, \ldots, n.$$

Put

$$T_{2n+1}^+ = \{\tau : \tau \in T_{2n+1}, \, \tau(x) \geq 0 \; (x \in R)\},$$

$$T_{2n+1}^- = \{\tau : \; \tau \in T_{2n+1}, \, \tau(x) \leq 0 \; (x \in R)\}$$

and denote by $T_{2n+1,m}$ the set of all polynomials $\tau \in T_{1n+1}$ having on period $[0, 2\pi)$ $m$ zeros with the respect of their multiplicities. Trigonometric polynomials from the set of $T_{2n+1,2n}$ are called by polynomials with complet complecte-total of zeros.

In this section we present some well-known properties of algebraic and trigonometric polynomials which repetedly will present furthermore.

Before all we present the statement which sometime is called as mane theorem of algebra .

**Theorem 2.1.1.** *Any algebraic polynomial in the form (2.1) has exactly $n$ zeros (complex in general case), if every from these zeros is calculated so much times how multiplicity it has.*

**Proof.** Show at first that polynomial $p(x)$ of the form (2.1) has at least one zero.

Let

$$a = \max\{|a_i| : i = 1, 2, \ldots, n\}$$

and

$$p_*(x) = \sum_{k=1}^{n} a_k x^{n-k}.$$

Then for $|x| > 1$

$$|p_*(x)| \leq a \sum_{k=1}^{n} |x|^k = \frac{a(|x|^n - 1)}{|x| - 1} \leq \frac{a|x|^n}{|x| - 1}.$$

Therefore when $|x| > 1$

$$|p(x)| = |a_0 x^n + p_*(x)| \geq |a_0 x^n| - |p_*(x)| \geq$$

$$|a_0|\,|x|^n - \frac{a|x|^n}{|x| - 1} = |x|^n a_0 \left(1 - \frac{a}{|a_0|(|x| - 1)}\right).$$

¿From this inequality it follows that for any $M > 0$ there exists $N > 1$ such that for all $|x| > N$ inequality $|p(x)| > M$ holds true. From this it follows that for some $N_1 > 0$

$$\inf\{|p(x)| : x \in R\} = \inf\{|p(x)| : |x| < N_1\}.$$

Besides this function $|p(x)|$ is continuous on the set $\{x : |x| \leq N_1\}$ that in view of Weierstrass's theorem it achieves the minimum on this set. Let

$$|p(x_*)| = \min\{|p(x)| : |x| < N_1\} = \min\{|p(x)| : x \in R\}.$$

Show that $p(x_*) = 0$. Suppose that opposite fact is true, i.e that $p(x_*) \neq 0$. Select index $k$ from the conditions

$$p^{(\nu)}(x_*) = 0, \;\; \nu = 1, 2, \ldots, k - 1, \;\; p^{(k)}(x_*) \neq 0.$$

Such index exists, because

$$p(x_*) = n!\, a_0 \neq 0.$$

¿From the Taylor's formula

$$p(x_* + h) = p(x_*) + \sum_{\nu=k}^{n} \frac{h^\nu}{\nu!} p^{(\nu)}(x_*)$$

and, hence,

$$\frac{p(x_* + h)}{p(x_*)} = 1 + ch^k + ch^k \sum_{\nu=1}^{n-k} b_\nu h^\nu,$$

where

$$c = \frac{p^{(k)}(x_*)}{p(x_*)k!}$$

and

$$b_\nu = \frac{p^{(k+\nu)}(x_*)}{cp(x_*)(k+\nu)!}, \quad \nu = 1, 2, \ldots, n-k.$$

Let $h$ is any number, which satisfies the conditions

$$arg\, h = (\pi - arg\, c)/k \tag{2.4}$$

$$|ch^k| < 1 \tag{2.5}$$

and

$$\left| \sum_{\nu=1}^{n-k} b_\nu h^\nu \right| < \frac{1}{2}. \tag{2.6}$$

Then in viev of (2.4) and (2.5) we have

$$|1 + ck^k| = 1 - |c|\,|h^k|$$

and therefore

$$\left| \frac{p(x_* + h)}{p(x_*)} \right| \leq |1 + ch^k| + \frac{1}{2}|ch^k| = 1 - \frac{1}{2}|c||h|^k < 1,$$

i.e. in such choice $h$

$$|p(x_* + h)| < |p(x_*)|.$$

This fact contradicts to previous fact. Since the point $x_*$ is a point of minimum for function $|p(x)|$.

Thus, polynomial $p(x)$ has at least one zero $x_*$. But then from Taylor's formula

$$p(x) = (x - x_*)p_1(x),$$

where

$$p_1(x) = \sum_{\nu=1}^{n-1} \frac{1}{(\nu+1)!} p^{(\nu+1)}(x_*)(x - x_*)^\nu$$

is a polynomial of order $n-1$. And now the statement of theorem it's easy to obtain with the help of induction.

**Theorem 2.1.2.** *Any polynomial $p \in P_n$ we may represent in the form*

$$p(x) = a \prod_{\nu=1}^{m} (x - \alpha_\nu) \prod_{\mu=1}^{k} ((x - \beta_\mu)^2 + \gamma_\mu^2), \quad m + 2k \le n,$$

*where $\alpha_\nu$, $\nu = 1, 2, \ldots, m$ are real zeros of polynomial $p(x)$ and $\beta_\mu \pm i\gamma_\mu$, $\mu = 1, 2, \ldots, k$ are complex zeros of polynomial $p(x)$.*

To prove this statement it is sufficiently to note that if $p(x)$ is a polynomial with real coefficients and $z_\mu = \beta_\mu + i\gamma_\mu$ is a complex zero of it multiplicity $l$ of $p(x)$, then number $\overline{z}_\mu = \beta_\mu - i\gamma_\mu$ is also a zero of multiplicity $l$ of $p(x)$ and

$$(x - z_\mu)(x - \overline{z}_\mu) = (x - \beta_\mu)^2 + \gamma_\mu^2.$$

The next statement presents us the analog of theorem 2.1.1 for trigonometric polynomials.

**Theorem 2.1.3.** *Any trigonometrical polynomial of order to $n$ has exactly $2n$ zeros (complex in general) in zone $0 \le Re\, z < 2\pi$, if every from these zeros is calculated so much times how multiplicity it has.*

**Proof.** Let polynomial $\tau(x)$ define by the equality (2.3). Consider the fraction

$$R(z) = \sum_{k=-n}^{n} c_k z^k. \tag{2.7}$$

Since

$$R((\overline{z})^{-1}) = \overline{R}(z),$$

then all zeros of fraction $R(z)$ are symmetric relatively to the sircumference $z = e^{ix}$. Besides that due to theorem 2.1.1 fraction $R(z)$ has $2n$ zeros with respect of their multiplicities. Thus there exist numbers $z_k$, $|z_k| < 1$, $k = 1, 2, \ldots, m$ and $|w_k| = 1$, $k = 1, 2, \ldots, 2n - 2m$, such that

$$R(z) = \frac{c_n}{z^n} \prod_{k=1}^{m} (z - z_k) \left( z - \frac{1}{\overline{z}_k} \right) \prod_{k=1}^{2n-2m} (z - w_k). \tag{2.8}$$

Supposing now that $z = e^{ix}$, we obtain the equality

$$\tau(x) = \frac{c_n}{z^n} e^{ix} \prod_{k=1}^{m} (e^{ix} - z_k) \left( e^{ix} - \frac{1}{\overline{z}_k} \right) \prod_{k=1}^{2n-2m} (e^{ix} - w_k) \tag{2.9}$$

from which the statement of theorem 2.1.3 easely follows.

**Corollary 2.1.4.** *For any trigonometrical polynomial $\tau \in T_{2n+1}$ there exist numbers $a \in R$ and $0 \leq x_1 \leq x_2 \leq \ldots \leq x_{2n} < 2\pi$ such that*

$$\tau(x) = a \prod_{k=1}^{2n} \sin \frac{x - x_k}{2}.$$

This statement immediately follows from (2.9) and from the fact that for $\tau \in T_{2n+1,2n}$ in the representation (2.9) under $m = 0$.

Present now representations for nonnegative algebraic and trigonometric polynomials.

**Theorem 2.1.5.** *Any polynomial $\tau \in T^{+}_{2n+1}$ we may represent in the form*

$$\tau(x) = \left| \sum_{k=0}^{l} \gamma_k e^{ikx} \right|^2, \quad l \leq n,$$

*where function*

$$g(x) = \sum_{k=0}^{l} \gamma_k e^{ikx}$$

*has no zeros at lower semiplane.*

**Proof.** Let polynomial $\tau \in T^{+}_{2n+1}$ has the form (2.3) and fraction $R(z)$ is defined by equality (2.7). Then from (2.8) and inequality $\tau(x) \geq 0$ it follows that all zeros $w_k$ of fraction $R(z)$ are zeroz of odd multiplicity. Fraction $R(z)$ we may represent in such way

$$R(z) = R(z) = \frac{c_n}{z^n} \prod_{k=1}^{m} (z - z_k)(z^{-1} - \overline{z}_k) \prod_{k=1}^{n-m} (z - w_k)^2 =$$

$$R(z) = c_* \prod_{k=1}^{m} (z - z_k)(z^{-1} - \overline{z}_k) \prod_{k=1}^{n-m} (z - w_k)(z^{-1} - \overline{w}_k).$$

Supposing that $z = e^{ix}$ we obtain

$$\tau(x) = R(e^{ix}) = c_* \prod_{k=1}^{m} \left|e^{ix} - z_k\right|^2 \prod_{k=1}^{n-m} \left|e^{ix} - w_k\right|^2 =$$

$$\left| \sqrt{c_*} \prod_{k=1}^{m} (e^{ix} - z_k) \prod_{k=1}^{n-m} (e^{ix} - w_k) \right|^2.$$

It remaines to note that $|z_k| < 1$, $k = 1, 2, \ldots, m$ and $|w_k| = 1$, $k = 1, 2, \ldots, n - m$.

**Corollary 2.1.6.** *Let $\tau \in T^{+}_{2n+1}$ is odd trigonometrical polynomial. Then there exist real numbers $\gamma_k$, $k = 0, 1, \ldots, l$, $l \leq n$ such that*

$$\tau(x) = \left| \sum_{k=0}^{l} \gamma_k e^{ikx} \right|^2,$$

*i.e. if*

$$\tau(x) = \gamma_0 + \sum_{k=1}^{l} \gamma_k \cos kx,$$

*then*

$$\gamma_0 = \sum_{\nu=0}^{l} \gamma_\nu^2,$$

$$\gamma_k = 2 \sum_{\nu=0}^{l-k} \gamma_\nu \gamma_{\nu+k}, \quad k = 1, 2, \ldots, l.$$

Really, in this case all coefficients $c_k$ in representation (2.9) are real and, consequently, all complex zeroz of function $R(z)$ are pair conjugate.

**Theorem 2.1.7.** *Let $p \in P_n^{+}$ and $n = 2m$, then there exist some polynomials $A \in P_m$ and $B \in P_{m-1}$ such that*

$$p(x) = A^2(x) + (1 - x^2) B^2(x), \tag{2.10}$$

*and if $n = 2m + 1$ then there exist some polynomials $A \in P_m$ and $B \in P_m$ such that*

$$p(x) = (1 - x) A^2(x) + (1 + x) B^2(x). \tag{2.11}$$

Consider the problem about the best approximation element in the case of appproximation for function by algebraic and trigonometric polynomials. This problem is decided by the next Chebyshev's theorem.

**Theorem 2.1.8.** *The algebraic polynomial $p \in P_n$ gets the best approximation on segment $[a, b]$ to function $f \in C[a, b]$ among all polynomials of power no less then $n$ if and only if there exist $n + 2$ points such that*

$$a \leq x_1 < x_2 < \ldots < x_{n+2} \leq b,$$

*in which the difference*

$$\Delta(x) = f(x) - p(x)$$

*accepts values $\pm \|\Delta\|_{C[a,b]}$ in turn changing the sign.*

The points $\{x_i\}_{i=1}^{n+2}$ are called as Chebychev's alternans of function $f$.

**Proof.** It is clear that difference $\Delta(x)$ has zeros, as in opposite case $p(x)$ doesn't get to function $f(x)$ the best approximation. Denote by $e$ the set of all points from segment $[a, b]$, in which $\Delta(x) = \pm A$, where $A = E(f, P_n)_{C[a,b]} = ||\Delta||_{C[a,b]}$. We shall call by $e_+$-points such points $x$ from the set $e$ in which $\Delta(x) = A$, and by $e_-$-points such points $x$ from the set $e$ in which $\Delta(x) = -A$. The set $e$ contains both $e_+$-points and $e_-$-points. On segment $[a, b]$ we may select points

$$a = z_0 < z_1 < \ldots < z_m < b = z_{m+1},$$

thus on every segment $[z_k, z_{k+1}]$, $(k = 0, 1, \ldots, m)$ is at least one $e$-point, all $e$-points from segment $[z_k, z_{k+1}]$ have same form, that is all they are either $e_+$ or $e_-$-points, $\Delta(z_k) = 0$, $k = 1, 2, \ldots, m$, besides that, the signes of $e$-points from segment $[z_k, z_{k+1}]$ and $[z_{k+1}, z_{k+2}]$, $k = 0\ 1, \ldots, m - 1$ are different. Prove, that $m \geq n + 1$. Assume, that $m \leq n$ and construct the algebraic polynomial

$$p_m(x) = \alpha \prod_{k=1}^{m} (x - z_k)$$

(it is clear, that $p_m \in P_m \subset P_n$), where $\alpha$ is selected, such that, if the segment $[z_m, z_{m+1}]$ contains $e_+$-points, then $\alpha > 0$, and if this segment contains $e_-$-points, then $\alpha < 0$. Let for clearness $\alpha > 0$.

Due to definition of segments $[z_k, z_{k+1}]$, $k = 0, 1, \ldots, m$, there exists number $h > 0$ such that

$$-A + h \leq \Delta(x) \leq A, x \in [z_{m-2\nu}, z_{m-2\nu+1}],$$

and

$$-A \leq \Delta(x) \leq A - h, x \in [z_{m-2\nu-1}, z_{m-2\nu}].$$

Select now $\varepsilon > 0$ such that

$$||\varepsilon p_m||_{C[a,b]} < \frac{h}{2}.$$

Then we shall have

$$||f - p - \varepsilon p_m||_{C[a,b]} < A - \frac{h}{2} = ||f - p||_{C[a,b]} - \frac{h}{2},$$

but this contradicts to the fact that $p(x)$ is a polynomial of the best approximation for $f(x)$ by the set $P_n$ in the metric of space $C[a, b]$.

Sufficiency to conditions of Chebyshev's theorem is established more simply. It's need to prove that for all polynomials $p_* \in P_n$ the inequality

$||f - p_*||_{C[a,b]} > ||f - p||_{C[a,b]} = A$ holds true. Suppose that this is not true, i.e. that for some polynomial $p_0 \in P_n$

$$||f - p_0||_{C[a,b]} < A. \tag{2.12}$$

Consider the algebraic polynomial $p_1(x) = p(x) - p_0(x) \neq 0$, which we may write in the form

$$p_1(x) = (f(x) - p_0(x)) - (f(x) - p(x)).$$

Under conditions of theorem the difference in second bracets in turn accepts values $\pm A$ at points $x_k$,$k = 1, 2, \ldots, n + 2$. From this and from (2.12) it follows that $p_1(x)$ in turn changes sign at points $x_k$, $k = 1, 2, \ldots, n + 2$, and, consequently, it has no less then $n + 1$ zeros on segment $[a, b]$, that is impossible.

The next statement almost immediately follows from theorem 2.1.8.

**Corollary 2.1.9.** *For any function $f \in C[a, b]$ the best uniform approximation polinomial in $P_n$ is unique.*

Really, if for function $f \in C[a, b]$ there exist two different polynomials $p, p_* \in P_n$ of the best approximation, then in view of theorem 2.1.8 there exist points

$$a \leq x_1 < x_2 < \ldots < x_{n+2} \leq b,$$

such that

$$f(x_i) - p(x_i) = (-1)^i \varepsilon A, |\varepsilon| = 1, i = 1, 2, \ldots, n + 2,$$

where

$$A = ||f - p||_{C[a,b]} = E(f, P_n)_{C[a,b]}.$$

But $p_*$ also is a polynomial of the best approximation and, consequently,

$$-A \leq f(x_i) - p_*(x_i) \leq A$$

and then $p(x_i) = p_*$, $i = 1, 2, \ldots, n + 2$, that is impossible.

There are many generalizations of Chebyshev's alternans theorem. In particular, this theorem is true, if space $P_n$ we substitute by arbitrary Chebyshev's system of functions. There are sufficient numbers of alternans theorems under the approximation of functions by splines.

We will need futhermore in such generalization of Chebyshev's theorem.

With the help of repetition of theorem 2.1.9 the truth of following statement is established.

**Theorem 2.1.10.** *Let $f \in C[-1,1]$. In order to the polynomial $p_0 \in P_n^0 := \{p : p \in P_n, p(-1) = f(-1)\}$ gets the best uniform approximation for function $f$ by the set of $P_n^0$ it is necessary and sufficiently that there exist points*

$$-1 < x_1^0 < x_2^0 < \ldots < x_{n+1}^0 \leq 1,$$

*such that*

$$p_0(x_k^0) - f(x_k^0) = (-1)^k \varepsilon \|f - p_0\|_{C[a,b]}, |\varepsilon| = 1, k = 1, 2, \ldots, n+1.$$

*The polynomial of the best uniform approximation for $f \in C[-1,1]$ by the set of $P_n^0$ exists and is unique.*

*In order to the polynomial*

$$p_{00} \in P_{00} := \{p : p \in P_n, p(\pm 1) = f(\pm 1)\}$$

*gets the best uniform approximation of $f$ by the set of $P_n^{00}$, it is necessary and sufficiently that there exist points*

$$-1 < x_1^{00} < x_2^{00} < \ldots < x_n^{00} < 1,$$

*such that*

$$p_{00}(x_k^{00}) - f(x_k^{00}) = (-1)^k \varepsilon \|f - p_{00}\|_{C[-1,1]},$$

$$|\varepsilon| = 1,\ k = 1, 2, \ldots, n+1.$$

Let us give the analog of theorem 1.2.8 for the case of approximations $2\pi$-periodic function by trigonometric polynomials.

**Theorema 2.1.11.** *Trigonometric polynomial $\tau \in T_{2n+1}$ gets to function $f \in C$ the best uniform approximation among all polynomials from the set of $T_{2n+1}$ if and only if when on period $[0, 2\pi)$ there exist $2n+2$ points*

$$0 \leq t_1 < t_2 < \ldots < t_{2n+2} < 2\pi,$$

*for which*

$$\Delta(z_k) = f(t_k) - \tau(t_k) = (-1)^i \varepsilon A, |\varepsilon| = 1, k = 1, 2, \ldots, 2n+2,$$

*where*

$$A = \|f - \tau\|_C.$$

*The polynomial of the best approximation in the set of $T_{2n+1}$ is unique.*

**Proof.** Prove the necessity to conditions of theorem. Let $\tau(t)$ is the best approximation polynomial of the function $f(t)$, i.e.

$$||f-\tau||_C = E(f,T_n)_C = d.$$

On interval $[0,2\pi)$ we select the points $z_k$

$$0 \le z_1 < z_2 < \ldots < z_{2m} < 2\pi \le z_{2m+1} = z_1 + 2\pi$$

or

$$0 < z_1 < z_2 < \ldots < z_{2m-1} < 2\pi \le z_{2m} < z_{2m+1} = z_1 + 2\pi,$$

such that the next conditions will be true: $\Delta(z_k) = 0, k = 1,2,\ldots,m$, Every from segments $[z_{2k-1}, z_{2k}]$ contains at least one $e_+$-point and doesn't contain $e_-$-points. The analogous fact is also true to the segment $[z_{2k}, z_{2k+1}]$. It is necessary to prove, that $m \ge n+1$. Let it is not true i.e. let $m \le n$ and

$$\tau_m(t) = \alpha \prod_{k=1}^{2m} \sin\frac{t-z_k}{2}, \ \alpha < 0.$$

So as

$$2\sin\frac{t-\beta}{2}\sin\frac{t-\gamma}{2} = \cos\frac{\beta-\gamma}{2} - \cos t,$$

then

$$\tau_m \in T_{2m+1} \subset T_{2n+1}.$$

Polynomial $\tau_m(t)$ is positive on the intervals $(z_{2k-1}, z_{2k})$ and negative on the intervals $(z_{2k}, z_{2k+1})$.

Select $h > 0$ so small that

$$-d + h \le \Delta(t) \le d, \ t \in [z_{2k-1}, z_{2k}]$$

and

$$-d \le \Delta(t) \le d - h, \ t \in [z_{2k}, z_{2k+1}].$$

If $\alpha < 0$ is selected such that $||\tau_m||_C < h/2$ then inequality

$$||f - (\tau + \tau_m)||_C \le ||f - \tau||_C - h/2.$$

will be true. This contradicts to the fact that $\tau(t)$ is the best approximation polynomial of the function $f(t)$.

Sufficientcy of conditions and uniqueness are proved thus as in algebraic case.

## 2.2 Least deviation from zero in $C$ and $L_1$

Function

$$T_n(x) = \cos n(arc\cos x) \tag{2.13}$$

is algebraic polynomial of order to $n$. Really, if

$$\cos nt = \frac{1}{2}(e^{int} + e^{-int}) = \frac{1}{2}((\cos t + i\sin t)^n + (\cos t - i\sin t)^n), \tag{2.14}$$

then

$$\cos nt = \cos^n t - C_n^2 \cos^{n-2} t sin^2 t + C_n^4 \cos^{n-4} t \sin^4 t + \dots$$
$$+(-1)^{[n/2]} \begin{cases} \sin^n t,\ n = 2k, \\ \cos t \sin^{n-1} t,\ n = 2k-1. \end{cases}$$

Changing $\sin^2 t$ on $1 - \cos^2 t$ we otain

$$\cos nt = 2^{n-1}\cos^n t - \frac{n}{1}2^{n-3}\cos^{n-2} t + \frac{n}{2}C_n^1 2^{n-5}\cos^{n-4} t -$$
$$\frac{n}{3}C_n^2 2^{n-7}\cos^{n-6} t + \dots + (-1)^{[n/2]} \begin{cases} 1,\ n - 2k, \\ n\cos t,\ n = 2k-1. \end{cases}$$

Therefore

$$T_n(x) = 2^{n-1}x^n - \frac{n}{1}2^{n-3}x^{n-2} + \frac{n}{2}C_n^1 2^{n-5}x^{n-4} - \frac{n}{3}C_n^2 2^{n-7}x^{n-6} + \dots$$
$$+ (-1)^{[n/2]} \begin{cases} 1,\ n - 2k, \\ nx,\ n = 2k-1. \end{cases} = \frac{n}{2}\sum_{k=0}^{[n/2]} \frac{(-1)^k}{n-k} C_{n-k}^k (2x)^{n-2k}. \tag{2.15}$$

Polynomial (2.13) is called as Chebyshev's polynomial of first order. Some of its wonderful properties we present in this section.

¿From (2.14) it follows that for $|x| > 1$ polynomial $T_n(x)$ we may represent in the form

$$T_n(x) = \frac{1}{2}\left((x + \sqrt{x^2-1})^n + (x - \sqrt{x^2-1})^n\right). \tag{2.16}$$

¿From the definition (2.13) and from the equality

$$\cos(n+2)\theta + \cos n\theta = 2\cos\theta\cos(n+1)\theta$$

the next recurrent correlations follow for Chebyshev's polynomials:

$$T_{n+2}(x) = 2xT_{n+1}(x) - T_n(x) \tag{2.17}$$

As $T_0(x) = 1$ and $T_2(x) = x$ then

$$T_2(x) = 2x^2 - 1,$$

$$T_3(x) = 4x^3 - 3x,$$

$$T_4(x) = 8x^4 - 8x^2 + 1,$$

$$T_5(x) = 16x^5 - 20x^3 + 5x,$$

$$T_6(x) = 32x^6 - 48x^4 + 8x^2 - 1,$$

$$T_7(x) = 64x^7 - 112x^5 + 56x^3 - 7x, \ldots.$$

¿From (2.13) it follows that Chebyshov's polynomial $T_n(x)$ has exactly $n$ zeros at points

$$x_k = -\cos\frac{2k-1}{2n}\pi, \quad k = 1, 2, \ldots, n, \tag{2.18}$$

and

$$-1 < x_1 < x_2 < \ldots < x_n < 1.$$

Moreover, $T_n(x)$ has alternans property. Really if

$$\theta_k = \cos(n-k)\pi/n, \quad k = 0, 1, \ldots, n, \tag{2.19}$$

then

$$-1 = \theta_0 < \theta_1 < \ldots < \theta_n = 1$$

and in view of (2.13)

$$T_n(\theta_k) = (-1)^{n-k} = (-1)^{n-k}||T_n||C, \quad k = 0, 1, \ldots, n. \tag{2.20}$$

¿From (2.20) and (2.15) and from theorem 2.1.8 it follows that

$$E(T_n, P_{n-1})_{C[-1,1]} = 2^{n-1}E(x^n, P_{n-1})_{C[-1,1]} = 2^{n-1}$$

and polynomial of the best approximation for function $T_n(x)$ (and, consequently, for $x^n$ ) in $P_n$ in the metric of space $C[-1,1]$ is unique. In equivalent form this fact we may formulate so.

**Theorem 2.2.1.** *Among all polynomials $p \in P_n(1)$ the unique polynomial $2^{1-n}T_n(x)$ has the least norm in the space $C[-1,1]$. In other terms*

$$min\{||p||_{C[-1,1]} : p \in P_n(1)\} = 2^{1-n}||T_n||_{C[-1,1]} = 2^{1-n}.$$

Chebyshev's polynomials have the numbers of other extremal properties. First we note, that from equations (2.20) and from theorem 2.1.10 the next statement will be obtained in similar way.

**Theorem 2.2.2.** *Among all polynomials $p \in P_n^0(1)$ the unique polynomial*

$$T_{n,0}(x) = 2^{1-n} \cos^{-2n} \frac{\pi}{4n} T_n \left( \sin^2 \frac{\pi}{4n} + x \cos^2 \frac{\pi}{4n} \right),$$

*has the least norm in the space $C[-1,1]$. In other terms*

$$min\{||p||_{C[-1,1]} : p \in P_n^0(1)\} = 2^{1-n} ||T_{n,0}||_{C[-1,1]} = 2^{1-n} \cos^{-2n} \frac{\pi}{4n}.$$

*Among all polynomials $p \in P_n^{00}(1)$ the unique polynomial*

$$T_{n,00}(x) = 2^{1-n} \cos^{-2} \frac{\pi}{2n} T_n \left( x \cos \frac{\pi}{2n} \right),$$

*has the least norm in the space $C[-1,1]$. In other terms*

$$min\{||p||_{C[-1,1]} : p \in P_n^{00}(1)\} = 2^{1-n} ||T_{n,00}||_{C[-1,1]} = 2^{1-n} \cos^{-n} \frac{\pi}{2n}.$$

Furthermore we will need in the next property of Chebyshev's polynomials.

**Proposition 2.2.3.** *For $n = 1, 2, \ldots$, $k = 1, 2, \ldots, n$ and $x \in [-1, 1]$*

$$|T_n^{(k)}(x)| \le T_n^{(k)}(1) = \frac{n^2(n^2 - 1^2)(n^2 - 2^2)\ldots(n^2 - (k-1)^2)}{(2k-1)!!}. \tag{2.21}$$

*(the first correlation holds true and when $k = 0$ ).*

**Proof.** Supposing that $\theta = arc \cos x$, we obtain

$$T_n(x) = n \frac{\sin n\theta}{\sin \theta} = 2n \sum_{k=0}^{[(n-1)/2]} T_{n-1-2k}(x). \tag{2.22}$$

With the help of induction we find

$$T_n^{(k)}(x) = 2^{k-1}(k-1)! \left( \frac{1}{2} C_{n-1}^{k-1} + \sum_{\nu=1}^{[(n-1)/2]} C_{n+1+\nu}^{k-1} C_{n-1-\nu}^{k-1} \right) T_{n-k-\nu}(x).$$

¿From here it immedietely follows that for $x \in [-1, 1]$

$$|T_n^{(k)}(x)| \le 2^{k-1}(k-1)! \left( \frac{1}{2} C_{n-1}^{k-1} + \sum_{\nu=1}^{[(n-1)/2]} C_{n+1+\nu}^{k-1} C_{n-1-\nu}^{k-1} \right) =$$

$$T_n^{(k)}(1). \tag{2.23}$$

It remains to prove that

$$T_n^{(k)}(1) = \frac{n^2(n^2-1^2)(n^2-2^2)\dots(n^2-(k-1)^2)}{(2k-1)!!}. \tag{2.24}$$

Wiht the help of direct verification it is easy to convinced that

$$(1-x^2)T"_n(x) - xT_n'(x) + n^2T_n(x) = 0. \tag{2.25}$$

By differentiation of this $(k-1)$-times, we obtain

$$(1-x^2)T_n^{(k+1)}(x) - (2x-1)T_n^{(k)}(x) + (n^2-(k-1)^2)T_n^{(k-1)}(x) = 0. \tag{2.26}$$

When $x = 1$ the equality (2.26) has such form

$$(2k-1)T_n^{(k)}(1) = (n^2-(k-1)^2)T_n^{(k-1)}(1).$$

¿From this equality, with the help of induction, we can easy obtain the equality (2.24).

**Proposition 2.2.4.** *When $y > 1$ for each $p \in P_n$ the inequality*

$$|p(y)| \le ||p||_{C[-1,1]}T_n(y) \tag{2.27}$$

*holds true. This inequality becomes an equality if and only if $p(x) = aT_n(x)$.*

**Proof.** Let

$$R(x) = \frac{p(y)}{T_n(y)}T_n(x) - p(x).$$

Then in view of (2.19) and (2.20)

$$R(\theta_k) = (-1)^{n-k}\frac{p(y)}{T_n(y)} - p(\theta_k),\ k = 0, 1, \dots, n.$$

If (2.27) is not true, then

$$sgn\ R(\theta_k) = \varepsilon(-1)^k,\ |\varepsilon| = 1,\ k = 0, 1, \dots, n,$$

and on every from intervals $(\theta_k, \theta_{k+1})$, $k = 0, 1, \dots, n$ there exists at least one zero of function $R(x)$. Moreover, $R(y) = 0$. Therefore $R(x) \equiv 0$, i.e.

$$p(x) = p(y)T_n(x)/T_n(y)$$

and

$$||p||_{C[-1,1]} \ge |p(1)| = |p(y)|/T_n(y)$$

that contradicts to assumption that (2.27) is not true.

The next statement is proved by analogy.

**Proposition 2.2.5.** *Let $|y| > 1$. Then for any $p \in P_n$ for which $p(\pm 1) = 0$ the inequality*

$$|p(y)| \leq \|p\|_{C[-1,1]} \cos^{-n} \frac{\pi}{2n} \left|T_n\left(y \cos \frac{\pi}{2n}\right)\right|$$

*holds true. This inequality becomes an equality if and only if*

$$p(x) = aT_n\left(x \cos \frac{\pi}{2n}\right)$$

Denote by $A_{n,k}$ the coefficient near $x^k$ for polynomial $T_n(x)$.

**Theorem 2.2.5.** *Among all polynomials $p \in P_n$ with equal to 1 coefficient near $x^k$, $k \leq n$, the polynomial $A_{n,k}^{-1} T_n(x)$ has the least norm in the space $C[-1,1]$, if $k$ and $n$ has identical oddness and the polynomial $A_{n-1,k}^{-1} T_n(x)$ if $k$ and $n$ has distinguish oddness.*

**Proof.** First we establish that if $\gamma_1 < \gamma_2 < \ldots < \gamma_m$, then the function

$$f(x) = \sum_{k=0}^{m} a_k x^{\gamma_k}, \quad a_k \neq 0, \tag{2.28}$$

has no more then $m$ positive zeros. When $m = 1$ this fact is clear. Suppose that this fact holds true for $m = l$, but it is not true for $m = l + 1$. Then there exists the function

$$f(x) = \sum_{k=0}^{l+1} a_k x^{\gamma_k}$$

such that it has more then $l + 1$ positive zeros. All its zeros will be also the zeros of function $f(x)x^{-\gamma_0}$. But then in view of Rolle's theorem the derivative of this function has no more $l$ positive zeros and has such form

$$f(x) = \sum_{k=0}^{l+1} a_k x^{\gamma_k - \gamma_0 - 1}(\gamma_k - \gamma_0)$$

that is impossible.

Let the numbers $n$ and $k$ are even and there exists the polynomial $p \in P_n$ which coefficient near $x^k$ is equal to one and such that

$$\|p\|_{C[-1,1]} \leq \|T_n / A_{n,k}\|_{C[-1,1]} = A_{n,k}^{-1}.$$

Then the even polynomial $(p(x)+p(-x))/2$ also has this property, i.e.

$$\|p(\cdot)+p(-\cdot)\|_{C[-1,1]} \le 2A_{n,k}^{-1}. \tag{2.29}$$

Let

$$R(x) = 2T_n(x)A_{n,k}^{-1} - p(x)_p(-x),$$

then $R(x)$ is even polynomial from $P_n$ which coefficient near $x^k$ equals to zero. Therefore polynomial $R(\sqrt{x})$ has the form (2.28), where $m \le (n-1)/2$, and, consequently, have no more $(n-1)/2$ zeros. But if the number $n$ is even, then the number of zeros of this polynomial no more $(n-2)/2$. Then $R(x)$ has no more $(n-2)/2$ positive zeros.

Moreover, in view of (2.20)

$$R(\theta_l) = 2(-1)^{k-l}A_{n,k}^{-1} - p(\theta_l) - p(-\theta_l), \quad l = 0, 1, \ldots, n.$$

Consequently, $R(x)$ has no less $n$ zeros on interval $(-1,1)$, and in view of eventness of $R(x)$, no less $n/2$ zeros on interval $(0,1)$, that contradicts previous.

The cases when $n$ and $k$ are odd and $n$ and $k$ have distinguish by analogy.

The next statement immediately follows from theorem 2.2.5 and equality (2.15).

**Corollary 2.2.6.** *Let*

$$p(x) = \sum_{\nu=0}^{n} c_\nu x^\nu.$$

*Then, if $n-k$ is even, then*

$$|c_k| \le 2^{k-1}\frac{n}{k!}\frac{((n+k)/2-1)!}{((n-k)/2)!}\|p\|_{C[-1,1]} \tag{2.30}$$

*and if $n-k$ is odd, then*

$$|c_k| \le 2^{k-1}\frac{n-1}{k!}\frac{((n+k-1)/2-1)!}{((n-k-1)/2)!}\|p\|_{C[-1,1]}. \tag{2.31}$$

*Inequalities (2.30) and (2.31) are exact.*

Now consider the algebraic polynomial of power to $n$

$$U_n(x) = \frac{1}{n+1}T'_{n+1}(x), \tag{2.32}$$

which is called Chebyshev's polynomial of order to two. Many their properties follow from properties of Chebyshev's polynomiasls $T_n(x)$.

Thus, from (2.13) it follows that

$$U_n(x) = \frac{\sin((n+1)arc\cos x)}{\sqrt{1-x^2}} \tag{2.33}$$

¿From this representations, in particular, it follows that points

$$y_k = \cos\frac{n+1-k}{n+1}\pi,\ k=1,\ldots,n, \tag{2.34}$$

and only these points are zeroz of polynomial $U_n(x)$.

Moreover, from definition (2.32) and from correlations (2.15), (2.17), (2.21) and (2.28) it follows that

$$2^{-n}U_n \in P_n(1), \tag{2.35}$$

$$U_0(x)=1,\ U_1(x)=2x,$$

$$U_{n+2}(x) = 2xU_{n+1}(x) - U_n(x),\ n=0,1,\ldots, \tag{2.36}$$

$$(1-x^2)U_n''(x) - 3xU_n'(x) + n(n+2)U_n(x) = 0 \tag{2.37}$$

and for all $x \in [-1,1]$, $k=0,1,\ldots,n$

$$|U_n^{(k)}(x)| \le U_n^{(k)}(1)| \le \frac{((n+1)^2-0^2)((n+1^2)-1^2)\ldots((n+1^2)-k^2)}{(n+1)(2k+1)!!}. \tag{2.38}$$

Besides that the correlations

$$\int_{-1}^{1} U_k(x)sgn\,U_n(x)dx = 0,\ k=0,1,\ldots,n-1. \tag{2.39}$$

Really

$$\int_{-1}^{1} U_k(x)sgn\,U_n(x)dx =$$

$$-\frac{1}{n+1}\int_{-1}^{1}\sin((k+1)arc\cos x)sgn\,\sin((n+1)arc\cos x)d(arc\cos x) =$$

$$\int_0^{\pi}\frac{1}{n+1}\sin((k+1)t)\,sgn\sin((n+1)t)dt =$$

$$\frac{1}{2n+2}\int_{-\pi}^{\pi}\sin((k+1)t)sgn\,\sin((n+1)t)dt. \tag{2.40}$$

But

$$sgn\,\sin t = \frac{4}{\pi}\sum_{\nu=1}^{\infty}\frac{\sin(2\nu-1)t}{2\nu-1}$$

and therefore

$$sgn\ \sin((n+1)t) = \frac{4}{\pi}\sum_{\nu=1}^{\infty}\frac{\sin((2\nu-1)(n+1)t)}{(2\nu-1)}.$$

¿From this and from (2.40) we immediately obtain equality (2.39).

The equalities (2.39) we may write in equivalent form

$$\int_{-1}^{1} x^k\, sgn\, U_n(x)dx = 0,\ k = 0,1,\ldots,n-1. \tag{2.41}$$

¿From this and from theorem 1.2.5 (in turn the fact that any doesn't equal to zero polynomial from $P_n$ has no more $2n$ zeros) it follows that

$$E(U_n, P_{n-1})_1 = ||U_n||_1 = \frac{2}{n+1}$$

and the best $L_1$-approximation polynomial of function $U_n$ by the set $P_{n-1}$ is unique. Taking into account now (2.35) we obtain

$$E(x^n, P_{n-1})_1 = 2^{-n}||U_n||_1 = \frac{2^{1-n}}{n+1}$$

and polynomial of the best $L_1$-approximation by the set $P_{n-1}$ is unique.

In equivalent form this fact we may write in such way.

**Theorem 2.2.7.** *Among all polynomials $p \in P_n(1)$ the least norm in $L_1[-1,1]$ has the unique polynomial $2^{-n}U_n(x)$ and*

$$min\{||p||_{L_1[-1,1]} : p \in P_n^{(1)}\} = 2^{-n}||U_n||_{L_1[-1,1]} = \frac{2^{1-n}}{n+1}.$$

The series another extremal properties of Chebyshev's polynomials will be presented in sections 4.3 and 4.4 and, also in next chapters this book when we will obtain the estimates of norms derivatives of agebraic polynomials.

## 2.3 Least deviation from zero in $L_q[-1,1]$

In this section we present some general properties for polynomials of least deviation from zero in the spaces $L_q[-1,1]$ for $1 < q < \infty$. For brevity we will assume that $||\cdot||_q = ||\cdot||_{L_q[-1,1]}$.

The polynomial $p_{n,q} \in P_n(1)$ for which

$$||p_{n,q}||_q = \min\{||p||_q : p \in P_n(1)\},$$

is called as a polynomial of least deviation from zero in the metric of the space $L_q[-1,1]$.

It's clear that

$$||p_{n,q}||_q = E(x^n, P_{n-1})_q.$$

¿From this equality and from criterion of the best approximation element in spaces $L_q[a,b]$ for $1 < q < \infty$ (look the theorems 1.2.5 and 1.2.6 ) it follows that the polynomial $p_{n,q}(x)$ is single-valuedly characterized by the next equalities

$$\int_{-1}^{1} |p_{n,q}|^{q-1} sgn\, p_{n,q}(x) x^k \, dx = 0. \tag{2.42}$$

¿From the fact of strict normity of the space $L_q$ for $q \in (1,\infty)$ it follows that polynomial of the least deviation from zero in the space $L_q$, $q \in (1,\infty)$ is unique.

**Proposition 2.3.1.** *Let $q \in [1,\infty)$. Then for even $n$ the polynomial $p_{n,q}(x)$ is even function and for odd $n$ is odd one.*

Really, if $n$ is even number and

$$p^*_{n,q} = \frac{1}{2}(p_{n,q}(x) + p_{n,q}(-x)),$$

then $p^*_{n,q} \in P_n(1)$ is even function and

$$||p^*_{n,q}||_q = \frac{1}{2}||p_{n,q}(\cdot) + p_{n,q}(-\cdot)||_q \leq$$

$$\frac{1}{2}||p_{n,q}(\cdot)||_q + \frac{1}{2}||p_{n,q}(-\cdot)||_q = ||p_{n,q}||_q.$$

¿From this and from uniqueness of the polynomial of least deviation from zero it follows that $p_{n,q}(x) = p^*_{n,q}(x)$. For odd $n$ the proof is analogical.

**Proposition 2.3.2.** *For $n = 1, 2\ldots$ and $q \in [1,\infty)$ the following equalities*

$$p_{n,q}(1) = (-1)^n p_{n,q}(-1) = ((nq+1)/2)^{1/q} ||p_{n,q}||_q$$

*hold true.*

**Proof.** So for any $\alpha > 0$ the polynomial $a^{-n}p_{n,q}(ax)$ belongs to $P_n(1)$, then for function

$$\phi(a) = a^{-nq}\int_{-1}^{1}|p_{n,q}(ax)|^q dx$$

the correlation

$$\min\{\phi(a) : a > 0\} = \phi(1)$$

will be true and, therefore, $\phi'(1) = 0$. Assuming $f(x) = p_{n,q}(x)$ we obtain

$$\phi'(a) = -nqa^{-nq-1}\int_{-1}^{1}|f(ax)|^q dx +$$

$$qa^{-nq}\int_{-1}^{1}|f(ax)|^{q-1} sgn\, f(ax) f'(ax) dx$$

and

$$\phi'(1) = -nq\int_{-1}^{1}|f(x)|^q dx + \int_{-1}^{1} x d(|f(x)|^q).$$

By means of integration by parts we obtain

$$\int_{-1}^{1} x d(|f(x)|^q) = x d(|f(x)|^q)|_{-1}^{1} - \int_{-1}^{1}|f(x)|^q dx =$$

$$|f(1)|^q + |f(-1)|^q - \int_{-1}^{1}|f(x)|^q dx$$

and taking into account the proposition 2.3.1 and necessary minimum condition we obtain next equations

$$0 = \phi'(1) = 2|f(1)|^q - (nq+1)\int_{-1}^{1}|f(x)|^q dx,$$

$$f(1) = (-1)^n f(-1),$$

that are equivalent to statement of proposition 2.3.2.

Note also that from conditions (2.42) it follows that the function

$$g(x) = |p_{n,q}(x)|^{q-1} sgn\, p_{n,q}(x)$$

changes sign on $(-1,1)$ no less $n$ times. Really, let this fact is not true and $-1 < x_1 < x_2 < \ldots < x_m < 1$, $m < n$ are all points where $g(x)$ changes sign and , consequently, $p_{n,q}$ does it too. If

$$\psi(x) = \prod_{k=1}^{m}(x - x_k),$$

then $\psi \in P_m$ and

$$\int_{-1}^{1} g(x)\psi(x)dx \neq 0,$$

that contradicts to (2.42).

Thus we proved the next statement.

**Proposition 2.3.3.** *All $n$ zeros of polynomial $p_{n,q}(x)$ are real, different and belong to the interval $(-1,1)$.*

Just one property of zeros of polynomial $p_{n,q}(x)$ give us the next proposition.

**Proposition 2.3.4.** *Let $n = 1, 2, \ldots$, $q \in [1, \infty)$ and*

$$-1 < x_1 < x_2 < \ldots < x_n < 1$$

*are all zeros of polynomial $p_{n,q}(x)$. Then*

$$\int_{-1}^{1} \frac{|p_{n,q}(x)|^q}{x - x_i} dx = 0, \quad i = 1, 2, \ldots, n.$$

**Proof.** For any $\varepsilon$ the polynomial

$$p_\varepsilon(x) := \frac{|p_{n,q}(x)|^q}{x - x_i}(x - x_i + \varepsilon)$$

belongs to the set $P_n(1)$ and , therefore, if

$$\psi(\varepsilon) = \int_{-1}^{1} |p_\varepsilon(x)|^q dx$$

then $\psi'(0) = 0$.

It ramains to account that

$$\psi'(0) = q \int_{-1}^{1} \frac{|p_{n,q}(x)|^q}{x - x_i} dx.$$

## 2.4 Legandre polynomials

Let in Gilbert space $H$ the elements $\varphi_k$, $k = 0, 1, \ldots$ construct ? mormolised basis the i.e

$$(\varphi_k, \varphi_i) = \begin{cases} 1, & k = i, \\ 0, & k \neq i \end{cases} \tag{2.43}$$

and $H_n$ is the set of all elements $\varphi$ of the form

$$\varphi = \sum_{k=0}^{n-1} c_k \varphi_k.$$

The proposition 1.2.2 asserts that if $\varphi(\psi) \in H_n$, $\psi \in H$ and

$$E(\psi, H_n)_H = \|\psi - \varphi(\psi)\|_H,$$

i.e. if $\varphi$ is the best approximation element for $\psi$ in $H$, then

$$\varphi(\psi) = \sum_{k=0}^{n-1} (\varphi_k, \psi)\varphi_k.$$

¿From this and from (2.43) it follows that $\varphi(\varphi_n) = 0$. This fact will be conveniently writen in such form.

**Proposition 2.4.1.** *Among all elements of the form*

$$\varphi = \varphi_n + \sum_{k=0}^{n-1} c_k \varphi_k$$

*the unique element $\varphi = \varphi_n$ has the least norm in the space $H$ and at that time*

$$\min_{c_k} \left\| \varphi - \sum_{k=0}^{n-1} c_k \varphi_k \right\|_H = \|\varphi_n\|_H = 1.$$

Consider some particular cases of this statement. The algebraic polynomial of order $n = 1, 2, \ldots$

$$L_n(x) = \frac{n!}{(2n)!} \frac{d^n}{dx^n} (x^2 - 1)^n \tag{2.44}$$

is called by Legandre polynomial. From the equality

$$(x^2 - 1)^n = \sum_{k=0}^{n} (-1)^k C_n^k x^{2n-2k}$$

and from the definition (2.44) it follows that

$$L_n(x) = \frac{1}{(2n)!} \sum_{k=0}^{[n/2]} (-1)^k C_n^k (2n-2k)! x^{n-2k} \tag{2.45}$$

and hence $L_n \in P_n(1)$.

**Lemma 2.4.2.** *For $m \neq n$ the following relations*

$$\int_{-1}^{1} L_n(x) L_m(x) dx = 0 \tag{2.46}$$

*and*

$$\int_{-1}^{1} L_m^2(x) dx = \frac{2}{2m+1} \left( \frac{2^m (m!)^2}{(2m)!} \right)^2 \tag{2.47}$$

*hold true.*

**Proof.** Without the loss of generality we may assume that $m > n$.

For any function $f \in L_1^m[-1,1]$ with the help of integrating by parts we receive

$$\int_{-1}^{1} f(x) L_m(x) dx = \frac{m!}{(2m)!} \int_{-1}^{1} f(x) \frac{d^m}{dx^m} (x^2-1)^m dx =$$

$$= \frac{m!}{(2m)!} \int_{-1}^{1} f^{(m)}(x) (1-x^2)^m dx,$$

hence

$$\int_{-1}^{1} f(x) L_m(x) dx = \frac{m!}{(2m)!} \int_{-1}^{1} f^{(m)}(x) (1-x^2)^m dx. \tag{2.48}$$

If we take into consideration now that $L_n \in P_n$ and $m > n$ , i.e. $L_n^{(m)}(x) = 0$ then we immediately obtain (2.46). Moreover, if $L_m \in P_m$ then $L_m^{(m)}(x) = m!$. ¿From this and from (2.48) we can get

$$\int_{-1}^{1} L_m^2(x) dx = \frac{(m!)^2}{(2m)!} \int_{-1}^{1} (1-x^2)^m dx =$$

$$= \frac{(m!)^2}{(2m)!} \frac{2}{2m+1} \frac{4^m (m!)^2}{(2m)!} = \frac{2}{2m+1} \left( \frac{(m!)^2 2^m}{(2m)!} \right)^2.$$

¿From lemma 2.4.2 the next statement immediately follows.

**Proposition 2.4.3.** *Algebraic polynomials*

$$L_n^*(x) = \sqrt{\frac{2n-1}{2}} \frac{(2n)!}{2^n (n!)^2} L_n(x) = \sqrt{\frac{2n+1}{2}} \frac{2^n n!}{(2n)!} \frac{d^n}{dx^n}(x^2-1) \qquad (2.49)$$

*construct the orthonormed system in the space* $L_2[-1,1]$.

¿From the propositions 2.4.1 and 2.4.3 it follows that among all polynomials of the form

$$L_n^*(x) + \sum_{k=0}^{n-1} c_k L_k^*(x)$$

the polynomial $L_n^*(x)$ has the least norm (that is equal to one) in the space $L_2[-1,1]$ . In accordance with (2.49) we conclude that among all polynomials of the form

$$L_n(x) + \sum_{k=0}^{n-1} c_k L_k(x)$$

the polynomial $L_n(x)$ has the least norm in the space $L_2[-1,1]$ which is equel to

$$\left(\sqrt{\frac{2n+1}{2}} \frac{(2n)!}{2^n (n!)^2}\right)^{-1}.$$

As any polynomial $p \in P_n(1)$ may be single-valuedly represented in the form

$$p(x) = L_n(x) + \sum_{k=0}^{n-1} c_k L_k(x),$$

then the next statement is valid.

**Theorem 2.4.4.** *Among all polynomials* $p \in P_n(1)$ *the polynomial* $L_n(x)$ *has the least norm in the space* $L_2[-1,1]$ . *In addition to this fact*

$$\|L_n\|_{L_2[-1,1]} = E(x^n, P_{n-1})_{L_2[-1,1]} =$$

$$= \min_{c_k} \left( \int_{-1}^{1} \left( x^n - \sum_{k=0}^{n-1} c_k x^k \right)^2 dx \right)^{1/2} = \sqrt{\frac{2}{2n+1} \frac{2^n (n!)^2}{(2n)!}}. \qquad (2.50)$$

Note that thanks to Stirling's formula we can get such relations

$$n! = \sqrt{2\pi n}(n/e)^n e^{\theta/12n},\ \theta \in (0,1),$$

and

$$(2n)! = 2\sqrt{\pi n}(2n/e)^{2n} e^{\theta_1/24n},\ \theta_1 \in (0,1).$$

Then

$$\|L_n\|_{L_2[-1,1]} = \sqrt{\frac{2\pi n}{2n+1}\frac{1}{2^n}e^{\mu/n}},\ \mu \in (-1/24, 1/6),$$

and when $n \to \infty$

$$\|L_n\|_{L_2[-1,1]} = \frac{\pi}{2^n} + O\left(\frac{1}{n2^n}\right).$$

Both the Legandre's polynomials and the Chebyshev's polynomials, find applications in many sections of mathematic analysis. Adduce some properties of Legandre's polynomials.

**Proposition 2.4.5.** *The Legandre's polynomial $L_n(x)$ satisfies for the next differential equation*

$$(1-x^2)L_n''(x) - 2xL_n'(x) + n(n+1)L_n(x) = 0. \tag{2.51}$$

Really, if $u = (x^2-1)^n$ then $u'(x^2-1) = 2nxu$ and according to Leibnitx formula

$$\sum_{k=1}^{n+1} C_{n+1}^k u^{n+2-k}(x^2-1)^k = 2n \sum_{k=0}^{n} C_{n+1}^k u^{n+1-k} x^k.$$

After the reductions we obtain

$$u^{n+2}(x^2-1) + (n+1)u^{n+1}2x + \frac{n(n+1)}{2}2u^{(n)} =$$

$$= 2n(u^{n+1}x + u^{(n)}(n+1))$$

or

$$(1-x^2)u^{n+2} - 2xu^{n+1} + n(n+1)u^{(n)} = 0$$

that is equivalent to (2.51).

**Proposition 2.4.6.** *The Legandre's polynomials $L_n(x)$ satisfy the reccurent correlations*

$$L_{n+2}(x) = xL_{n+1}(x) - \frac{(n+1)^2}{(2n+3)(2n+1)}L_n(x). \tag{2.52}$$

**Proof.** Polynomial

$$L^0_{n+2}(x) = xL_{n+1}(x) - \alpha_n L_n(x), \quad \alpha_n = \frac{(n+1)^2}{(2n+3)(2n+1)},$$

belongs to the set $P_{n+2}$. Therefore to prove the equation (2.52) it is sufficiently to show that

$$a_k = \int_{-1}^{1} x^k L^0_{n+2}(x)dx = 0, \quad k = 0, 1, 2, \ldots, n+1.$$

But in view of (2.48)

$$a_k = \int_{-1}^{1} x^{k+2} L_{n+1}(x)dx - \alpha_n \int_{-1}^{1} x^k L_n(x)dx =$$

$$= \frac{(n+1)!}{(2n+2)!}\int_{-1}^{1} (x^{k+1})^{(n+1)}(1-x^2)^{n+1}dx -$$

$$\alpha_n \frac{n!}{(2n)!}\int_{-1}^{1} (x^k)^{(n)}(1-x^2)^n dx.$$

Thus, it is clear that $a_k = 0$, $k = 0, 1, \ldots, n-1$. Moreover

$$a_n = \frac{((n+1)!)^2}{(2n+2)!}\int_{-1}^{1} (1-x^2)^{n+1}dx - \alpha_n \frac{(n!)^2}{(2n)!}\int_{-1}^{1} (1-x^2)^n dx = 0$$

and

$$a_{n+1} = \frac{n+2}{C^{n+2}_{2n+2}}\int_{-1}^{1} x(1-x^2)^{n+1}dx - \alpha_n \frac{n+1}{C^n_{2n}}\int_{-1}^{1} x(1-x^2)^n dx = 0$$

According to the proposition 2.4.6 we can easy write some first Legandre's polynomials

$$L_0(x) = 1, \; L_1(x) = x, \; L_2(x) = x^2 - \frac{1}{3}$$

$$L_3(x) = x^3 - \frac{3}{5}x,$$

$$L_4(x) = x^4 - \frac{6}{7}x^2 + \frac{3}{35},$$

$$L_5(x) = x^5 - \frac{70}{63}x^3 + \frac{5}{21}x, \ldots .$$

**Proposition 2.4.7.** *All zeros of polynomial $L_n(x)$ are real and belong to the interval* $(-1, 1)$

This statement is easy obtained with the help of Rolle's theorem and from the definition (2.44).

**Proposition 2.4.8.** *Let $x_k$, $k = 1, 2, \ldots, n$*

$$-1 < x_1 < x_2 < \ldots < x_n < 1$$

*are all zeros of Legandre's polynomials $L_n(x)$*

$$a_k = \int_{-1}^{1} \frac{L_n(x)}{(x - x_k)L_n'(x_k)} dx.$$

*Then for any algebraic polynomial $p \in P_{2n-1}$*

$$\int_{-1}^{1} p(x)dx = \sum_{k=1}^{n} a_k p(x_k).$$

**Proof.** As the polynomials

$$l_{k,m}(x) = \frac{L_m(x)}{(x - x_k)L_m(x_k)}$$

satisfy the following conditions

$$l_{k,m}(x_i) = \begin{cases} 1, & k = i, \\ 0, & k \neq i \end{cases}$$

then the algebraic polynomial of order $\leq n - 1$

$$p_{n-1}(f, x) = \sum_{k=1}^{n} f(x_k) l_{k,m}(x), \; f \in C[-1, 1]$$

interpolated the $f(x)$ at points $x_k$, $k = 1, 2, \ldots, n$. Therefore for any algebraic polynomial $p(x)$ of order at most $2n-1$ there exists the polynomial $q \in P_{n-1}$ such that

$$p(x) - p_{n-1}(x) = L_n(x)q(x)$$

and, consequently,

$$\int_{-1}^{1} p(x)dx - \sum_{k=1}^{n} a_k p(x_k) = \int_{-1}^{1} (p(x) - p_{n-1}(x))dx =$$

$$\int_{-1}^{1} L_n(x)q(x)dx. \tag{2.53}$$

That in view of (2.46) the last integral in (2.53) is equal to zero.

The proposition 2.4.8 show that quadrature

$$\int_{-1}^{1} f(x)dx = \sum_{k=1}^{n} a_k f(x_k) + R(f) \tag{2.54}$$

is exact for all polynomials from $P_{2n-1}$.

The quadrature (2.52) is called Gauss quadratue or (due to proposition 2.4.8) quadrature of the highest algebraic power exactness. This formula has the great number of applications both in the theory of mechanic quadratures and in other sections of calculation mathematic.

The next easy statement adds the proposition 2.4.8.

**Proposition 2.4.9.** *Let $\omega(t) \geq 0$, $t \in (-1, 1)$ and*

$$mes\{t : t \in (-1,1), \omega(t) > 0\} > 0.$$

*Then there is no the quadrature formula in the form*

$$\int_{-1}^{1} f(x)\omega(x)dx = \sum_{k=1}^{m} c_k f(x_k) + R(f),$$

*where $m \leq n$ and*

$$-1 \leq x_1 < x_2 < \ldots < x_m \leq 1$$

*which is exact on the set $P_{2n}$.*

Really, if

$$p_*(x) = \prod_{k=1}^{m} (x - x_k)^2,$$

then $p_* \in P_{2n}$ and $p_*(x_k) = 0$, $k = 1, \ldots, m$ and hence

$$\int_{-1}^{1} p_*(x)\omega(x)dx - \sum_{k=1}^{m} c_k p_*(x_k) = \int_{-1}^{1} p_*(x)\omega(x)dx > 0.$$

Present now some extremal properties of Chebyshev polynomials $T_n(x)$ and $U_n(x)$ in the space of $L_2$ with the some special weight.

**Proposition 2.4.10.** *For $m \neq n$*

$$\int_{-1}^{1} T_n(x)T_m(x)\frac{dx}{\sqrt{1-x^2}} = 0$$

*and*

$$\int_{-1}^{1} T_n^2(x)\frac{dx}{\sqrt{1-x^2}} = \frac{\pi}{2}.$$

Really due to (2.13)

$$\int_{-1}^{1} T_n(x)T_m(x)\frac{dx}{\sqrt{1-x^2}} =$$

$$-\int_{-1}^{1} \cos(n\arccos x)\cos(m\arccos x)d\arccos x =$$

$$\int_{0}^{\pi} \cos nt \cos mt dt = \begin{cases} 0, & m \neq n, \\ \pi/2, & m = n. \end{cases}$$

Let $L_{2,\omega} = L_{2,\omega}[-1,1]$, $\omega(x) > 0$, $x \in (-1,1)$ is the space of all measurable on $[-1,1]$ functions $f$ for which

$$||f||_{2,\omega} := \left(\int_{-1}^{1} f^2(x)\omega(x)dx\right)^{1/2} < \infty.$$

Analogicaly to proving of the theorem 2.4.4 and by means the proposition 2.4.10 instead of lemma 2.4.2 we obtain the next statement.

**Theorem 2.4.11.** *Among all polynomials $p \in P_n(1)$ the unique polynomial $2^{1-n}T_n(x)$ has the least norm in the space $L_{2,\omega}$ for $\omega(x) = (1-x^2)^{-1/2}$. In addition*

$$2^{1-n}\|T_n\|_{2,\omega} = E(x^n, P_{n-1})_{2,\omega} =$$

$$\min_{c_k}\left(\int_{-1}^{1}\left(x^n - \sum_{k=0}^{n-1} c_k x^k\right)^2 \frac{dx}{\sqrt{1-x^2}}\right)^{1/2} = 2^{1-n}\sqrt{\frac{\pi}{2}}.$$

With the help of proposition 2.4.10 it is easy to prove the next analog of proposition 2.4.8.

**Proposition 2.4.12.** *Let*

$$x_k = \cos\frac{2k-1}{2n}\pi,\ k = 1, 2, \ldots, n,$$

*are zeros of Chebyshev polynomial $T_n(x)$ and*

$$a_k = \int_{-1}^{1} \frac{T_n(x)}{(x-x_k)T_n'(x_k)}\frac{dx}{\sqrt{1-x^2}} = \frac{\pi}{n}, \quad k = 1, 2, \ldots, n.$$

*Then quadrature*

$$\int_{-1}^{1} f(x)\frac{dx}{\sqrt{1-x^2}} = \frac{\pi}{n}\sum_{k=1}^{n} f(x_k) + R(f)$$

*is exact on the set $P_{2n-1}$.*

In a similar way the next statements may be proved.

**Theorem 2.4.13.** *Among all polynomials $p \in P_n(1)$ the unique polynomial $U_n(x)$ has the least morm in the space $L_{2,\omega}$ for $\omega = \sqrt{1-t^2}$. At that time*

$$\|U_n\|_{2,\omega} = \min_{c_k}\left(\int_{-1}^{1}\left(x^n - \sum_{k=0}^{n-1} c_k x^k\right)^2 \sqrt{1-x^2}dx\right)^{1/2} =$$

$$\frac{1}{n+1}2^{1-n}\sqrt{\frac{\pi}{2}}.$$

**Proposition 2.4.14.** *Let*

$$x_k = \cos\frac{k}{n+1}\pi, \quad k = 1, 2, \ldots, n,$$

*are zeros of Chebyshev polynomial* $U_n(x)$ *and*

$$a_k = \int_{-1}^{1} \frac{U_n(x)}{(x - x_k)U_n'(x_k)}\sqrt{1 - x^2}\,dx =$$

$$\frac{\pi}{n+1}\sin^2\frac{k\pi}{n+1}, \quad k = 1, 2, \ldots, n.$$

*Then quadrature*

$$\int_{-1}^{1} f(x)\sqrt{1 - x^2}\,dx = \frac{\pi}{n+1}\sum_{k=1}^{n}\sin^2\frac{k\pi}{n+1} f\left(\cos\frac{k\pi}{n+1}\right) + R(f)$$

*is exact on the set* $P_{2n-1}$.

## 2.5 Jacoby polynomials

At previous section we proved that Legandre poly- nomials $L_n(x)$ and Chebyshav polynomials $T_n(x)$ and $U_n(x)$ are construct systems of functions in the spaces of $L_{2,\omega}[-1, 1]$ when

$$\omega(x) = 1, \quad \omega(x) = (1 - x^2)^{-1/2}$$

and

$$\omega(x) = (1 - x^2)^{1/2}$$

correspondingly.

The natural generalization of such polynomials is given by orthogonal in the space $L_{2,\omega}[-1, 1]$ polynomials when

$$\omega(x) = (1 - x)^\alpha(1 + x)^\beta, \quad \alpha, \beta > -1. \tag{2.55}$$

These polynomials are called Jacoby polynomials. Here we present some properties of Jacoby's polynomials, which we shall use in further.

Let

$$||f||_{L_p^{\alpha,\beta}} = \left(\int_{-1}^{1} |f(x)|^p(1 - x)^\alpha(1 + x)^\beta dx\right)^{1/p}.$$

Consider the function

$$y_n^{\alpha,\beta}(x) = (1 - x)^{n+\alpha}(1 + x)^{n+\beta} \tag{2.56}$$

and

$$J_n^{\alpha,\beta} = (-1)^n (2^n n!)^{-1} (1-x)^{-\alpha} (1+x)^{-\beta} \frac{d^n}{dx^n} y_n^{\alpha,\beta}(x). \tag{2.57}$$

According to Leibnitz formula we have

$$J_n^{\alpha,\beta}(x) = \sum_{k=0}^{n} d_{n,k} (1-x)^{n-k} (1+x)^k, \tag{2.58}$$

where

$$d_{n,k} = \frac{\Gamma(n+\alpha+1)\Gamma(n+\beta+1)}{2^n k!(n-k)!\Gamma(n+\alpha+1-k)\Gamma(\beta+1+k)} \tag{2.59}$$

and

$$\Gamma(x) = \int_0^1 e^{-t} t^{x-1} dt$$

is the Γ-function.

Thus $J_n^{\alpha,\beta}(x)$ is an algebraic polynomial of order $n$, which in further we shall called the Jacoby's polynomial.

¿From the equalities (2.58) and (2.59), in particular,it follows that

$$J_n^{\alpha,\beta}(1) = \frac{\Gamma(n+\alpha+1)}{n!\Gamma(\alpha+1)} \tag{2.60}$$

and

$$J_n^{\alpha,\beta}(-1) = (-1)^n \frac{\Gamma(n+\beta+1)}{n!\Gamma(\beta+1)}. \tag{2.61}$$

¿From binomial Newton's formula it follows that for $x > 1$

$$(x-1)^{n+\alpha} = x^{n+\alpha} \left(1 - \frac{1}{x}\right)^{n+\alpha} = x^{n+\alpha} - (n+\alpha) x^{n+\alpha-1} + \dots$$

and

$$(x+1)^{n+\beta} = x^{n+\beta} + (n+\beta) x^{n+\beta-1} + \dots$$

and therefore for $x > 1$

$$(x-1)^{n+\alpha}(x+1)^{n+\beta} = x^{2n+\alpha+\beta} + (\beta-\alpha) x^{2n+\alpha+\beta-1} + \dots.$$

Moreover

$$(x-1)^{-\alpha}(x+1)^{-\beta} = x^{-\alpha-\beta} + (\alpha-\beta) x^{-\alpha-\beta-1} + \dots$$

Thus

$$J_n^{\alpha,\beta} = K_n(\alpha,\beta) x^n + K_n^*(\alpha,\beta) x^{n-1} + \dots,$$

where

$$K_n(\alpha,\beta) = (2n+\alpha+\beta)\dots(n+\alpha+\beta+1)(2^n n!)^{-1} =$$

$$\frac{\Gamma(2n+\gamma)}{2^n n!\Gamma(n+\gamma)}; \tag{2.62}$$

$$K_n^*(\alpha,\beta) = (\beta-\alpha)(2n+\alpha+\beta-1)\dots(n+\alpha+\beta)(2^n n!)^{-1} =$$

$$\frac{(\beta-a)\Gamma(2n+\gamma)}{2^n n!\Gamma(n+\gamma-1)} \tag{2.63}$$

and

$$\gamma = \alpha+\beta+1. \tag{2.64}$$

¿From this

$$(K_n(\alpha,\beta))^{-1} J_n^{\alpha,\beta} \in P_n(1). \tag{2.65}$$

The next statement proves, that Jacoby's polynomials $J_n^{\alpha,\beta}$ establish orthogonal system of polynomials in the space $L_{2,\omega}[-1,1]$ with the weight (2.55).

**Theorem 2.5.1.** *If*

$$S_{n,m}(\alpha,\beta) = \int_{-1}^{1} J_n^{\alpha,\beta}(x) J_m^{\alpha,\beta}(x)(1-x)^\alpha(1+x)^\beta dx, \tag{2.66}$$

*then*

$$S_{n,m}(\alpha,\beta) = 0, \quad m \neq n \tag{2.67}$$

*and*

$$S_{n,n}(\alpha,\beta) = \|J_n^{\alpha,\beta}\|_{L_2^{\alpha,\beta}} = \frac{2^\gamma\Gamma(n+\alpha+1)\Gamma(n+\beta+1)}{(2n+\gamma)n!\Gamma(n+\gamma)}. \tag{2.68}$$

**Proof.** Note before all that without the loss of generality we may think that $m \leq n$. As

$$\frac{d^\nu}{dx^\nu} J_n^{\alpha,\beta}(x)|_{x=\pm1} = 0, \quad \nu = 0,1,\dots,n-1,$$

and

$$2^n n! J_n^{\alpha,\beta}(x) = (-1)^n (y_n^{\alpha,\beta}(x))^{(n)},$$

then, with the help of integration by parts, we obtain

$$S_{n,m}^{\alpha,\beta} = \frac{(-1)^n}{2^n n!} \int_{-1}^{1} J_n^{\alpha,\beta}(x)(y_n^{\alpha,\beta}(x))^{(n)} dx =$$

$$\frac{1}{2^n n!}\int_{-1}^{1}(J_n^{\alpha,\beta}(x))^{(n)}y_n^{\alpha,\beta}(x)dx. \qquad (2.69)$$

Now equality (2.67) follows from fact that for $m < n$ the identity

$$(J_n^{\alpha,\beta}(x))^{(n)} \equiv 0$$

holds true. Moreover from (2.69) and (5.65) it follows that

$$S_{n,n}^{\alpha,\beta} = \frac{K_n(\alpha,\beta)}{2^n}\int_{-1}^{1} J_n^{\alpha,\beta}(x)dx =$$

$$\frac{K_n(\alpha,\beta)}{2^n}2^{2n+\gamma}\int_0^1 x^{n+\alpha}(1-x)^{n+\beta}dx =$$

$$2^{n+\gamma}K_n(\alpha,\beta)B(n+\alpha+1, n+\beta+1),$$

where

$$B(p,q) = \int_0^1 x^{p-1}(1-x)^{q-1}dx = \frac{\Gamma(p)\Gamma(q)}{\Gamma(p+q)}.$$

¿From here and from (2.62) we obtain the equality (2.68).

¿From the theorem 2.5.1 it follows that if

$$\tilde{J}_n^{\alpha,\beta}(x) = (S_{n,m}(\alpha,\beta))^{-1/2}J_n^{\alpha,\beta}(x), \qquad (2.70)$$

then

$$\int_{-1}^{1}\tilde{J}_m^{\alpha,\beta}(x)\tilde{J}_n^{\alpha,\beta}(x)(1-x)^{\alpha}(1+x)^{\beta}dx = \begin{cases} 0, & m \neq n, \\ 1, & m = n \end{cases}$$

and therefore $\tilde{J}_n^{\alpha,\beta}(x)$, $n = 0, 1, \ldots$ is called orthonormed Jacoby's polynomials.

¿From theorem 2.5.2, proposition 2.4.1 and from the correlations (2.65), (2.62) and (2.68) the next statement immediately follows.

**Theorem 2.5.2.** *If $\alpha, \beta > -1$ and $n = 0, 1, \ldots$ then among all polynomials $p \in P_n(1)$ the unique polynomial*

$$\overline{J}_n^{\alpha,\beta}(x) = (K_n(\alpha,\beta))^{-1}J_n^{\alpha,\beta}(x) \qquad (2.71)$$

*has the least norm in the space $L_2^{\alpha,\beta}[-1,1]$. Moreover*

$$\min\{||p||_{L_2^{\alpha,\beta}} : \ p \in P_n(1)\} = ||\overline{J}_n^{\alpha,\beta}||_{L_2^{\alpha,\beta}} =$$

$$\left(\frac{2^{2n+\gamma}n!\Gamma(n+\gamma)\Gamma(n+\alpha+1)\Gamma(n+\beta+1)}{\Gamma(2n+\gamma)\Gamma(2n+\gamma+1)}\right)^{1/2}.$$

¿From this statement and from theorems 2.4.11 and 2.4.13 it follows that

$$J_n^{-1/2,-1/2}(x) = \frac{(2n-1)!}{n!(n-1)!}T_n(x), \tag{2.72}$$

$$J_n^{0,0}(x) = \frac{(2n)!}{n!(n-1)!}L_n(x) \tag{2.73}$$

and

$$J_n^{1/2,1/2}(x) = 2\frac{(2n+1)!}{n!(n+1)!}U_n(x). \tag{2.74}$$

In a similar way as we proved proposition 2.4.6, we can infer the next recurrent formulas for the Jacoby's poynomials

$$\overline{J}_0^{\alpha,\beta}(x) = 1$$

$$\overline{J}_1^{\alpha,\beta}(x) = \frac{1}{\gamma+1}\left((1+\alpha)(x+1)+(1+\beta)(x-1)\right); \tag{2.75}$$

and

$$\overline{J}_{n+2}^{\alpha,\beta}(x) = (x-\lambda_n(\alpha,\beta))\overline{J}_{n+1}^{\alpha,\beta}(x) - \mu_n(\alpha,\beta)\overline{J}_n^{\alpha,\beta}(x), \quad n=0,1,\ldots, \tag{2.76}$$

where

$$\lambda_n^{\alpha,\beta} = \frac{\beta^2-\alpha^2}{(2n+\gamma+1)(2n+\gamma+3)} \tag{2.77}$$

and

$$\mu_n^{\alpha,\beta} = 2\frac{(n+\gamma)(n+\alpha+1)(n+\beta+1)(n+1)}{(2n+\gamma)(2n+\gamma+1)^2(2n+\gamma+2)}. \tag{2.78}$$

Now with the help of induction it is easy to verify that the polynomials

$$g_n(x) = \frac{1}{n+1}(\overline{J}_n^{\alpha,\beta}(x))'$$

satisfy the correlations

$$g_0(x) = 1, \quad g_1(x) = \frac{1}{\gamma+3}((\alpha+2)(x+1)+(\beta+2)(x-1)),$$

$$g_{n+2}(x) = (x-\lambda_n(\alpha+1,\beta+1))g_{n+1}(x) - \mu_n(\alpha+1,\beta+1)g_n(x), \quad n=0,1,\ldots$$

and as the polynomials $\overline{J}_n^{\alpha,\beta}(x)$ are single-valuedly determined by correlations (2.75)-(2.78) then

$$g_n(x) = \overline{J}_n^{\alpha+1,\beta+1}(x).$$

Thus

$$(\overline{J}_{n+1}^{\alpha,\beta}(x))' = (n+1)\overline{J}_n^{\alpha+1,\beta+1}(x).$$

and

$$((\overline{J}_{n+1}^{\alpha,\beta}(x))')^2 = \frac{(n+1)^2 S_{n,n}(\alpha+1,\beta+1)K_{n+1}^2(\alpha,\beta)}{S_{n+1,n+1}(\alpha,\beta)K_n^2(\alpha+1,\beta+1)}(\overline{J}_n^{\alpha+1,\beta+1}(x))^2.$$

Then

$$((\overline{J}_{n+1}^{\alpha,\beta}(x))')^2 = (n+1)(n+1+\gamma)(\overline{J}_n^{\alpha+1,\beta+1}(x))^2$$

and, consequently, for all $k = 0, 1, \ldots, n$

$$((\overline{J}_{n+1}^{\alpha,\beta}(x))^{(k)})^2 = \frac{n!\Gamma(n+\gamma+1)}{(n-k)!\Gamma(n+\gamma+1-k)}(\overline{J}_n^{\alpha+k,\beta+k}(x))^2 \tag{2.79}$$

Moreover, from (2.60), (2.61), (2.70) and (2.68) it follows that

$$(\overline{J}_n^{\alpha,\beta}(1))^2 = \frac{(2n+\gamma)\Gamma(n+\gamma)\Gamma(n+\alpha+1)}{2^\gamma\Gamma^2(\alpha+1)\Gamma(n+\beta+1)\Gamma(n+1)} \tag{2.80}$$

and

$$(\overline{J}_n^{\alpha,\beta}(-1))^2 = \frac{(2n+\gamma)\Gamma(n+\gamma)\Gamma(n+\beta+1)}{2^\gamma\Gamma^2(\beta+1)\Gamma(n+\alpha+1)\Gamma(n+1)} \tag{2.81}$$

and hence

$$\xi_{n,k}^+(\alpha,\beta) := ((\overline{J}_n^{\alpha,\beta}(x))^{(k)}|_{x=1})^2 =$$

$$\frac{(2n+\gamma)\Gamma(n+k+\gamma)}{2^{\gamma+2k}\Gamma^2(\alpha+k+1)}\frac{\Gamma(n+\alpha+1)}{\Gamma(n+\beta+1)\Gamma(n-k+1)} \tag{2.82}$$

and

$$\xi_{n,k}^-(\alpha,\beta) := ((\overline{J}_n^{\alpha,\beta}(x))^{(k)}|_{x=-1})^2 =$$

$$\frac{(2n+\gamma)\Gamma(n+k+\gamma)}{2^{\gamma+2k}\Gamma^2(\beta+k+1)}\frac{\Gamma(n+\beta+1)}{\Gamma(n+\alpha+1)\Gamma(n-k+1)}. \tag{2.83}$$

In further the values $\xi_{n,k}^{\pm}(\alpha,\beta)$ will play a great role into the inequalities between the norms for derivatives of polynomials and the norms of polynomials. The next statement gives us the possibility to get the estimates of values $\xi_{n,k}^{\pm}(\alpha,\beta)$ for big $n$.

**Lemma 2.5.3.** *If $\alpha, \beta > -1$, $n = 1, 2, \ldots$ then at $n \to \infty$*

$$\xi^{+}_{n,k}(\alpha,\beta) = \frac{n^{2k+2\alpha+1}}{2^{2k+\gamma-1}\Gamma^2(\beta+k+1)}\left(1+O\left(\frac{1}{n}\right)\right). \tag{2.84}$$

*and*

$$\xi^{-}_{n,k}(\alpha,\beta) = \frac{n^{2k+2\beta+1}}{2^{2k+\gamma-1}\Gamma^2(\beta+k+1)}\left(1+O\left(\frac{1}{n}\right)\right). \tag{2.85}$$

**Proof.** It is known that

$$\ln\Gamma(x) = \left(x-\frac{1}{2}\right)\ln x - x + \ln\sqrt{2\pi} + \frac{1}{12x} + \frac{\theta_0(x)}{360x^3},$$

where $0 < \theta_0(x) < 1$. Therefore for $y > -1$ we have

$$\ln\frac{\Gamma(x+y)}{\Gamma(x)} = \left(x+y-\frac{1}{2}\right)\ln\left(\frac{y}{x}+1\right) - y + y\ln x + \theta_0,$$

where

$$\theta_0 = -\frac{y}{12x^2} + \frac{\theta_1}{x^3};\ |\theta_1| < \frac{1}{10}.$$

Using now the correlation

$$\ln\left(1+\frac{y}{x}\right) = \frac{y}{x} - \frac{y^2}{2x^2} + \frac{y^3}{3x^3} + \frac{\theta_2 y^4}{4x^4},\ |\theta_2| \le 1,\ |y| < x,$$

we obtain

$$\ln\frac{\Gamma(x+y)}{\Gamma(x)} = y\ln x + \frac{y(y-1)}{2x} + \theta_3\frac{y(y-1)(y^2+1)}{4x^2},\ |\theta_3| \le 1,$$

and if now $x > y^2+1$ then

$$\frac{\Gamma(x+y)}{\Gamma(x)} = x^y exp\frac{y(y-1)}{2x} + \theta_3\frac{y(y-1)(y^2+1)}{4x^2} =$$

$$x^y + \frac{1}{2}y(y-1)x^{y-1} + \theta(y)x^{y-2},\ |\theta(y)| \le |y(y-1)|(y^2+1). \tag{2.86}$$

¿From the correlation (2.86) it follows that if $n \to \infty$ then

$$\frac{\Gamma(n+\gamma+2k)}{\Gamma(n+\gamma)} = n^{2k}\left(1+\frac{2k(2k-1)}{2n}+O\left(\frac{1}{2n}\right)\right),$$

$$\frac{\Gamma(n+\alpha+1)}{\Gamma(n+1-k)} = n^{k+\alpha}\left(1 + \frac{(k+\alpha)(k+\alpha-1)}{2n} + O\left(\frac{1}{n^2}\right)\right)$$

and

$$\frac{\Gamma(n+\gamma+k)}{\Gamma(n+\beta+1)} = n^{k+\alpha}\left(1 + \frac{(k+\alpha)(k+\alpha-1)}{2n} + O\left(\frac{1}{n^2}\right)\right),$$

that together with (2.82) and (2.83) gives us the correlations (2.84) and (2.85).

**Lemma 2.5.4.** *If $\alpha, \beta > -1$ and $n = 1, 2, \ldots$, then Jacoby's polynomial $J_n^{\alpha,\beta}$ satisfies the next differential equation*

$$(1-x^2)y'' + (\beta - \alpha - (\gamma+1)x)y' + n(n+\gamma)y = 0.$$

**Proof.** If function $J_n^{\alpha,\beta}$ is defined by equation (2.56) and

$$u(x) = \frac{(-1)^n}{2^n n!} y_n^{\alpha,\beta}(x),$$

then

$$(1-x^2)u' = (\beta - \alpha - (2n+\gamma-1)x)u.$$

Differentiating this equation $(n+1)$ times, we obtain

$$(1-x^2)u^{(n+2)} - 2(n+1)xu^{(n+1)} - n(n+1)u^{(n)} =$$

$$= (\beta - \alpha + (2n+\gamma+1))u^{(n+1)} - (2n+\gamma-1)u^{(n)}$$

or

$$(1-x^2)u^{(n+2)} + (\alpha - \beta - (\gamma-3)x)u^{(n+1)} + (n+1)(n+\gamma-1)u^{(n)} = 0. \quad (2.87)$$

Besides that, in view of (2.57)

$$u^{(n)}(x) = (1-x)^\alpha (1+x)^\beta J_n^{(\alpha,\beta)}(x).$$

Expressing from here $u^{(n+1)}$ and $u^{(n+2)}$ and substituting these functions in (2.87) we are easy convinced of lemma justice.

**Lemma 2.5.5.** *If $\alpha \geq \beta > -1$, $\alpha \geq -1/2$ and $\gamma = \alpha + \beta + 1$, then for all $x \in [-1, 1]$*

$$(J_n^{\alpha,\beta}(x))^2 + \frac{1-x^2}{n(n+\gamma)}((J_n^{\alpha,\beta}(x))')^2 \leq (\tilde{J}_n^{\alpha,\beta}(1))^2 \tag{2.88}$$

*and consequently*

$$|J_n^{\alpha,\beta}(x)| \leq J_n^{\alpha,\beta}(1),$$

*and if $\beta \geq \alpha > -1$ and $\beta \geq -1/2$ then for all $x \in [-1, 1]$*

$$(J_n^{\alpha,\beta}(x))^2 + \frac{1-x^2}{n(n+\gamma)}((J_n^{\alpha,\beta}(x))')^2 \leq (J_n^{\alpha,\beta}(-1))^2$$

*and hence*

$$|J_n^{\alpha,\beta}(x)| \leq (-1)^n J_n^{\alpha,\beta}(-1).$$

**Proof.** Let $\alpha \geq \beta > -1$ and $\alpha \geq -1/2$.Put now $y(x) = J_n^{\alpha,\beta}(x)$ and

$$\psi(x) = y^2 + \frac{1-x^2}{n(n+\gamma)}(y')^2.$$

Then due to lemma 2.5.4, if $\gamma = 0$ then

$$\psi'(x) = \frac{2(\alpha-\beta)}{n^2}(y')^2 \geq 0$$

and if $\gamma \neq 0$ then

$$\psi'(x) = \frac{2\gamma}{n(n+\gamma)}(x - x_0)(y')^2, \; x_0 = (\alpha - \beta)/\gamma.$$

If $\alpha \geq \beta \geq -1/2$ then $\gamma > 0$ and $-1 \leq x_0 \leq 1$. Thus the function $\psi(x)$ decreases on $[-1, x_0]$ and increases on $[x_0, 1]$ and, consequently, for

$$\psi(x) \leq \max\{\psi(1), \psi(-1)\} = \max\{y^2(1), y^2(-1)\}.$$

If $\alpha \geq 1/2$ and $\beta < -1/2$ then $\psi'(x) > 0$ for $x \in [-1, 1]$ and consequently for $x \in [-1, 1]$ it will be $\psi(x) \leq \psi(1) \leq y^2(1)$. Now to complete the proof of lemma 2.5.5 it is remain to note that if $\alpha > \beta > -1$ then in view of (2.60) and (2.61)

$$y(1) = \frac{\Gamma(n+\alpha+1)}{n!\Gamma(\alpha+1)} = \frac{1}{n!}(n+a)\ldots(\alpha+1) >$$

$$\frac{1}{n!}(n+\beta)\ldots(\beta+1) = (-1)^n y(-1).$$

The next statements immediately follow from lemma 2.5.5.

**Lemma 2.5.6.** *If $\alpha \geq \beta > -1$ and $\alpha \geq -1/2$ then for all $x \in [-1, 1]$*

$$\sum_{k=0}^{n} (\tilde{J}_k^{\alpha,\beta}(x))^2 + \frac{1-x^2}{n(n+\gamma)} \sum_{k=1}^{n} ((\tilde{J}_k^{\alpha,\beta}(x))')^2 \leq \theta_{n,0}^{+}(\alpha, \beta) \tag{2.89}$$

*where*

$$\theta_{n,0}^{+}(\alpha, \beta) = \sum_{k=0}^{n} (\tilde{J}_k^{\alpha,\beta}(1))^2 = \sum_{k=0}^{n} \xi_{k,0}^{+}(\alpha, \beta) \tag{2.90}$$

*If $\beta \geq \alpha > -1$ and $\beta \geq -1/2$ then for all $x \in [-1, 1]$*

$$\sum_{k=0}^{n} (\tilde{J}_k^{\alpha,\beta}(x))^2 + \frac{1-x^2}{n(n+\gamma)} \sum_{k=1}^{n} ((\tilde{J}_k^{\alpha,\beta}(x))')^2 \leq \theta_{n,0}^{-}(\alpha, \beta), \tag{2.91}$$

*where*

$$\theta_{n,0}^{-}(\alpha, \beta) = \sum_{k=0}^{n} (\tilde{J}_k^{\alpha,\beta}(-1))^2 = \sum_{k=0}^{n} \xi_{k,0}^{-}(\alpha, \beta) \tag{2.92}$$

¿From here and from (2.79) it immediately follows that

**Lemma 2.5.7.** *If $n \in N$, $m = 1, 2, \ldots, n-1$ and $x \in [-1, 1]$ then for $\alpha \geq \beta > -1$*

$$\sum_{k=m}^{n} ((\tilde{J}_k^{\alpha,\beta}(x))^{(m)})^2 + \frac{1-x^2}{n(n+m+\gamma)} \sum_{k=m+1}^{n} ((\tilde{J}_k^{\alpha,\beta}(x))^{(m+1)})^2 \leq \theta_{n,m}^{+}(\alpha, \beta)$$

*where*

$$\theta_{n,m}^{+}(\alpha, \beta) = \sum_{k=m}^{n} ((\tilde{J}_k^{\alpha,\beta}(x))^{(m)}|_{x=1})^2 = \sum_{k=m}^{n} \xi_{k,m}^{+}(\alpha, \beta). \tag{2.93}$$

*If $\beta \geq \alpha > -1$ and $\beta \geq -1/2$ then for all $x \in [-1, 1]$*

$$\sum_{k=m}^{n} ((\tilde{J}_k^{\alpha,\beta}(x))^{(m)})^2 + \frac{1-x^2}{n(n+m+\gamma)} \sum_{k=m+1}^{n} ((\tilde{J}_k^{\alpha,\beta}(x))^{(m+1)})^2 \leq \theta_{n,m}^{-}(\alpha, \beta)$$

*where*

$$\theta_{n,m}^{-}(\alpha, \beta) = \sum_{k=m}^{n} (\tilde{J}_k^{\alpha,\beta}(x))^{(m)}|_{x=-1})^2 = \sum_{k=m}^{n} \xi_{k,0}^{-}(\alpha, \beta).$$

Note also that from lemma 2.5.3 and from the definition of values $\theta_{n,k}^{\pm}(\alpha, \beta)$ we may easy get the next statement.

**Lemma 2.5.8.** *If $\alpha, \beta > -1$, $n \in N$ and$k = 1, 2, \ldots$ then $n \to \infty$*

$$\theta^{+}_{n,k}(\alpha, \beta) = \frac{n^{2(k+\alpha+1)}}{2^{2k+\gamma}(k+\alpha+1)\Gamma^2(k+\alpha+1)}\left(1 + O\left(\frac{1}{n}\right)\right)$$

*and*

$$\theta^{-}_{n,k}(\alpha, \beta) = \frac{n^{2(k+\beta+1)}}{2^{2k+\gamma}(k+\beta+1)\Gamma^2(k+\beta+1)}\left(1 + O\left(\frac{1}{n}\right)\right).$$

For the exact calculation of values $\theta^{\pm}_{n,k}(\alpha, \beta)$ we shall need in some supplementary constructions.At first we prove Cristofel-Darby identity which is very important in the orthogonal series theory. Formulate it in common situation.

Let $\omega(x)$ is arbitrary nonnegative, summable on the segment $[-1, 1]$ function and $\{p_k\}_{k=0}^{\infty}$ is orthonormed in the space $L_{2,\omega}[-1, 1]$ system of polynomials i.e. $p_k \in P_n$ and

$$\int_{-1}^{1} p_n(x)p_m(x)\omega(x)dx = 0, \quad n \neq m \tag{2.94}$$

and

$$\int_{-1}^{1} p_n^2(x)\omega(x)dx = 1. \tag{2.95}$$

It is clear that if weight $\omega(x)$ is defined by equation (2.55), then

$$p_k(x) = \tilde{J}_k^{\alpha,\beta}, \quad k = 0, 1, \ldots.$$

**Theorem 2.5.9.** *If $a_k$ is the coefficient near $x^k$ at polynomial $p_k(x)$ then for all $x, t \in R$*

$$(t - x)\sum_{k=0}^{n} p_k(t)p_k(x) = \frac{a_n}{a_{n+1}}\left(p_{n+1}(t)p_n(x) - p_n(t)p_{n+1}(x)\right). \tag{2.96}$$

**Proof.** As $xp_k(x) \in P_{k+1}$ then

$$xp_k(x) = \sum_{\nu=0}^{k+1} c_{k,\nu}p_\nu(x),$$

where

$$c_{k,\nu} = \int_{-1}^{1} xp_k(x)p_\nu(x)\omega(x)dx, \quad \nu = 0, 1, \ldots, k+1.$$

For $\nu = 0, 1, \ldots, k-2$ polynomial $xp_\nu(x)$ belongs to $P_{k-1}$ and, hence,

$$c_{k,\nu} = 0, \quad \nu = 0, 1, \ldots, k-2.$$

Moreover, there exist the polynomials $q_m \in P_m$, $m = k-1, k$ such that

$$xp_{k-1}(x) = \frac{a_{k-1}}{a_k} p_k(x) + q_{k-1}(x)$$

and

$$xp_k(x) = \frac{a_k}{a_{k+1}} p_{k+1}(x) + q_k(x).$$

Therefore

$$c_{k,k-1} = \int_{-1}^{1} p_k(x) \frac{a_{k-1}}{a_k} p_k(x)\omega(x)dx == \frac{a_{k-1}}{a_k}$$

and

$$c_{k,k+1} = \int_{-1}^{1} p_{k+1}(x) \frac{a_k}{a_{k+1}} p_{k+1}(x)\omega(x)dx = \frac{a_k}{a_{k+1}}.$$

Thus

$$xp_k(x) = \frac{a_{k-1}}{a_k} p_{k-1}(x) + c_{k,k} p_k(x) + \frac{a_k}{a_{k+1}} p_{k+1}(x)$$

and

$$xp_k(x)p_k(t) = \frac{a_{k-1}}{a_k} p_{k-1}(x)p_k(t) +$$
$$c_{k,k} p_k(x)p_k(t) + \frac{a_k}{a_{k+1}} p_{k+1}(x)p_k(t).$$

Replacing $x$ by $t$ and $t$ by $x$ we can get following equality

$$tp_k(t)p_k(x) = \frac{a_{k-1}}{a_k} p_{k-1}(t)p_k(x) +$$
$$c_{k,k} p_k(t)p_k(x) + \frac{a_k}{a_{k+1}} p_{k+1}(t)p_k(x),$$

which together with previous equality allows us to obtain for $k = 1, 2, \ldots, n$ the next relation

$$(t-x)p_k(x)p_k(t) = \frac{a_k}{a_{k+1}} \left(p_{k+1}(t)p_k(x) - p_{k+1}(x)p_k(t)\right) -$$
$$\frac{a_{k-1}}{a_k} \left(p_{k-1}(x)p_k(t) - p_k(x)p_{k-1}(t)\right). \tag{2.97}$$

It's also clear that

$$(t-x)p_0(x)p_0(t) = \frac{a_0}{a_1} \left(p_1(t)p_0(x) - p_0(t)p_1(x)\right). \tag{2.98}$$

Now to complete the proof of the theorem it remains to add all equations (2.98) and (2.99).

**Corollary 2.5.10.** *For $n \in N$ and $x \in R$ the next equations*

$$\sum_{k=0}^{n} p_k^2(x) = \frac{a_n}{a_{n+1}} \left(p'_{n+1}(x)p_n(x) - p_{n+1}(x)p'_n(x)\right), \tag{2.99}$$

$$\sum_{k=0}^{n} ((p_k(x))')^2 = \frac{a_n}{2a_{n+1}} \left(p''_{n+1}(x)p'_n(x) - p'_{n+1}(x)p''_n(x)\right) +$$

$$\frac{a_n}{6a_{n+1}} \left(p'''_{n+1}(x)p_n(x) - p_{n+1}(x)p'''_n(x)\right) \tag{2.100}$$

*are valid.*

**Proof.** Puting in theorem 2.2.9 $t = y + \delta$ and $x = y - \delta$ we obtain

$$\sum_{k=0}^{n} p_k(y-\delta)p_k(y+\delta) =$$

$$\frac{a_n}{a_{n+1}} \frac{p_{n+1}(y+\delta)p_n(y-\delta) - p_n(y+\delta)p_{n+1}(y-\delta)}{2\delta} \tag{2.101}$$

and therefore when $\delta \to 0$

$$\sum_{k=0}^{n} p_k^2(y) - \delta^2 \left(\sum_{k=2}^{n} p''_k(y)p_k(y) - \sum_{k=1}^{n} (p'_k(y))^2\right) + O(\delta^3) =$$

$$\frac{a_n}{a_{n+1}} \left(p'_{n+1}(y)p_n(y) - p_{n+1}(y)p'_n(y)\right) +$$

$$\delta^2 \frac{a_n}{6a_{n+1}} \Big(3\left(p''_{n+1}(y)p_n(y) - p_{n+1}(y)p''_n(y)\right) +$$

$$\left(p'_{n+1}(y)p''_n(y) - p''_{n+1}(y)p'_n(y)\right)\Big) + O(\delta^3),$$

and thus the functions, which are both in right and left parts (2.101) are polynomials (of $\delta$), then we have

$$\sum_{k=0}^{n} p_k^2(y) = \frac{a_n}{a_{n+1}} \left(p'_{n+1}(y)p_n(y) - p_{n+1}(y)p'_n(y)\right),$$

and

$$\left(\sum_{k=2}^{n} p''_k(y)p_k(y) - \sum_{k=1}^{n} (p'_k(y))^2\right) =$$

$$\frac{a_n}{6a_{n+1}}\Big(3\left(p'''_{n+1}(y)p_n(y)-p_{n+1}(y)p'''_n(y)\right)+$$
$$\left(p'_{n+1}(y)p''_n(y)-p''_{n+1}(y)p'_n(y)\right)\Big). \tag{2.102}$$

By analogy from theorem 2.5.9 it follows that

$$\sum_{k=0}^{n}p_k(y+\delta)p_k(y)=\frac{a_n}{a_{n+1}}\frac{p_{n+1}(y+\delta)p_n(y)-p_n(y+\delta)p_{n+1}(y)}{\delta}$$

therefore when $\delta\to 0$

$$\sum_{k=0}^{n}p_k^2(y)+\delta\sum_{k=1}^{n}p_k(y)p'_k(y)+\frac{\delta^2}{2}\sum_{k=2}^{n}p_k(y)p''_k(y)+O(\delta^3)=$$
$$\frac{a_n}{a_{n+1}}\left(p'_{n+1}(y)p_n(y)-p_{n+1}(y)p'_n(y)\right)+$$
$$\delta\frac{a_n}{2a_{n+1}}\left(p''_{n+1}(y)p_n(y)-p_{n+1}(y)p''_n(y)\right)+$$
$$\delta^2\frac{a_n}{6a_{n+1}}\left(p'''_{n+1}(y)p_n(y)-p_{n+1}(y)p'''_n(y)\right)+O(\delta^3).$$

¿From here, in particular, it follows that

$$\sum_{k=2}^{n}p_k(y)p''_k(y)=\frac{a_n}{3a_{n+1}}\left(p'''_{n+1}(y)p_n(y)-p_{n+1}(y)p'''_n(y)\right)$$

but this fact with (2.102) gives us the equality (2.100).

**Corollary 2.5.11.** *If the values $\theta^{\pm}_{n,k}(\alpha,\beta)$ are defined by equalities (2.90), (2.92) and (5.2.93) then*

$$\theta^{+}_{n,0}(\alpha,\beta)=\frac{(n+\gamma)\Gamma(n+\alpha+2)\Gamma(n+\gamma+1)}{2^{\gamma+1}\Gamma(n+\beta+1)\Gamma^2(\alpha+1)\Gamma^2(\alpha+2)(n+1)!}, \tag{2.103}$$

*and*

$$\theta^{-}_{n,0}(\alpha,\beta)=\frac{(n+\gamma)\Gamma(n+\beta+2)\Gamma(n+\gamma+1)}{2^{\gamma+1}\Gamma(n+\alpha+1)\Gamma^2(\beta+1)\Gamma^2(\beta+2)(n+1)!}. \tag{2.104}$$

In fact in this case due to (2.79)

$$K(x)=(\tilde{J}^{(\alpha,\beta)}_{n+1})'\tilde{J}^{(\alpha,\beta)}_{n}-(\tilde{J}^{(\alpha,\beta)}_{n})'\tilde{J}^{(\alpha,\beta)}_{n+1}=$$

$$\sqrt{(n+1)(n+1+\gamma)}\tilde{J}_n^{(\alpha+1,\beta+1)}\tilde{J}_n^{(\alpha,\beta)}$$

$$\sqrt{n(n+\gamma)}\tilde{J}_{n-1}^{(\alpha+1,\beta+1)}\tilde{J}_{n+1}^{(\alpha,\beta)}.$$

¿From here and from (2.80) and (2.99) we immediately obtain the equalities (2.93) and (2.104). Using (2.100) instead of (2.99) by analogy we may obtain the meaning of values $\theta_{n,1}^{\pm}(\alpha,\beta)$.

## 2.6 Hermit and Lagerre polynomials

Now we present the analog of the results from section 2.4 on axis and semiaxis.

Consider the function

$$H_n(x) = (-1)^n e^{x^2/2} \frac{d^n}{dx^n} e^{-x^2/2}. \tag{2.105}$$

With the help of induction it is easy to verify that $H_n(x)$ is the algebraic polynomial of order $n$ with the coefficietn at $x^n$ that is equal to one, i.e. $H_n \in P_n(1)$.

The polynomials $H_n(x)$ are called Hermit polynonials.

As

$$A_{n,m} := \int_{-\infty}^{\infty} e^{-x^2/2} H_n(x) H_m(x) dx = (-1)^n \int_{-\infty}^{\infty} H_m(x) \frac{d^n}{dx^n} e^{-x^2/2} dx$$

and for any polynomial $Q$

$$\lim_{x\to\infty} Q(x) \frac{d^n}{dx^n} e^{-x^2/2} = 0,$$

then

$$A_{n,m} = \int_{-\infty}^{\infty} e^{-x^2/2} H_m^{(n)}(x) dx.$$

¿From here, in particular, it follows that

$$\int_{-\infty}^{\infty} e^{-x^2/2} H_n(x) H_m(x) dx = 0, \quad m \neq n, \tag{2.106}$$

and

$$A_{n,n} = \int_{-\infty}^{\infty} e^{-x^2/2} H_n^2(x) dx = n! \int_{-\infty}^{\infty} e^{-x^2/2} dx = n!\sqrt{2\pi}. \tag{2.107}$$

The next simple statement immediately follows from equalities (2.107) and (2.108) and from proposition 2.4.1.

**Theorema 2.6.1.** *Among all polynomials $p \in P_n(1)$ the unique polynomial $H_n(x)$ has the least norm in the space $L_{2,\omega}(R)$, $(\omega(x) = e^{-x^2/2})$. And*

$$\min\{||p||_{L_{2,\omega}(R)} : \ p \in P_n(1)\} = ||H_n||_{L_{2,\omega}(R)} = (n!\sqrt{2\pi})^{1/2}.$$

¿From (2.107) it follows that for all $k = 1, 2, \ldots, n-1$

$$\int_{-\infty}^{\infty} e^{-x^2/2} H2_n(x) H'_{n+1}(x) dx = (-1)^k \int_{-\infty}^{\infty} \frac{d^n}{dx^n} \left(e^{-x^2/2}\right) dH_{n+1}(x) =$$

$$(-1)^{k+1} \int_{-\infty}^{\infty} \frac{d^{n+1}}{dx^{n+1}} \left(e^{-x^2/2}\right) H_{n+1}(x) dx =$$

$$\int_{-\infty}^{\infty} e^{-x^2/2} H_{n+1}(x) H_{k+1}(x) dx = 0.$$

Taking into account now that $H_n \in P_n(1)$ and $(n+1)^{-1} H'_{n+1} \in P_n(1)$ we obtain

$$H'_{n+1}(x) = (n+1) H_n(x). \tag{2.108}$$

Moreover, from (2.105) it follows that

$$H'_n(x) = x H_n(x) - H_{n+1}(x).$$

¿From here and from (2.105) we receive

$$H_{n+2}(x) = x H_{n+1}(x) - (n+1) H_n(x). \tag{2.109}$$

In particular,

$$h_0(x) = 1, \ H_1(x) = x, \ H_2(x) = x^2 - 1,$$

$$H_3(x) = x^3 - 3x,$$

$$H_4(x) = x^4 - 6x^2 + 3,$$

$$H_5(x) = x^5 - 10x^3 + 15x, \ldots.$$

Besides that, from equality (2.109) it follows that Hermit polynomial $H_n(x)$ satisfies to differential equation

$$H''_n(x) - x H'_n(x) + n H_n(x) = 0. \tag{2.110}$$

Put now

$$H(x,t) = e^{xt - \frac{t2}{2}}.$$

Now we write the decomposition of this function in M'c Loren series (by t)

$$H(t,x) = \sum_{k=0}^{\infty} \frac{G_k(x)}{k!} t^k.$$

It is easy to see that $G_0(x) = 1$ and $G_1(x) = 0$. We can directly verify also that function $H(t,x)$ satisfies to differential equation

$$\frac{\partial H}{\partial t} = (x - t)H.$$

¿From here it follows, that

$$\frac{1}{(n+1)!} G_{n+2}(x) = \frac{x}{(n+1)!} G_{n+1}(x) - \frac{1}{n!} G_n(x),$$

i.e.

$$G_{n+2}(x) = xG_{n+1}(x) - (n+1)G_n(x).$$

¿From here and from (2.110) it follows that for all $n = 1, 2, \ldots$ the equality $G_n(x) = H_n(x)$ holds true. Thus

$$e^{xt - \frac{t^2}{2}} = \sum_{k=0}^{\infty} \frac{H_k(x)}{k!} t^k,$$

i.e. $H(x,t) = e^{xt - \frac{t^2}{2}}$ is giving birth function for polynomials $H_n(x)/n!$.

Consider now the functions

$$L_n^{\alpha}(x) = (-1)^n x^{-\alpha} e^x \frac{d^n}{dx^n} \left( x^{n+\alpha} e^{-x} \right), \quad \alpha > -1.$$

With the help of induction it is easy to verify, that $L_n^{\alpha} in P_n(1)$.

The functions $L_n^{\alpha}(x)$ are called Lagerre polynomials.

For each algebraic polynomial $Q$ and for any $m = 0, 1, \ldots, n-1$ the following equality is valid

$$\lim_{x \to \infty} Q(x) \frac{d^m}{dx^m} \left( x^{n+\alpha} e^{-x} \right) = 0.$$

Besides that for any $m = 0, 1, \ldots, n-1$

$$\frac{d^m}{dx^m} \left( x^{n+\alpha} e^{-x} \right) |_{x=0} = 0.$$

¿From these correlations we obtain

$$\int_0^\infty x^\alpha e^{-x} \left(L_m^\alpha\right)^{(n)}(x)dx = \int_0^\infty x^{n+\alpha} e^{-x} L_m^\alpha L_n^\alpha dx,$$

and, consequently,

$$\int_0^\infty x^\alpha e^{-x} L_m^\alpha(x) L_n^\alpha(x) dx = 0, \quad n \neq m,$$

and

$$\int_0^\infty x^\alpha e^{-x} \left(L_n^\alpha(x)\right)^2 dx == n! \int_0^\infty x^{n+\alpha} e^{-x} dx = n!\Gamma(n+\alpha+1).$$

The next statement follows from previous correlations and from proposition 2.4.1.

**Theorem 2.6.2.** *Among all polynomials $p \in P_n(1)$ the unique polynomial $L_n^\alpha$ has the least norm in the space $L_{2,\omega[0,\infty]}$, $\omega(x) = x^\alpha e^{-x}$, $\alpha > -1$. In other words*

$$\min\{||p||_{L_{2,\omega[0,\infty]}};\, p \in P_n(1)\} = ||L_n^\alpha||_{L_{2,\omega[0,\infty]}} = \sqrt{n!\Gamma(n+\alpha+1)}.$$

## 2.7 Constant sign polynomials

Let we have continuous nondecreasing on $[0, \infty)$ function $\Phi$ such that $\Phi(0) = 0$ and measurable nonnegative and almost everywhere different from zero on segment $[-1, 1]$ function $\omega(x)$.

Denote by $\Im_{\Phi,\omega}$ the set of all measurable on $[-1, 1]$ functions $f$ for which

$$||f||_{\Im_{\Phi,\omega}} := \int_{-1}^1 \Phi(|f(x)|\omega(x))dx < \infty. \qquad (2.111)$$

In this section we will consider the problem of constant sign on $[-1, 1]$ polynomials deviating least from zero in the space of $\Im_{\Phi,\omega}$, i.e. polynomials $p \in P_n(1)$ minimizing the value (2.111). The exact organization of this problem is: we are to find polynomial

$$p_n^\pm = p_n^\pm(\Phi, \omega; x) \in P_n^\pm(1)$$

such that

$$\min\{||p||_{\Im_{\Phi,\omega}} : p \in P_n^\pm(1)\} = ||p||_{\Im_{\Phi,\omega}}. \qquad (2.112)$$

**Proposition 2.7.1.** *For each $n \in N$ the polynomial $p_n^{\pm}(x)$ exists.*

**Proposition 2.7.2.** *The polynomial $p_n^{\pm}(x)$ has only real zeros.*

**Proof.** Suppose, that $x = \alpha + i\beta$, $\beta \neq 0$ is a zero of polynomial $p_n^{+}(x)$. Then in view of corollary 2.1.2 this polynomial we may represent in the form

$$p_n^{+}(x) = \left((x-\alpha)^2 + \beta^2\right) q(x), \quad q \in P_{n-2}^{+}(1),$$

and, consequently, for all $x \in [-1, 1]$ it will be

$$p_n^{+}(x) > (x-\alpha)^2 q(x) \geq 0.$$

But then

$$\|(\cdot - \alpha)^2 q(\cdot)\|_{\Im_{\Phi,\omega}} < \|p_n^{+}\|_{\Im_{\Phi,\omega}},$$

that contradicts to the definition of polynomial $p_n^{+}(x)$.

**Proposition 2.7.3.** *All zeros of the polynomial $p_n^{\pm}(x)$ belong to segment $[-1, 1]$.*

Really, if $\alpha > 1$ is zero of polynomial $p_n^{+}(x)$ then in view of corollary 2.1.2 the polynomial $p_n^{+}(x)$ we may represent in the form

$$p_n^{+}(x) = (x-\alpha) q(x), \quad q \in P_n^{+}(1).$$

But then

$$p_n^{+}(x) > (x-1) q(x)$$

and therefore

$$\|(\cdot - 1) q(\cdot)\|_{\Im_{\Phi,\omega}} < \|p_n^{+}\|_{\Im_{\Phi,\omega}},$$

that contradicts to the definition of $p_n^{+}(x)$. In a similar way we obtain the contradiction in the case when $\alpha < -1$ is zero of polynomial $p_n^{+}(x)$.

**Proposition 2.7.4.** *Let all zeros of polynomial $p \in P_n(1)$ are in a segment $[-1, 1]$. Then if $p \in P_{2n}^{+}(1)$ there exists the polynomial $q \in P_n(1)$ such that*

$$p(x) = q^2(x); \tag{2.113}$$

*if $p \in P_{2n-1}^{+}(1)$ then there exists the polynomial $q \in P_{n-1}$ such that*

$$p(x) = (x+1) q^2(x); \tag{2.114}$$

*if $p \in P_{2n-1}^{-}(1)$ then there exists the polynomial $q \in P_{n-1}$ such that*

$$p(x) = (x-1) q^2(x); \tag{2.115}$$

*if* $p \in P^-_{2n}(1)$ *then there exists the polynomial* $q \in P_{n-1}$ *such that*

$$p(x) = (x^2 - 1)q^2(x). \tag{2.116}$$

**Proof.** For $n = 1$ the statement is obvious. Suppose that statement holds true for $n = 1, 2, \ldots, m$ and prove its justice for $n = m + 1$. If $m = 2k + 1$, $p \in P^+_{2k+2}$ and points $x = 1$ and $x = -1$ are not odd multiplicity zeros of polynomial , all is obvious, as zeros of $p(x)$ which are in segment $(-1, 1)$ have even multiplicity. If, for example, $x = -1$ is odd multiplicity zero of polynomial $p(x)$ then there exists the polynomial $q_1 \in P^+_{2k+1}(1)$ such that all its seros are in segment $[-1, 1]$ and

$$p(x) = (x + 1)q_1(x).$$

But due to assumption of induction there exists the polynomial $q_2 \in P_m(2)$ such that

$$q_1(x) = (x + 1)q_2^2(x).$$

Therefore

$$p(x) = (x + 1)^2 q_2^2(x),$$

and now it is sufficiently to put $q(x) = (x + 1)q_2(x)$.

If $x = -1$ is odd multiplicity zero of polynomial $p(x)$, then there exists the polynomial $q_3 \in P^-_{2m+1}(1)$ such that all its zeros are in segment $[-1, 1]$ and

$$p(x) = (x - 1)q_3(x).$$

But then in view of induction supposition there exists the polynomial $q_4 \in P_k(1)$ such that

$$q_3(x) = (x - 1)q_4^2(x).$$

Therefore

$$p(x) = (x - 1)^2 q_4^2(x).$$

The case $m = 2k + 1$, $p \in P^+_{m+1}(1)$ is fully settled. The analogous proof is given in the cases $m = 2k$, $p \in P^+_{m+1}(1)$ , $m = 2k + 1$, $p \in P^-_{m+1}(1)$ and $m = 2k$, $p \in P^-_{m+1}(1)$.

Denote by $p_n(\Phi, \omega; x)$ the polynomial from the set $P_n(1)$ deviating least from zero in the space $\Im_{\Phi,\omega}$, i.e. $p_n(\Phi, \omega) \in P_n(1)$ and

$$\min\{||p||_{\Im_{\Phi,\omega}};\ \} = ||p_n(\Phi, \omega, \cdot)||_{\Im_{\Phi,\omega}}.$$

The next statement immediately follows from propositions 2.7.3 and 2.7.4.

**Theorem 2.7.5.** *For $n = 1, 2, \ldots$ the next equalities*

$$p^+_{2n}(\Phi, \omega; x) = \left(p_n(\Phi^2, \sqrt{\omega}; x)\right)^2;$$

$$p^+_{2n+1}(\Phi, \omega; x) = (x+1)\left(p_{2n}(\Phi^2, \sqrt{(1+\cdot)\omega(\cdot)}; x)\right)^2;$$

$$p^-_{2n+1}(\Phi, \omega; x) = (x-1)\left(p_n(\Phi^2, \sqrt{(1-\cdot)\omega(\cdot)}; x)\right)^2;$$

$$p^-_{2n+2}(\Phi, \omega; x) = (x^2-1)\left(p_{2n}(\Phi^2, \sqrt{(1-(\cdot)^2)\omega(\cdot)}; x)\right)^2$$

*are valid.*

Puting in theorem 2.7.5 $\omega(x) = (1-x)^\alpha(1+x)^\beta$ and taking into account the statement of theorem 2.5.2 we obtain

**Corollary 2.7.6.** *Let $n \in N$, $\alpha, \beta > -1/2$ and*

$$C^\pm_n(\alpha, \beta) = \min_{p \in P^\pm_n(1)} \pm \int_{-1}^1 p(x)(1-x)^\alpha(1+x)^\beta\, dx.$$

*Then*

$$C^+_{2n}(\alpha, \beta) = \int_{-1}^1 \left(\overline{J}^{\alpha,\beta}_n\right)^2 (1-x)^\alpha(1+x)^\beta\, dx$$

$$C^+_{2n+1}(\alpha, \beta) = \int_{-1}^1 \left(\overline{J}^{\alpha,\beta+1}_n\right)^2 (1-x)^\alpha(1+x)^{\beta+1}\, dx$$

$$C^-_{2n+1}(\alpha, \beta) = \int_{-1}^1 \left(\overline{J}^{\alpha+1,\beta}_n\right)^2 (1-x)^{\alpha+1}(1+x)^\beta\, dx$$

*and*

$$C^-_{2n+2}(\alpha, \beta) = \int_{-1}^1 \left(\overline{J}^{\alpha+1,\beta+1}_n\right)^2 (1-x)^{\alpha+1}(1+x)^{\beta+1}\, dx$$

*where $\overline{J}^{\alpha,\beta}_n$ are Jacoby's polynomials.*

Let now $P^+_{n,*}(1)$ is the set of all nonnegative on semiaxis $[0, \infty)$ polynomials $p \in P_n(1)$ and measurable nonnegative and almost everywhere finite on $[0, \infty)$ function $\omega(x)$ is such that

$$\int_0^\infty \Phi((x^n+1)\omega(x))dx < \infty.$$

Put now

$$||f||_{\Im^*_{\Phi,\omega}} = \int_0^\infty \Phi(|f(x)|)\omega dx$$

and denote by $p_n^*(\Phi, \omega; x)$ and $p_n^{+*}(\Phi, \omega; x)$ the algebraic polynomials from the sets $P_n(1)$ and $P_{n,*}^+(1)$ correspondingly, such that

$$\min\{||p||_{\Im^*_{\Phi,\omega}}; p \in P_n(1)\} = ||p_n^*(\Phi, \omega; \cdot)||_{\Im^*_{\Phi,\omega}}$$

and

$$\min\{||p||_{\Im^*_{\Phi,\omega}}; p \in P_{n,*}^+(1)\} = ||p_n^{+*}(\Phi, \omega; \cdot)||_{\Im^*_{\Phi,\omega}}.$$

By analogy we may prove the next statement

**Theorem 2.7.7.** *For $n = 1, 2, \ldots$ the next equalities*

$$p_{2n}^{+*}(\Phi, \omega; x) = \left(p_n^*(\Phi^2, \sqrt{\omega}; x)\right)^2$$

*and*

$$p_{2n+1}^{+*}(\Phi, \omega; x) = x\left(p_n^*(\Phi^2, \sqrt{\cdot\omega(\cdot)}; x)\right)^2.$$

*are valid*

The next result immediately follows from theorems 2.7.7 and 2.7.5.

**Corollary 2.7.8.** *Let $n = 1, 2, \ldots$, $\alpha > -1$ and*

$$Q_n(\alpha) = \min_{p \in P_n^+(1)} \int_0^\infty x^\alpha e^{-x} p(x) dx.$$

*Then*

$$Q_{2n}^+(\alpha) = \int_0^\infty x^\alpha e^{-x} \left(L_n^{(\alpha)}\right)^2 dx$$

*and*

$$Q_{2n+1}^+(\alpha) = \int_0^\infty x^{\alpha+1} e^{-x} \left(L_n^{(\alpha)}\right)^2 dx$$

## 2.8 Polynomials with two fixed coefficients

Denote by $p_n(1, \sigma)$ the set of all algebraic polynomials $p \in P_n$ of the form

$$p(x) = x^n - \sigma x^{n-1} + a_2 x^{n-1} + \ldots + a_n,$$

where $\sigma \in R$ is given and $a_2, \ldots, a_n$ are arbitrary.

The first Zolotarev's problem consists in finding among all polynomials $p \in P_n(1, \sigma)$ a polynomial $Z_{n,\infty,\sigma} \in P_n(1, \sigma)$ such that

$$||Z_{n,\infty,\sigma}||_{C[-1,1]} = inf\{||p||_{C[-1,1]} : \ p \in P_n(1, \sigma)\}.$$

The most general problem is: to find for given $1 \leq q \leq \infty$ and $\sigma \in R$ the polynomial $Z_{n,q,\sigma} \in P_m(1\sigma)$ such that

$$||Z_{n,q,\sigma}||_{L_q[-1,1]} = inf\{||p||_{L_q[-1,1]} : \ p \in P_n(1, \sigma)\}.$$

The alalogous problems we may consider in other (in weights, for example) spaces also.

First we note that for each $1 \leq q \leq \infty$ and $\sigma$ the next almost obvious correlation

$$Z_{n,q,-\sigma}(x) = (-1)^n Z_{n,q,\sigma}(-x)$$

is true.That is why, in further,studying the problem of finding the polynomials $Z_{n,q,\sigma}(x)$, when it is convenient, we shall assume that $\sigma > 0$.

Crossing to elucidation of the results we begin from the case $q = \infty$.

**Theorem 2.8.1.** *a) For any $\sigma$ the polynomial $Z_{n,\infty,\sigma}(x)$ is single-valuedly characterized so that it has on segment $[-1, 1]$ $n$-alternans, i.e. at $n$ points $-1 \leq x_1 < x_2 < \ldots < x_n \leq 1$ this polynomial takes alternating signs values,that equal by modulus to its norm in the space $C[-1, 1]$.*

*b)If $0 \leq \sigma \leq \tan^2(\pi/(2n))$, then*

$$Z_{n,\infty,n\sigma}(x) = (\sigma + 1)^n \overline{T}_n \left( \frac{x - \sigma}{\sigma + 1} \right),$$

*where $\overline{T}_n(x)$ is the Chebyshev polynomial of first kind normed by conditions $\overline{T}_n \in P_(1)$. In this case*

$$||Z_{n,\infty,n\sigma}||_{C[-1,1]} = 2^{1-n}(\sigma + 1)^n.$$

**Proof.** It is clear that

$$Z_{n,\infty,n\sigma}(x) = x^n - \sigma x^{n-1} - p_{n-2}(x),$$

where $p_{n-2}(x)$ is the best approximation polynomial of function $x^n - \sigma x^{n-1}$ in the space $C[-1, 1]$. But in view of Chebyshev alternans theorem (theorem 2.1.8) the polynomial $p_{n-2}(x)$ is uniformly characterized by fact,that difference $x^n - \sigma x^{n-1} - p_{n-2}(x)$ has $n$-alternans on segment $[-1, 1]$ The statement a) is proved.

Furthermore, let

$$y(x) = (\sigma + 1)^n \overline{T}_n \left( \frac{x - \sigma}{\sigma + 1} \right).$$

It is easy to verify that $y \in P_n(1, n\sigma)$. If $x \in [-1, 1]$ then

$$-1 \leq \frac{x - \sigma}{\sigma + 1} \leq \frac{1 - \sigma}{1 + \sigma}.$$

So far as the alternans points of the polynomial $\overline{T}_n$ are the points $x_k = \cos(k\pi/n)$, $k = 0, 1, \ldots, n$ we see that if

$$\frac{1 - \sigma}{1 + \sigma} \geq \cos \frac{\pi}{n},$$

or, that is equivalent,

$$0 \leq \sigma \leq \tan^2 \frac{\pi}{2n}, \tag{2.117}$$

then the polynomial $y(t)$ has $n$-alternans on segment $[-1, 1]$ (at points $\cos(k\pi/n)$, $k = n, n-1 \ldots, 1$). Thus, if the condition (8.117) holds true, then

$$Z_{n,\infty,n\sigma}(x) = y(x) = (\sigma + 1)^n \overline{T}_n \left( \frac{x - \sigma}{\sigma + 1} \right).$$

**Remark.** The finding problem of evident form for polynomial $Z_{n,\infty,n\sigma}(x)$ when $\sigma > \tan^2 \frac{\pi}{2n}$ is more difficult. Zolotarev himself reduces it to differential equations. For example, if searching polynomial $y(x)$ takes alternating signs value $||y||_{C[-1,1]} = L$ at points

$$-1 = x_1 < x_2 < \ldots < x_{n-1} < x_n = 1,$$

then $y'(x)$ we may represent in the form

$$y'(x) = n(x - x_2) \ldots (x - x_{n-1})(x - c),$$

and difference $L^2 - y^2$ may be written in such form

$$L^2 - y^2(x) = (1 - x)(1 + x)(x - x_2)^2 \ldots (x - x_{n-1})(x^2 + px + q).$$

¿From here we obtain the next differential equation

$$\frac{y'}{\sqrt{L^2 - y^2}} = \frac{n(x - c)}{\sqrt{(1 - x^2)(x^2 + px + q)}}$$

for definition of polynomial $y$, its norm $L$ and parameters $x, p$ and $q$.

The other approach connecting with the using of elliptical functions, one can find in Ahiezer's monograph [6, p. 314-319].

Consider now the case $q = 2$. In this case the solution of the problem is very simple.

**Theorem 2.8.2.** *For all $n = 1, 2, \ldots$ and any $\sigma \in R$*

$$Z_{n,2,\sigma}(x) = \overline{L}_n(x) - \sigma \overline{L}_{n-1}, \tag{2.118}$$

*where $\overline{L}_n(x)$ is the Legandre polynomial normed by condition $L \in P_n(1)$.*

**Proof.** The polynomials $L(x)$ are orthogonal on segment $[-1, 1]$ thus for any $j = 0, 1, \ldots, n-2$ we have

$$\int_{-1}^{1} Z_{n,2,\sigma}(x) x^j \, dx = 0$$

for polynomial (2.118). Therefore, in view of theorem 1.1.10 the best approximation polynomial from $P_{n-2}$ for $Z_{n,2,\sigma}$ in the space $L_2[-1, 1]$ identically equal to zero. To complete the proof it remains to account that $Z_{n,2,\sigma} \in P_n(1)$ as for even $n$ the coefficients of L (near odd powers of $x$) equal to zero and for odd $n$ the coefficients of L (near even powers of $x$) are equel to zero.

The theorem 2.8.2 admits the next generalisation. Let the positive on $[-1, 1]$ weigth $\omega(x)$ is given and the sequence of polynomials

$$p_1, p_2, \ldots, p_n, \ldots$$

on $[-1, 1]$ with the weight $\omega(x)$ and normed by condition $p_n \in P_n(1)$, $n = 1, 2, \ldots$.

Let also $Z_{n,2,\omega,}$ is the polynomial deviating least from zero in the space $L_{2\omega}[-1, 1]$. Likewise to the theorem 2.8.2 we can prove the following statement.

**Theorem 2.8.3.** *For all $n = 1, 2, \ldots$*

$$Z_{n,2,\omega,\sigma} = p_n(x) - \left( \sigma - \frac{p_n^{(n-1)}(0)}{(n-1)!} \right) p_{n-1}(x).$$

Let furthermore $\omega_1(x) = e^{x^2/2}$ and $Z_{n,2,\omega_1,\sigma}$ is the polynomial from $p_n(1,\sigma)$ deviating least from zero in the space $L_{2,\omega_1}(-\infty,\infty)$. Let also $\omega_2(x) = x^\alpha e^x$ $(\alpha > -1)$ and $Z_{n,2,\omega_2,\sigma}$ is the polynomial deviating least from zero in the space $L_{2,\omega_2}[0,\infty)$.

By analogy to the theorems 2.8.2 an 2.8.3 the next statement is established.

**Theorem 2.8.4.** *For all $n = 1, 2, \ldots$*

$$Z_{n,2,\omega_1,\sigma}(x) = H_n(x) - \sigma H_{n-1}(x),$$

*where $H_n(x)$ is the Hermite polynomial.*
*Moreover, if $\alpha > -1$, then*

$$Z_{n,2,\omega_2,\sigma}(x) = L_n^{(\alpha)}(x) - \left(\sigma + \frac{(L_{n-1}^{(\alpha)})^{(n-1)}(0)}{(n-1)!}\right) L_{n-1}^{(\alpha)}(x),$$

*where $L_n^{(\alpha)}(x)$ is the Lagerre polynomial.* medskip
Cross to the case $q = 1$.

**Theorem 2.8.5.** *Let $n = 1, 2, \ldots$. If $|\sigma| \geq 1$ then*

$$Z_{n,1,\sigma}(x) = (x-\sigma)\overline{U}_{n-1}(x) \tag{2.119}$$

*where $\overline{U}_n(x)$ is the Chebyshev polynomial of second kind normed by condition $\overline{U}_n \in P_n(1)$. If $|\sigma| < 1$ then*

$$Z_{n,1,\sigma}(x) = \overline{U}_n(x) - \sigma\overline{U}_{n-1}(x) + \frac{\sigma^2}{4}\overline{U}_{n-2}(x). \tag{2.120}$$

**Proof.** It is clear that polynomials defined by relations (2.119) and (2.120) belong to the set $P_n(1,\sigma)$ and to prove the theorem in both cases it is sufficiently to establish the following relation

$$E(Z_{n,1,\sigma}, P_{n-2})_{L_1[-1,1]} = ||Z_{n,1,\sigma}||_{L_1[-1,1]},$$

which, in view of theorem 1.1.9 is equivalent to the next conditions

$$\int_{-1}^{1} x^j Z_{n,1,\sigma}(x)dx = 0, \ j = 0, 1, \ldots, n-2. \tag{2.121}$$

For $|\sigma| \geq 1$ we have

$$sgn\, Z_{n,1,\sigma}(x) = \pm sgn \overline{U}_{n-1}(x).$$

Moreover (see (2.41))

$$\int_{-1}^{1} x^j\, sgn \overline{U}_{n-1}(x) dx = 0, \; j = 0, 1, \ldots, n-2.$$

Therefore for $|\sigma| \geq 1$ the correlations (2.121) held true and the first statement of theorem proved.

The second statement is proving more difficult. First of all note that in this case all zeros of polynomial $Z_{m,1,\sigma}(x)$ lie on segment $[-1,1]$. Really if it is not true, then the polynomial $Z_{m,1,\sigma}(x)$ (and consequently the function $sgn\, Z_{m,1,\sigma}(x)$) would have at most $n-1$ change of sign. But if the function $g(x)$ takes on segment $[-1,1]$ only values 1 and $-1$, has at most $n-1$ change of sign and is orthogonal on this segment to polynomials of order no more then $n-2$, then $g(x) = \pm sgn \overline{U}_{n-1}(x)$. But then it will be

$$Z_{m,1,\sigma}(x) = (x - c)\overline{U}_{n-1},$$

where $|c| \geq 1$ (the unique zero of $Z_{m,1,\sigma}(x)$ that isn't on [-1,1]). Taking into account that $Z_{m,1,\sigma}(x) \in P_n(1,\sigma)$ we see that $|\sigma| = |c| \geq 1$ but at that time we have $|\sigma| < 1$.

Suppose, that $n = 2m$ and denote by $\xi_1, \eta_1, \ldots, \xi_m, \eta_m$ the roots of polynomial $Z_{m,1,\sigma}(x)$ being in $[-1,1]$ and numbered in order of increase by $\xi_1, \eta_1, \ldots, \xi_m, \eta_m$ Then the conditions (2.121) may be written in the form

$$\int_{-1}^{\xi_1} x^k dx - \int_{\xi_1}^{\eta_1} x^k dx + \int_{\eta_1}^{\xi_2} x^k dx - \ldots - \int_{\xi_m}^{\eta_m} x^k dx + \int_{\eta_m}^{1} x^k dx = 0,$$
$$k = 0, 1, \ldots, 2m-2.$$

This correlations after the integration may be reduced to following equalities

$$\sum_{i=1}^{m} \xi_i^{k+1} + \frac{1}{2}(1 - (-1)^{k+1}) = \sum_{i=1}^{m} \eta_i^{k+1}, \; k = 0, 1, \ldots, 2m-2. \qquad (2.122)$$

So far as

$$\prod_{i=1}^{m} (x - \xi_i)(x - \eta_i) = x^{2m} - \sigma x^{2m-1} + \ldots,$$

we obtain

$$\sum_{i=1}^{m} (\xi_i + \eta_i) = \sigma. \qquad (2.123)$$

**Lemma 2.8.6.** *For any of numbers $c_1, \ldots, c_k$ there exist the numbers $\alpha_1, \ldots, \alpha_k$ such that*

$$\sum_{i=1}^{k} \alpha_i^j = c_j, \quad j = 1, \ldots, k.$$

Really for $k = 1$ it is obvious. Let for $k = n - 1$ lemma is proved. Prove it for $k = n$.

$$s_m = s_m(\alpha_1, \ldots, \alpha_{n-1}) = \alpha_1^m + \ldots + \alpha_{n-1}^m.$$

Put

$$s_m = s_m(\alpha_1, \ldots, \alpha_{n-1}) = \sum_{k=1}^{n-1} \alpha_k^m.$$

It is known that value $s_n$ is polynomially expressed by $s_1, \ldots, s_{n-1}$, i.e. $s_n = P(s_1, \ldots, s_{n-1})$. But $s_j = c_j - \alpha_n^j$, i.e.

$$s_n(c_1 - \alpha_n, \ldots, c_{n-1} - \alpha_n^{n-1}) = Q(\alpha_n).$$

Now it remains to find the root $\overline{\alpha}_n$ of equation

$$Q((\alpha_n) = c_n - \alpha_n^n,$$

and then get the solutions $\overline{\alpha}_1, \ldots, \overline{\alpha}_{n-1}$ of equationts $s_j = c_j - \alpha_n^j$, $j = 1, \ldots, n-1$, which exist in view of induction supposition.

Find now numbers $\alpha_1, \ldots, \alpha_{2m-1}$ such that

$$\sum_{i=1}^{2m-1} \alpha_i^{k+1} = \frac{-1 + (-1)^{k+1}}{2}, \quad k = 0, 1, \ldots, 2m-2.$$

Donote by $\xi_{m+k} = \alpha_k$, $k = 1, \ldots, 2m-1$ and represent the system (2.122) in the form

$$\sum_{i=1}^{3m-1} \xi_i^j = \sum_{i=1}^{m} \eta_i^j, \; j = 1, \ldots, 2m-1, \tag{2.124}$$

or

$$s_j(\xi_1, \ldots, \xi_{3m-1}) = s_j(\eta_1, \ldots, \eta_m).$$

By Viet's theorem

$$Z_{2m,1,\sigma}(t) = \prod_{i=1}^{m} (t - \xi_i)(t - \eta_i) = t^{2m} - (\sigma_1 + \tau_1)t^{2m-1} +$$

$$+(\sigma_2 + \sigma_1 t_1 + \tau_2)t^{2m-2} + \ldots + \sigma_m \tau_m,$$

where for $k = 1, 2, \ldots, m$

$$\sigma_k(\eta_1, \ldots, \eta_n) = \sum_{1 \le i_1 < \ldots < i_k \le m} \eta_{i_1} \ldots \eta_{i_k} \tag{2.125}$$

and

$$\tau_k(\xi_1, \ldots, \xi_n) = \sum_{1 \le i_1 < \ldots < i_k \le m} \xi_{i_1} \ldots \xi_{i_k}. \tag{2.126}$$

In addition to (2.125) and (2.126) we consider that $\sigma_0 = \tau_0 = 1$ and $\sigma_k = 0$, $\tau_k = 0$ if $k$ is more than number of variables.

In view of (2.126) for $k = 1, 2, \ldots, 2m - 1$ we have

$$\sigma_k(\eta_1, \ldots, \eta_m) = \tau_k(\xi_1, \ldots, \xi_{3m-1}) =$$

$$\sum_{i=0}^{k} \tau_i(\xi_1, \ldots, \xi_m)\tau_{k-i}(\xi_{m+1}, \ldots, \xi_{3m-1}) =$$

$$\sum_{i=0}^{k} \tau_i(\xi_1, \ldots, \xi_m)\tau_{k-i}(\alpha_1, \ldots, \alpha_{2m-1}) =$$

$$\sum_{i=0}^{k} \alpha_{ki}\tau_i(\xi_1, \ldots, \xi_m)$$

or

$$\sigma_k = \sum_{i=0}^{k} \alpha_{ki}\tau_i, \; k = 1, 2, \ldots, 2m - 1. \tag{2.127}$$

Taking into account (2.125) we obtain also

$$\sigma_1 + \tau_1 = \sigma. \tag{2.128}$$

¿From relations (2.127) and (2.128) it follows that the numbers $\tau_1, \ldots, \tau_m$ and $\sigma_1, \ldots, \sigma_m$ satisfy to the system of $2m$ linear equations with $2m$ unknown values, moreover $\sigma$ enters only in one equation as the free member.

But then all $\tau_k$ and $\sigma_k$, $k = 1, 2, \ldots, m$ linearly depend of $\sigma$ and therefore the coefficients of polynomial $Z_{2m,1,\sigma}$ will be quadratically dependent of $\sigma$. By analogy we can establish that for odd $n$ the coefficients of polynomial $Z_{n,1\sigma}$ quadratically depend of $\sigma$.

Thus, if (for each $n$)

$$Z_{n,1\sigma}(x) = x^n - \sigma x^{n-1} + \sum_{k=0}^{n-2} a_k(\sigma)x^k,$$

then

$$a_k(\sigma) = \alpha_k\sigma^2 + \beta_k\sigma + \gamma_k,\ k = 0, \ldots, n-2. \tag{2.129}$$

Now from correlations (2.129) knowing $a_k(\sigma)$ for $\sigma = 0, 1$ and $-1$ we may find $\alpha_k$, $\beta_k$, $\gamma_k$ and thus the polynomial $Z_{n,1,\sigma}$ for any $\sigma$ ($|\sigma| < 1$).

Bellow, by $A_{n,k}$ we denote the coefficient near $x^k$ of polynomial $\overline{U}_n(x)$. Taking into account that $Z_{n,1,0}(x) = \overline{U}_n(x)$ from (2.130) we obtain

$$\gamma_k = a_k(0) = A_{n,k}. \tag{2.130}$$

¿From (2.117) and (2.117) for $\sigma = 1$ we can get

$$\alpha_k + \beta_k + \gamma_k = A_{n-1,k-1} - A_{n-1,k} \tag{2.131}$$

and for $\sigma = -1$ we obtain

$$\alpha_k - \beta_k + \gamma_k = A_{n-1,k-1} + A_{n-1,k}. \tag{2.132}$$

Deciding the system of equations (2.130)-(2.132) about the unknown values $\alpha_k$, $\beta_k$, $\gamma_k$ and substituting given values in (2.119) we get that

$$a_k(\sigma) = (A_{n-1,k-1} - A_{n,k})\sigma^2 - \sigma A_{n-1,k} + A_{n,k}$$

Taking into account the reccurent formula

$$\overline{U}_n(x) = \overline{U}_{n-1}(x)x - \frac{1}{4}\overline{U}_{n-2}(x)$$

we infer the equality (2.120).

## 2.9 Trigonometric polynomials case

Let $T_{2n+1}(1)$ is the set of all trigonometric polynomials $\tau \in T_{2n+1}$ of the form

$$\tau(t) = \frac{1}{2}\rho_0 + \sum_{k=1}^{n} \rho_k \cos(kt + \theta_k), \tag{2.133}$$

such that $\rho_n = 1$.

**Proposition 2.9.1.** *Let the morm $\|\cdot\|$ is translate invariant (i.e. $\|f(\cdot+\lambda)\| = \|f\|$ for all $\lambda$), then for any trigonometric polynomial $\tau \in T_{2n+1}(1)$ the next inequality*

$$\|\tau\| \geq \|\cos(\cdot)\|$$

*holds true. In particular*

$$\cos n(\cdot) \prec \tau(\cdot) \tag{2.134}$$

*and for all $p \in [1,\infty]$*

$$\|\tau\|_p \geq \|\cos n(\cdot)\|_p = \|\cos(\cdot)\|_p, \tag{2.135}$$

*i.e. among all trigonometric polynomials $\tau \in T_{2n+1}(1)$ the polynomial $\tau(t) = \cos(nt+\theta)$ has least deviation from zero in any normed space of functions which have the translate invariant norm. Moreover, in the space $L_p$, $p \in [1,\infty]$ the polynomial deviating least from zero is unique with the exactness to shift of argument $\theta$.*

**Proof.** Put for each $2\pi$-periodic function $f$

$$\phi_n(f,t) = \frac{1}{n}\sum_{k=1}^{n} f\left(t + \frac{2k\pi}{n}\right).$$

Then

$$\|\phi_n(f)\| = \frac{1}{n}\left\|\sum_{k=1}^{n} f\left(\cdot + \frac{2k\pi}{n}\right)\right\| \leq \frac{1}{n}\sum_{k=1}^{n}\left\|f\left(\cdot + \frac{2k\pi}{n}\right)\right\| = \|f\|,$$

and so far as

$$\phi_n(\cos(k\cdot+\theta),t) = \begin{cases} \cos(kt+\theta), & k = nm, \\ 0, & k \neq nm. \end{cases}$$

That is why for any trigonometric polynomial $\tau$ of the form (2.133)

$$\phi_n(\tau,t) = \rho_0 + \rho_n\cos(nt+\theta_n).$$

Therefore

$$|\rho_n|\,\|\cos n(\cdot)\| = \|\rho_n \sin n(\cdot)\| = \frac{1}{2}\|\phi_n(\tau,\cdot) - \phi_n(\tau,\cdot+\pi)\| \leq$$

$$\frac{1}{2}(\|\phi_n(\tau)\| + \|\phi_n(\tau,\cdot+\pi)\|) \leq \|\phi_n(\tau)\| = \|\tau\|.$$

The uniqueness (with the exactness to shift of argument) of polynomial deviating least from zero in metric $L_p$ for $1 < p < \infty$ follows from fact that spaces $L_p$ for $1 < p < \infty$ are strictly normed. For $p = 1$ and $p = \infty$ the uniqueness is established without difficulty.

In the similar way two next statements may be proved.

**Proposition 2.9.2.** *If the norm $||\cdot||$ is translate invariant and such that from inequalities $f(t) \geq g(t) \geq 0$, $(t \in R)$ it follows that $||f|| \geq ||g||$, then for any nonnegative on all axis trigonometric polynomial $\tau \in T_{2n+1}$ an inequality*

$$||\tau|| \geq ||1 + \cos n(\cdot)||$$

*holds true. In particular*

$$||\tau||_p \geq ||1 + \cos n(\cdot)||_p = ||1 + \cos(\cdot)||_p, \quad p \in [1, \infty].$$

**Proposition 2.9.3.** *If the norm $||\cdot||$ is translate invariant and numbers $n$ and $k$ are mutually simple, then for any trigonometric polynomial $\tau$ of the form (2.133)*

$$||\tau|| \geq |\rho_k| \, ||\cos k(\cdot)||.$$

Note that equalities (2.135) may be easy infered from other considerations.

For example,the passage to limit when $k \to \infty$ in Zigmund's inequality (3.3.23)

$$||n^{-k}\tau^{(k)}||_p \leq ||\tau||_p$$

immediately leads us to inequality (2.135).

¿From propositions 2.9.1 and 2.9.2 we can easy obtain

**Corollary 2.9.4.** *If $\Phi(t)$ is arbitrary N-function, then for any algebraic polynomial $p \in P_n(1)$*

$$\int_{-1}^{1} \Phi(|p(t)|) \frac{dt}{\sqrt{1-t^2}} \geq \int_{-1}^{1} \Phi(2^{1-n}|T_n(t)|) \frac{dt}{\sqrt{1-t^2}}$$

*and for any algebraic polynomial $p \in P_n^+(1)$*

$$\int_{-1}^{1} \Phi(p(t)) \frac{dt}{\sqrt{1-t^2}} \geq \int_{-1}^{1} \Phi\left(2^{1-n}\left(1 + T_n(t)\right)\right) \frac{dt}{\sqrt{1-t^2}}$$

*The inequalities turn into the equalities only if*

$$p(t) = 2^{1-n}T_n(t) \quad or \quad p(t) = (1 + T_n(t))2^{1-n}$$

*correspondingly.*

Adduce still two extremal properties of nonnegative trigonometric polynomials.This properties are widely used in approximation theory and lead us to Fejer's and Korovkin's polynomials. Denote by $T_{2n+1}$ the set of all nonnegative on all axis trigonometric polynomials $\tau \in T_{2n+1}^+$.

**Proposition 2.9.5.** *Among all polynomials $\tau \in T^+_{2n-1}$ for which $\tau(0) = n$ the unique Fejer's polynomial*

$$F_n(t) = \frac{1}{2} + \sum_{k=1}^{n-1} \left(1 - \frac{k}{n}\right) \cos kt = \frac{1}{2n} \left(\frac{\sin nt/2}{\sin t/2}\right)^2,$$

*deviates least from zero in the metric of the space $L_1$, i.e.*

$$\min\{||\tau||_1 : \tau \in T^+_{2n+1},\ \tau(0) = n\} = ||F_n||_1 = \pi.$$

*A polynomial deviating least from zero is unique.*

**Proof.** From the theorem 2.1.5 it follows that for each $\tau \in T^+_{2n+1}$ there exist the numbers $x_1, \ldots, x_n \in R$ such that

$$\tau(t) = \left|\sum_{k=0}^{n-1} x_k e^{ikt}\right|^2 = \frac{1}{2}\rho_0 + \sum_{k=1}^{n-1} \rho_k \cos kt, \tag{2.136}$$

where

$$\rho_k = \sum_{\nu=1}^{n-k} x_\nu x_{\nu+k} \quad k = 1, 2, \ldots, n. \tag{2.137}$$

. But then in view of Cauchy-Bunyakowski's inequality

$$\tau(0) = \frac{1}{2}\rho_0 + \sum_{k=1}^{n-1} \rho_k = \frac{1}{2}\left(\sum_{k=1}^{n} x_k\right)^2 \leq \frac{1}{2} n \sum_{k=1}^{n} x_k^2 =$$

$$\frac{1}{2} n\rho_0 \leq \frac{n}{2\pi} \int_0^{2\pi} \tau(t)dt = \frac{n}{2\pi}||\tau||_1,$$

i.e.

$$n = \tau(0) \leq \frac{n}{2\pi}||\tau||_1,$$

and inequality turns into equality if and only if $x_k = 1/\sqrt{2},\ k = 1, 2, \ldots, n$, i.e. if and only if $\rho_0 = 1/2$ and $\rho_k = (1 - k/n)/2,\ k = 1, 2, n$, that is equivalent to the statement of proposition 2.9.5.

¿From the proposition 2.9.5 and from the fact that norm $||\tau||_\infty$ is translate invariant the next proposition immediately follows.

**Proposition 2.9.6.** *For any polynomial $\tau \in T^{+}_{2n-1}$ the inequality*

$$\|\tau\|_\infty \leq \frac{n}{2\pi}\|\tau\|_1 \tag{2.138}$$

*holds true. This inequality turns into equality if and only if*

$$\tau(t) = aF_n(t+t_0).$$

**Theorem 2.9.7.** *If $\tau \in T^{+}_{2n-1}$ then*

$$\frac{1}{\pi}\int_0^{2\pi} \tau(t)\cos t\, dt \leq \cos\frac{\pi}{n+1}. \tag{2.139}$$

*Besides that, the sign of equality is attained if and only if*

$$\tau(t) = \frac{1}{2} + \sum_{k=1}^{n-1} \rho_{k,n}\cos kt, \tag{2.140}$$

*where*

$$\rho_{k,n} = \frac{n-k}{n}\cos\frac{k\pi}{n+1} + \frac{1}{n}\sin\frac{k\pi}{n+1}\cot\frac{\pi}{n+1}.$$

**Proof.** Let

$$x_k^0 = \sqrt{\frac{2}{n}\frac{\sin k\pi/(n+1)}{\sin\pi/(n+1)}}, \quad k = 1, 2, \ldots, n. \tag{2.141}$$

Then

$$\rho_k^0 = \sum_{\nu=1}^{n-k} x_\nu^0 x_{\nu+k}^0 = \frac{1}{n\sin\pi/(n+1)}\sum_{\nu=1}^{n-k}\left(\cos\frac{k\pi}{n+1} - \cos\frac{2\nu+k}{n+1}\pi\right) =$$

$$\frac{n-k}{n}\cos\frac{k\pi}{n+1} + \frac{1}{n}\sin\frac{k\pi}{n+1}\cot\frac{\pi}{n+1},$$

and in particular $\rho_1^0 = \cos\pi/(n+1)$. From here and from (2.137) it follows that in order to prove the theorem it is sufficiently to show that extremal problem

$$\sum_{k=1}^{n-1} x_k x_{k+1} \to max, \quad \sum_{k=1}^{n-1} x_k^2 = 1 \tag{2.142}$$

has the unique solution $x_k = x_k^0,\ k = 1, 2, \ldots, n$.

Decide this problem using the rule of Lagrange's multipliers.Lagrange's function has the form (it's easy to prove that we may consider that $\lambda_0 = 1$)

$$\Im = \sum_{k=1}^{n-1} x_k x_{k+1} - \lambda \left( \sum_{k=1}^{n-1} x_k^2 - 1 \right).$$

The necessary extremum conditions $\partial\Im/\partial x_i = 0,\ k = 1, 2, \ldots, m$ of this problem acquire the form

$$\begin{gathered} x_2 - 2\lambda x_1 = 0 \\ x_{k-1} - 2\lambda x_k + x_{k+1} = 0, \quad k = 2, \ldots, n-1, \\ x_{n-1} - 2\lambda x_n = 0. \end{gathered} \tag{2.143}$$

¿From here it follows that if $x_k^*,\ k = 1, 2, \ldots, n$ and $\lambda^*$ are the solutions of extremal problem (2.142) then

$$x_k^* = p_k(\lambda^*) x_1^*, \quad k = 1, 2, \ldots, n; \tag{2.144}$$

$$x_1^* p_{n+1}(\lambda^*) = 0, \tag{2.145}$$

where

$$\begin{gathered} p_1(\lambda) = 1; \quad p_2(\lambda) = 2\lambda; \\ p_k(\lambda) = 2\lambda p_{k-1}(\lambda) - p_{k-2}(\lambda), \quad k = 3, \ldots, n+1. \end{gathered} \tag{2.146}$$

Therefore, in view of (2.24)

$$p_k(\lambda) = U_{k-1}(\lambda) \tag{2.147}$$

is the Chebyshav's polynomial of second kind.

Multiplying $k$-th equation of sustem on $x_k^*$ and adding its we obtain

$$\sum_{k=1}^{n-1} x_k^* x_{k+1}^* - \lambda^* \sum_{k=1}^{n-1} (x_k^*)^2 = 0,$$

and consequently

$$|\lambda^*| \sum_{k=1}^{n-1} (x_k^*)^2 \left| \sum_{k=1}^{n-1} x_k^* x_{k+1}^* \right| \leq \left( \sum_{k=1}^{n-1} (x_k^*)^2 \sum_{k=2}^{n-1} (x_k^*)^2 \right)^{1/2} \leq \sum_{k=1}^{n-1} (x_k^*)^2$$

i.e. $|\lambda^*| \leq 1$. From here and from (2.147) and (2.143) it follows that

$$p_k(\lambda) = \frac{\sin k\theta}{k \sin \theta},$$

that together with (2.135) gives us $\theta = \pi/(n+1)$. Thus

$$x_k^* = a\,\frac{\sin k\pi/(n+1)}{\sin \pi/(n+1)}, \quad k = 1, \ldots, n.$$

Now from condition

$$\sum_{k=1}^{n-1} (x_k^*)^2 = 1$$

it follows that $x_k^* = x_k^0$, $k = 1, 2, \ldots, n$ and this fact completes the proof of theorem. By analogy we can get the next generalization of theorem 2.9.7.

**Theorem 2.9.8.** *If $\tau \in T_{2n+1}^+$ then for all $k = 1, 2, \ldots, n$*

$$\int_0^{\pi} \tau(t) \cos kt\, dt \le \cos \frac{\pi}{[n/k]+2},$$

*where $[a]$ is integral part of $a$.*

## 2.10 Commentaries

Section 2.1. Theorem 2.5.1 was proved by M.Riesz [1].Theorem 2.1.7 is Markov-Lucach's theorem (see V.A.Markov [1] , Lucach [1]).

Theorems 2.1.8 and 2.1.1 are famous Chebyshev's theorems about alternans, being one of the corner-stones at the foundation of approximation theory.

Section 2.2.The problem of polynomials deviating least from zero is one from important problems of approximation theory. Theorems 2.2.1 and 2.2.7 were proved by P.L.Chebyshev [1]. Chebyshev's polynomials of first and second kind are extremal in the theorems 2.2.1 and 2.2.7 and besides that have a great number of remarkable properties and find numerous applications in mathematics (see,for example, the monograph of S.Pashkovski [1]).

Section 2.3. Although the evident shape of polynomials deviating least from zero in the spaces $L_p[-1,1]$ (when $p$ is not equal to $1, 2, \infty$) is unknown , nevertheless under the investigations of a great many important problems of approximation theory some their general properties, given in this section, are very useful.

Sections 2.4 - 2.6. At Hilbert spaces of functions the polynomials deviating least from zero are orthogonal on $[-1,1]$, on axis and semiaxis to all polynomials with lesser order.

The properties of important class orthogonal Legandr's polynomials are given in the section 2.4.

The section 2.5 is devoted to Jacoby's polynomials and section 2.6 is devoted to Hermit's and Lagerr's polynomials.Systematically orthogonal polynomials are studied in fundamental book G.Szego [3].

Section 2.7. Constant sign polynomials deviating least from zero in $L_1[-1,1]$ were found by R.Boyanic and R.De Vore [1].Theorem 2.7.5 was proved by V.F.Babenko and V.F.Kofanov [1] and its simple proof was suggested also by I.J.Tyrygin [1].

Section 2.8. Theorem 2.8.1 was proved by E.I.Zolotarev [1].Theorem 2.8.5 was proved by M.L.Geronimus [1] (see, also V.M.Tihomirov [7]).

Section 2.9. Proposition 2.9.5 is wellknown Fejer's result [1].Theorem 2.9.7 belongs to P.P.Korovkin (see [2]).

## 2.11 Supplementary results

**1.** Subspace $H_n \subset C[a,b]$ of $n$-th dimension is called by Chebyshev subspace, if any nonzero element $u \in H_n$ has on segment $[a,b]$ at most $n$ zeros.The next generalization of alternans theorem is valid.

Let $f \in C[a,b], Q \in [a,b]$ is arbitrary closed set, $H_n \subset C[a,b]$ is $n$-dimensional Chebyshev subspace. In order to an element $u \in H_n$ would be the best approximation element for $f$ in the space $C(Q)$ it is necessary and sufficiently that difference $f(x) - u(x)$ would attain maximum of its modulus on $Q$ with successive change of sign at least at $n+1$ points from this set. Bernstein S.N. [1, p. 16].

**2.** Let $\Omega(x)$ is continuous and strictly positive on the segment $[a,b]$ function; $C([a,b];\Omega)$ is linear space of all continuous on $[a,b]$ functions $f$ whith the norm

$$||f|| = \max_{t \in [a,b]} |f(t)|\Omega(t).$$

The algebraic polynomial $p \in P_n$ is the best approximation polynomial for $f$ in the space $C([a,b];\Omega)$ if and only if weighted difference $\Omega(t)|f(t) - p(t)|$ attains on $[a,b]$ the maximum of its modulus at least at $n+2$ different points with successive change of sign.

**3.** Let closed (finite or infinite) interval $[a,b] \subset R$ and two real, continuous on $[a,b]$ functions $f(x)$ and $s(x)$ are given. Let also

$$Q(x) = s(x)\frac{q_0 x^n + q_1 x^{n-1} + \ldots + q_n}{p_0 x_m + p_1 x_{m-1} + \ldots + p_m}, \tag{2.148}$$

where $n, n \in N$ and $q_0, \ldots, q_n, p_0, \ldots, p_m \in R$.

The function $p(x)$ of the form (2.148) given above which among all such function has the least deviation from $f(x)$ (in uniform metric) is characterized by the next property. If it may be represented in the form

$$p(x) = s(x)\frac{b_0x^{n-\nu} + b_1x^{n-\nu-1} + \ldots + b_{n-\nu}}{a_0x^{m-\mu} + a_1x^{m-\mu-1} + \ldots + a_{m-\mu}} = s(x)\frac{B(x)}{A(x)},$$

where $0 \leq \mu \leq m$; $0 \leq \nu \leq n; a_0 \neq 0$ and fraction $B(x)/A(x)$ is unreduced then the number of $N$ successive points from inferval $[a, b]$ in which the difference $f(x) - p(x)$ attains the maximum by modulus value with alternating signs is at most $m + n - d + 2$, where $d = \min\{\mu, \nu\}$ and if $p(x) = 0$, $x \in [a, b]$ then $N > n + 2$.

Chebyshev P.L., Ahiezer N.I. [6, p. 66].

**4.** Let $F_A$ is the set on the plane,restricted by the ellips with the focuses at points $(-1, 0)$ and $(1, 0)$ and with the sum of semiaxis $A > 1$. Then among all functions of the form

$$\sum_{k=0}^{n} a_k z^{n-k} + \sqrt{z^2 - 1}\sum_{k=0}^{n-1} b_k z^{n-1-k}$$

whith the complex coefficients $a_k$ and $b_k$, from which $a_0$ and $b_0$ are fixed and the rest are arbitrary, the least norm in the space $C(F_A)$, which is equal to

$$2^{-n}\left(A^n \max\{|a_0 + b_0|, |a_0 - b_0|\} + A^{-n} \min\{|a_0 + b_0|, |a_0 - b_0|\}\right)$$

is attained by the function

$$2^{1-n}\left(a_0 T_n(z) + b_0\sqrt{z^2 - 1}U_{n-1}(z)\right)$$

and only by this function.

Pashkovski S.[1, p. 65-66]

**5.** Let $n, m \in Z$ and $n \geq m + 2$, $0 \leq r \leq R$, $r > 0$ if $m < 0$. Among all functions of the form

$$\sum_{k=m}^{n} a_k z^k$$

with the complex coefficients $a_k$, where $a_m$ and $a_n$ are fixed and the rest are arbitrary, the sum $a_m z^m + a_n z^m$ and only it has the least norm in the space $C(F_{R.r})$ $(F_{R,r} = \{z : r \leq |z| \leq R\})$ which is equal to

$$max\{|a_m|r^m + |a_n|r^n, |a_m|R_m + |a_n|R^n\}.$$

Pashkovski S.[1, p. 65]

**6.** Let $z_1, \ldots, z_n \in C$, $|z_j| > 1$, $J = 1, 2, \ldots, k$ and $|z_j| \leq 1$, $j = k + 1, \ldots, n$. Then for $N > n$

$$\min_{A_1,\ldots,A_N} \max_{|z|=1} \frac{|x^N + A_1 z^{N-1} + \ldots + A_N|}{|(z - z_1)\ldots(z - z_n)|} = \frac{1}{|z_1 \ldots z_k|}$$

and for any $p > o$

$$\min_{A_1,\ldots,A_N} \frac{1}{2\pi} \int_{|z|=1} \left| \frac{x^N + A_1 z^{N-1} + \ldots + A_N}{(z - z_1)\ldots(z - z_n)} |^p |dz| \right| = \frac{1}{|z_1 \ldots z_k|^p}.$$

**7.** Let positive on segment $[-1, 1]$ polynomial $\omega(x)$ is given and

$$\omega(x) = \left(1 - \frac{x}{a_1}\right)\left(1 - \frac{x}{a_2}\right)\ldots\left(1 - \frac{x}{a_{2q}}\right).$$

Let also

$$x = \frac{1}{2}\left(v - \frac{1}{v}\right), \; a_k = \frac{1}{2}\left(c_k - \frac{1}{c_k}\right), \; |c_k| < 1, \; k = 1, 2, \ldots, 2q,$$

$$\Omega(v) = \prod_{k=1}^{2q} \sqrt{v - c_k}$$

and

$$\Im_m = \begin{cases} 2^{1-m} \prod_{k=1}^{2q} \sqrt{1 + c + k^2}, & m > q, \\ 2^{1-q} \frac{1}{1 + c_1 \ldots c_{2q}} \prod_{k=1}^{2q} \sqrt{1 + c + k^2}, & m = q. \end{cases}$$

Put now

$$T_m(x, \omega) = \frac{\Im_m}{2}\left(v^{2q-m}\frac{\Omega(1/v)}{\Omega(v)} + v^{m-2q}\frac{\Omega(v)}{\Omega(1/v)}\right)\sqrt{\omega(x)},$$

where $m \in N$, $m \geq q$. Then $T_m(x, \omega) \in P_m(1)$ and

$$\min_{A_1,\ldots,A_N} \max_{|z|=1} \frac{|x^m + A_1 z^{m-1} + \ldots + A_m|}{\sqrt{\omega(x)}} = \Im_m.$$

Moreover polynomial $T_m(x, \omega)$ is extremal.

Let

$$U_m(x, \omega) = \Im_{m+1}\left(v^{2q-m}\frac{\Omega(1/v)}{\Omega(v)} - v^{m-2q}\frac{\Omega(v)}{\Omega(1/v)}\right)\frac{\sqrt{\omega(x)}}{1/v - v}.$$

Then $U_m(\cdot,\omega) \in P_n(1)$ and for any $p \geq 1$

$$\min_{A_1,\ldots,A_m} \int_{-1}^{1} \left| (x^m + A_1 x^{m-1} + \ldots + A_m) \frac{\sqrt{1-x^2}}{\sqrt{\omega(x)}} \right|^p \frac{dx}{\sqrt{1-x^2}} =$$

$$\int_{-1}^{1} \left| U_m(x,\omega) \frac{\sqrt{1-x^2}}{\sqrt{\omega(x)}} \right|^p \frac{dx}{\sqrt{1-x^2}} = \frac{\Gamma(1/2)\Gamma((p+1)/2)}{\Gamma(1+p/2)} \Im_{m+1}^p .$$

Ahiezer N.I. [6, p. 290-292].

**8.** Let $p_m \in P_m$ is arbitrary positive on segment $[-1,1]$ polynomial and $T_{n.m} \in P_n(1)$ is the polynomial deviating least from zero in the space $C[-1,1]$ with the weigth $(p_m(x))^{-1/2}$ i.e.

$$\min\{\|p(p_m)^{-1/2}\|_{C[-1,1]} : p \in P_n(1)\} = \left\| T_{n,m}(p_m)^{-1/2} \right\|_{C[-1,1]} .$$

Then

$$T_{n,m}(x) = \sqrt{p_m(x)} \cos \sum_{k=1}^{2n} \phi_k(x),$$

where $\phi_k(x)$ is defined by conditions

$$\cos\phi_k = \left( \frac{(1+a_k)(1+x)}{2(1+xa_k)} \right)^{1/2}, \sin\phi_k = \left( \frac{(1-a_k)(1-x)}{2(1+xa_k)} \right)^{1/2},$$

$$-\pi < \phi_k \leq \pi, \quad k = 1,2,\ldots,2n.$$

In these equalities the numbers $a_1,\ldots,2n$ (some from these numbers may be zeros) such that

$$p_m(x) = p_m(0) \prod_{k=1}^{2n} (1 + x a_k).$$

Markov A.A. [1].

**9.** Let $\omega(t)$, $t \in [-\pi,\pi]$ is given and for it

$$\int_{-\pi}^{\pi} \omega(t)\, dt \geq 0$$

and let

$$G(\omega) = \begin{cases} \exp\left(\frac{1}{2\pi}\int_{-\pi}^{\pi} \ln \omega(t)\, dt\right), & if \ \ln \omega(\cdot) \in L_1, \\ 0, & if \ \ln \omega(\cdot) \notin L_1. \end{cases}$$

Then for any $p > 0$

$$\lim_{n\to\infty} \min_{A_k} \frac{1}{2\pi} \int_{-\pi}^{\pi} \left| z^n + \sum_{k=1}^{n} A_k x^{n-k} \right| \omega(t)\, dt = G(\omega), \quad z = e^{it}.$$

**10.** Let $m \in N$ and $n > 1$. Then for each algebraic polynomial $p \in P_n(1)$ with integral coefficients the next inequality

$$||p||_{C[-1/m,1/m]} \geq m^{-m}$$

is valid. This inequality turns into equality for polynomial $p(x) = x^n$.
Bernstein S.N.[1].

**11.** Among all of functionts $\theta(x)$ of the type

$$\theta(x) = \frac{1}{1-x} + \sum_{k=0}^{n} a_k x^k$$

the function

$$\theta_n(x) = \frac{1}{1-x} - \sum_{k=0}^{n} x^k.$$

has the least deviation from zero in the space $L_1[-1/2, 1/2]$. In this case

$$E\left(\frac{1}{1-x}, P_n\right)_{C[-1/2,1/2]} = ||\theta_n||_{C[-1/2,1/2]} = \frac{1}{2^n}.$$

Bernstein S.N.[1].

**12.** For $a > o$ and $n \geq 0$

$$E\left(\frac{1}{x^2 - a^2}, P_n\right)_{C[-1,1]} = \frac{1}{2a^2(a^2-1)(a+\sqrt{a^-1})^n}$$

and

$$E\left(\frac{x}{x^2 - a^2}, P_n\right)_{C[-1,1]} = \frac{1}{2a(a^2-1)(a+\sqrt{a^-1})^n}$$

Bernstein S.N. [3].

**13.** If $Q \in R$ is the closed set of measure $2h$ then for any polynomial $p \in P_n(1)$

$$||p||_{C(Q)} \geq 2(h/2)^n,$$

and moreover the equality is possible only if $Q$ is segment of length $2h$.

Polia G.(See Bernstein S.N. [1])

The problem of finding the polynomial deviating least from zero on the given closed set $Q$ is very difficult. In the case, when $Q$ consists of two unintersected segments this problem was establihsed by Ahiezer [1, 6, p. 320].If $Q = [-1, -\alpha] \cup [\alpha, 1]$ $(0 < \alpha < 1)$ and $n = 2m$ then polynomial has the form

$$p_*(x) = L_{2m} T_m \left( \frac{2x^2 - 1 - \alpha^2}{1 - \alpha^2} \right),$$

where

$$L_{2m} = ||P_*||_{C[-1,1]} = 2^{1-2m}(1 - \alpha^2)^m.$$

**14.** Let $M \subset C$ is closed set and

$$L_n = \min_{A_k} \left\| z^n + \sum_{k=1}^{n} A_k z^{n-k} \right\|_{C(M)}.$$

Then there exists

$$\lim_{n \to \infty} (L_n)^{1/n} = \tau = \tau(M).$$

The value $\tau(M)$ is called by transfinite diameter of the set $M$ and is equal to

$$\lim_{n \to \infty} (V_n)^{n/2},$$

where $V_n$ is the most value attained by the modulus of Vandermond's determinant $V(x_1, \ldots, x_n)$, when the points $x_1, \ldots, x_n$ run the set $M$. ¿From the results of item number 13 it immediately follows that if $M = [-1, -\alpha] \cup [\alpha, 1]$, then

$$\tau(M) = \frac{1}{2}\sqrt{1 - \alpha^2}.$$

Ahiezer N.I. [6, p. 32].

**15.** If $q(x)$ is the polynomial with real coefficients, having only complex zeros $\alpha_k \pm \beta_k$, $k = 1, 2, \ldots, n$ then the least deviation from zero of fraction

$$\frac{a_0 + x + \sum_{k=1}^{m} a_k x^{k+1}}{\sqrt{q(x)}}, \quad a_k \in R, \; m \geq n,$$

in the space $C(R)$ is equal to

$$\left(q(0)\left|\sum_{k=1}^{n}\frac{1}{\alpha_k - i\beta_k}\right|\right)^{-1}.$$

If $q \in P_n$ has only negative roots and

$$p(x) = \sum_{k=0}^{m} b_k x^k, \quad b_k \in R, \ k = 0, 1, \ldots, m,$$

then

$$\sum_{k=0}^{m} |b_k| \leq 2^{n/2-1}\sqrt{q(1)} \sup_{x\in R^+}\left|\frac{p(x)}{\sqrt{q(x)}}\right|,$$

and equality sign has place when

$$q(x) = (1+x)^n$$

and

$$p(x) = a\left(\left(1-\sqrt{-x}\right)^n + \left(1-\sqrt{-x}\right)^n\right).$$

Bernstein S.N. [1, p. 136-137].

**16.** If $V_{-1}^{1}$ is the variation of funcion $f$ on segment $[-1,1]$ then for any polynomial $p \in P_n(1)$

$$V_{-1}^{1}(p) \geq 2^{1-n}V_{-1}^{1}(T_n)$$

and inequality turns into equality if and only if $p(x) = 2^{1-n}T_n(x)$.

**17.** For $a \in (o, \pi)$ and for each even trigonometric polynomial $\tau \in T_{2n-1}$ the exact inequality

$$\|\cos n(\cdot) + \tau(\cdot)\|_{C[-a,a]} \geq \left(\sin\frac{a}{2}\right)^n$$

holds true.

Bernstein S.N. [1].

**18.** For all $n \in N$

$$\inf_{f\in C,\, f\perp \cos n(\cdot)} \|\cos n(\cdot) - f(\cdot)\|_\infty = \frac{\pi}{4}$$

and for any $p \in [1, \infty)$

$$\inf_{f\in C,\, f\perp \cos n(\cdot)} \|\cos n(\cdot) - f(\cdot)\|_p = \pi\|\cos(\cdot)\|_{p'}^{-1}, \quad \left(\frac{1}{p}+\frac{1}{p'}=1\right).$$

The first correlation was proved by Bernstein S.N. [1, p. 29] and second one was proved by Tajkov L.V. [1].

**19.** If $p \in (0,1)$ and

$$\|f\|_p = \left(\int_0^\pi |f(t)|^p dt\right)^{1/p},$$

then uniformly by $m$ and $n$, $m \le n$

$$\inf_{c_k} \|\tau(\cdot) - \sum_{|k|\le n,\, k\ne m} c_k e^{ik(\cdot)}\|_p \asymp (n-m+1)^{1-1/p},$$

and for any trigonometric polynomial $\tau$ of order $m$

$$\inf_{c_k} \left\| \tau(\cdot) - \sum_{m<|k|\le n} c_k e^{ik(\cdot)} \right\|_p \asymp n^{1-1/p}.$$

Ivanov V.I., Judin V.A. [1].

# Chapter 3

# Trigonometric polynomials

Extremal properties of trigonometric polynomials have been studied specially thoroughly. Specific features of the simplest triginometric functions, linear combination of which is a trigonometric polynomial, and independence of their norm from the shift of argument in the spaces $C$ and $L_p$ facilitates the investigation and allows to obtain exact inequalities with the help of elegant and quite simple transformations. Besides that,a solution of many problems of the analysis on classes of periodical functions is based on inequalities for trigonometric polynomial, moreover, inequalities that estimate the norm of the certain derivative of polynomial by norm (may be in another space) of the polynomial itself have special significance. First such inequality was obtained by S.N.Bernstein in 1912.

Major attention in this chapter is paid to the inequalities of Bernstein's type, and in the most importent, as it seems to us, situations we adduce detailed proofs and elucidate their messages. Apart from this in section 3.5 there are expounded results connected with the inequalities for norm of the polynomial itself in different spaces.

## 3.1 Bernstein's inequalities and its generalization

The classical Bernstein's inequality

$$||\tau'||_C \leq n||\tau||_C, \tag{3.1}$$

that holds true for any trigonometric polynomial $\tau(t)$ of order $n$ has an importent applications in the approximation theory and other branches of the mathematics. At present it is known many various proofs of this inequality wich give us the possibility of its generalisation in various directions (the entire functions, the $L_p$-spaces etc.). Below we shall present its proof which based on the proposition 1.4.2 about derivative's comparison and is probably the simplest of the known proofs. The considerations which follows from the proposition 1.4.2 allow us to prove a series useful generalisations of the inequality (3.1) and obtain some inequality which is sharper then inequality (3.1).

We set $||\cdot|| = ||\cdot||_C$ everywhere in this paragraph.

**Lemma 3.1.1.** *Any trigonometric polynomial $\tau(t)$ of order $n$ ($\tau(t) \in T_{2n+1}$) owns the $\mu$-property relatively to $2\pi/n$-right function $A\cos nt$ if $||\tau|| \leq A$.*

Realy, in the opposite case for some $\beta \in R$ and sufficiently small $\varepsilon > 0$ the difference

$$\delta(t) = A\cos nt - (1-\varepsilon)\tau(t+\beta)$$

will have no less $2n+2$ sign changes. But it is impossible because $\delta(t)$ is a trigonometric polynomial of order $n$.

**Theorema 3.1.2.** *For any trigonometric polynomial $\tau(t)$ of order $n$ the next inegauleties*

$$||\tau^{(k)}|| \leq n^k ||\tau||, \; k = 1, 2, \ldots, \tag{3.2}$$

*hold true. This inegualities become equalities for polynomials of the form $\tau(t) = a\cos n(t+\alpha)$.*

**Proof.** It is clear that inequality (3.2) will be proved if we shall prove this inequality for $k = 1$.

In view of lemma 3.1.1 we may aply the proposition 1.4.2 to polynomial $\tau(t)$ and $2\pi/n$-right function $q_n(t) = ||\tau||\cos nt$.

By proposition 1.4.2 we have that if

$$\tau(t) = q_n(y) = ||\tau||\cos ny \tag{3.3}$$

and $\tau'(x)q_n'(y) \geq 0$ then

$$||\tau'(x)|| \leq |q_n'(y)| = n||\tau|||\sin ny|. \tag{3.4}$$

Thus

$$|\tau'(t)| \leq n||\tau||, \; -\infty < t < \infty,$$

that is equivalent to (3.2) for $k = 1$.

It follows from theorem 3.1.2 that for each polynomial $\tau \in T_{2n+1}$ for all $k = 1, 2, \ldots$

$$\|\tau^{(k)}\| \leq n^k E_0(\tau)_C.$$

Thus if $\tau(t) \geq 0$ for all $t \in R$, then

$$\|\tau^{(k)}\| \leq \frac{1}{2} n^k \|\tau\|.$$

This inequalities are also exact and become equalities for polynomials of the form $a \cos n(t + t_0)$ and $a \cos n(t + t_0) + |a|$ respectively. For one of this polynomial it will be proved in this paragraph and in the sequel this will not be noticed in the statemens of the theorems.

Multiplying the inequality (3.3) by $n$, raising (3.3) and (3.4) to the second power and next adding the obtained inequalities we get the next inequality which is more precise then Bernstein's inequality.

**Theorem 3.1.3.** *For each trigonometric polynomial $\tau \in T_{2n+1}$ the next inequality*

$$\sqrt{(\tau'(t))^2 + n^2(\tau(t))^2} \leq n\|\tau\| \tag{3.5}$$

*which is exact for every point $t$ holds true.*

**Proof.** Since for an arbitrary $x, y \in R$

$$(x^2 + y^2)^{1/2} = \sup\{x \sin \alpha + y \cos \alpha : \alpha \in R\}$$

the inequality (3.5) we may write in the next equivalent form:

$$\|\tau(t) \sin \alpha + n\tau(t) \cos \alpha\| \leq n\|\tau\|, \quad \forall \tau \in R. \tag{3.6}$$

Applying the inequalities (3.6) to the polynomial

$$\tau(t) - (\max\{\tau(u) : u\} + \min\{\tau(u) : u\})/2$$

we obtain for every $\tau \in T_{2n+1}$ and all $t$ that

$$[(\tau'(t))^2 + n^2(\tau(t) - (\max\{\tau(u) : u\} + \min\{\tau(u) : u\})/2)^2]^{1/2} \leq n E_0(\tau)_C.$$

In particurar, if $\tau(t) \geq 0$ and $t \in R$ then

$$[(\tau'(t))^2 + n^2(\tau(t) - \max\{\tau(u) : u\}/2)^2]^{1/2} \leq n\|\tau\|/2.$$

Let now for $h > 0$ and $k = 1, 2, \ldots$

$$\Delta_h^k(f;t) = \sum_{i=0}^{k} (-1)^i \binom{k}{i} f(t + (k-2i)h),$$

and

$$\Delta_h^1(f;t) = \Delta_h(f;t).$$

The next theorem gives us the inequalities which are more precise then Bernstein's anes but in some other sence.

**Theorem 3.1.4.** *For each polynomial $\tau(t) \in T_{2n+1}$ $0 < h < \pi/n$ and $0 < h < \pi/(2n)$ respectively the exact inequalities*

$$||\tau^{(k)}|| \leq \left(\frac{n}{2\sin nh}\right)^k ||\Delta_h^k(\tau)|| \leq n^k ||\tau|| \tag{3.7}$$

*holds true*

**Proof.** It is sufficiently to prove the inequalities (3.7) for $k = 1$ and then use the induction on $k$.

Let $||\tau|| = M$ and let $M = |\tau'(t_0)|$. Without loss of generality we may assume, that $M = \tau'(t_0)$. In view of lemma 3.1.1 polynomial $\tau'(t)$ posseses the $\mu$-property relatively to the function $M \cos nt$. Then in view of lemma 1.4.1 we have

$$\tau'(t_0 + u) \geq M \cos nu, \quad |u| < \pi/n. \tag{3.8}$$

By integrating the inequality (3.8) in $u$ from $-h$ to $h$, $|h| \leq \pi/n$, we have

$$\tau(t_0 + h) - \tau(t_0 - h) \geq \frac{2M}{n} \sin nh.$$

Consequently

$$M = ||\tau|| \leq \frac{n}{2\sin nh} ||\Delta_h(\tau)||.$$

and the first inequality in (3.7) is proved for $k = 1$ .

To prove the second inequality we recall that polynomial $\tau(t)$ possesses the $\mu$-property relatively to $q_n(t) = ||\tau|| \cos nt$. Let

$$||\Delta_h(\tau)|| = |\tau(x_0 + h) - \tau(x_0 - h)|$$

and let for definitness the value standing under the sign of the modulus is positive.

Choose the point $y_0 \in R$ such that

$$\tau(x_0 + h) = q_n(y_0),\ \tau'(x_0 + h)q_n'(y_0) \geq 0.$$

Next on the interval of the monotonicity of the function $q_n(t)$ which includes the point $y_0$ we choose the point $y_1$ from condition $\tau(x_0 + h) = q_n(y_1)$. Then in view of lemma 1.4.3 we shall have $|y_1 - y_0| := 2h_0\tau \leq 2h$ and since

$$||\Delta_{h_0}(q_n)|| \leq ||\Delta_h(q_n)||$$

then for $0 < h_0 \leq h < \pi/(2n)$ we obtain

$$||\Delta_h(\tau)|| = |q_n(y_0) - q_n(y_1)| \leq ||\Delta_{h_0}(q_n)|| \leq ||\Delta_h(q_n)|| = 2||\tau|| \sin nh,$$

and the second inequality (3.7) is proved.

The theorem 3.1.4 is completely proved.

Setting in inequality (3.8) $h = \pi/(2n)$ we obtain the next statement.

**Corollary 3.1.5.** *For any trigonometric polynomial $\tau \in T_{2n+1}$ and for all $k = 1, 2, \ldots$ the exact inequalities*

$$||\tau^{(k)}|| \leq (n/2)^k ||\Delta^k_{\pi/(2n)}(\tau)|| \leq n^k ||\tau||$$

*holds true.*

Now we present without proof (it is based on the same ideas) analogous to (3.6) inequality which is more precise than the second inequality in (3.7).

**Theorema 3.1.6.** *For any $\tau \in T_{2n+1}$ $t, h \in (0, n^{-1} arc\cos(|\tau|/||\tau||))$ the exact inequality*

$$[(\tau(t)^2 4\sin^2 nh + (\Delta(\tau(t))^2]^{1/2} \leq 2||\tau|| \sin nh.$$

*holds true.*

Prove now one another strictness of the inequality (3.7).

**Theorem 3.1.7.** *For any $k, n \in N$, $0 < h \leq \pi/n$ and $\tau \in T_{2n+1}$*

$$||\tau^{(k)}|| \leq \left(\frac{2}{h}\right)^k \left(||\Delta_h^k \tau|| + \left(\left(\frac{nh/2}{\sin nh/2}\right)^k - 1\right) \frac{||\Delta_h^{k+2}\tau||}{\sin^2(nh/2)}\right)$$

*and in particular (for $h = \pi/n$)*

$$||\tau^{(k)}|| \leq n^k \left(\left(\frac{2}{\pi}\right)^k ||\Delta^k_{\pi/n}\tau|| + \left(1 - \left(\frac{2}{\pi}\right)^k\right) ||\Delta^{k+2}_{\pi/n}\tau||\right).$$

**Remark** If theorem 3.1.7 $||\Delta_h^{k+2}\tau||$ estimates by $||\Delta_h^k\tau||$ we obtain the corollary 3.1.5.

**Proof.** It is easy to see that

$$\tau'(x) - \frac{2}{h}\Delta_h^1\tau(x) = \int_0^{h/2}\left(\frac{2t}{h} - 1\right)\Delta_{2t}^1\tau''(x)dt.$$

Therefore

$$\left\|\tau' - \frac{2}{h}\Delta_h^1\tau\right\| + \left\|\frac{2}{h}\Delta_h^1\tau\right\| \leq \int_0^{h/2}\left(\frac{2t}{h} - 1\right)||\Delta_{2t}^1\tau''||dt + \left\|\frac{2}{h}\Delta_h^1\tau\right\|.$$

¿From this and from the Boas inequality

$$\frac{||\Delta_\delta^k\tau||}{\sin^k n\delta/2} \leq \frac{||\Delta_h^k\tau||}{\sin^k nh/2}, \quad 0 < \delta < h \leq \pi/n$$

we obtain

$$||\tau'|| \leq \frac{||\Delta_h^1\tau''||}{\sin(nh/2)}\int_0^{h/2}\left(1 - \frac{2t}{h}\right)\sin nt\,dt + \left\|\frac{2}{h}\Delta_h^1\tau\right\|$$

or

$$||\tau'|| \leq ||\Delta_h^1\tau''||\left(\frac{1}{n\sin(nh/2)} - \frac{2}{n^2h}\right) + \left\|\frac{2}{h}\Delta_h^1\tau\right\|.$$

¿From this and from corollary 3.1.5 we obtain the assertion of theorem for $k = 1$.

Let now this assertion is valid for $k = 1, 2, \ldots, m$. Then

$$||\tau^{(m+1)}|| \leq \left(\frac{2}{h}\right)^m\left(||\Delta_h^m\tau'|| + \left(\left(\frac{nh/2}{\sin nh/2}\right)^m - 1\right)\frac{||\Delta_h^{m+2}\tau'||}{\sin^2(nh/2)}\right)$$

and moreover in view of corollary 3.1.5

$$||\Delta_h^{m+2}\tau'|| \leq n\frac{||\Delta_h^{m+3}\tau||}{\sin(nh/2)}$$

and due to statement of theorem for $k = 1$

$$||\Delta_h^m\tau'|| \leq \frac{2}{h}\left(||\Delta_h^{m+1}\tau|| + \left(\frac{nh/2}{\sin(nh/2)} - 1\right)\frac{||\Delta_h^{m+3}\tau||}{\sin^2(nh/2)}\right).$$

Comparing three last inequalities we obtain demanding statement for $k = m + 1$ and finish proof by induction.

Finally we shall obtain the generalizations of inequalities (3.3) to the case of linear differential operators having constant coefficients and being more general than $d^k/dt^k$ linear differential operators having real coefficients.

Let $m = 1, 2, \ldots$ and

$$\Im_m(x) = a_0 x^m + a_1 x^{m-1} + \ldots + a_m \tag{3.9}$$

is an arbitrary polynomial of degree $m$ having real coefficients and

$$\Im_m(D) = a_0 D^m + a_1 D^{m-1} + \ldots + a_m . D = d/dt \tag{3.10}$$

by the corresponding differential operator.

**Theorem 3.1.7.** *Let the polynomial $\Im_m$ of the form (3.9) has only real zeros. Then for all trigonometric polynomial $\tau \in T_{2n+1}$ the exact inequality*

$$||\Im_m(D)\tau|| \leq |\Im_m(in)|||\tau||. \tag{3.11}$$

*holds true.*

**Proof.** If $\Im_m$ is a polynonial of the form (3.9) with real zeros $\beta_j$, $j = 1, \ldots, m$, then

$$\Im_m(x) = a_0 \prod_{j=1}^{m} (x - \beta_j)$$

and, consequently, operator $\Im_m(D)$ is a composition of the operators $D - \beta_j$ of the first order. For this reason it suffices to prove the inequality (3.11) for $m = 1$, i.e. for the operapor $\Im_1(D) - \beta$. Then at will be easy to prove the (3.11) for abetrary $m$ by induction. As it wos already notised the polynomial $\tau \in T_{2n+1}$ posseses the $\mu$-property relatively to $||\tau|| \cos nt$. So in view of proposition 1.4.5 we have

$$||\tau' - \beta\tau|| \leq ||\tau|| \, || - n \sin n(\cdot) - \beta \cos n(\cdot)|| = |in - \beta| \, ||\tau||.$$

If the polynomial $\Im_m(x)$ has not only real seros the next theorem is valid.

**Theorem 3.1.8.** *Let $n, m \in N$ and $\Im_m(x)$ be a polynomial of the form (3.9). Let $x_k$, $k = 1, 2, \ldots, m$ be its zeros. Then for any*

$$n > 2 \max\{|Im\, x_k| : k\}$$

*for each polynomial $\tau \in T_{2n+1}$ the exact inequality*

$$||\Im_m(D)\tau|| \leq |\Im_m(in)| \, ||\tau|| \tag{3.12}$$

*holds true.*

**Proof.** Let $x_1, \ldots, x_l$ are real zeros and $x_k = \gamma_k + i\alpha_k$, $k = l+1, \ldots, m-l$ are complex zeros of the polynomial $\Im_m(x)$. If $\gamma_k + i\alpha_k$ is a zero of the polynomial $\Im_m(x)$ then $\gamma_k - i\alpha_k$ is a zero of $\Im_m(x)$ too. So $Xi_m(x)$ we may write as

$$\Im_m(x) = a \prod_{j=1}^{l} (x - xj) \prod_{j=i+1}^{(m-l)/2} (x^2 - 2\gamma_j x + \gamma_j^2 + \alpha_j^2).$$

Therefore operator $\Im_m(D)$ is a superposition of the $l$ operators of the form $D - \beta$ and $(r-l)/2$ operators of the form

$$\Im(D) = D^2 - 2\gamma D + \gamma^2 + \alpha^2, \tag{3.13}$$

where $\alpha$ and $\beta$ are the real numbers.

In view of the theorem 3.1.7 we see that to prove the theorem 3.1.8 it is suffitient to establish the inequality)

$$||\Im(\tau)|| \leq |\Im(in)| \, ||\tau||, \; \tau \in T_{2n+1} \tag{3.14}$$

for differential operator $\Im(D)$ of the form (3.13) and $n > 2\alpha$.

A simple calculation shows us that for any function $f \in C^2$ the inequality

$$(D^2 - 2\gamma D + \gamma^2 + \alpha^2) f(t) =$$

$$\frac{\exp(-\gamma(t-a))}{\sin \alpha(t-a)} D \left( \sin^2 \alpha(t-a) D \left( \frac{\exp(-\gamma(t-a))}{\sin \alpha(t-a)} f(t) \right) \right)$$

holds true. From this it follows that if $b - a < \pi/\alpha$, $f(b) = f(a)$ and $f(y) \neq 0$ for $0 < t < b$ then there exists a point $\xi \subset (a, b)$ such that $\Im(D) f(\xi) sgn\, f(\xi) < 0$.

Thus if the lenght of the maximal interval of sign-constancy $f \in C^2$ is no more than $\pi/\alpha$ then

$$\mu(\Im(D) f) \geq \mu(f). \tag{3.15}$$

Let $f(t) = \tau(t)$ and $g(t) = ||\tau|| \cos nt$. Then for any $\lambda$, $|\lambda| < 1$ and every $\xi \in R$ we have

$$\mu(g(\cdot) - \lambda f(\cdot + \xi)) \geq 2n$$

and the lenght of the maximal interval of the constant sign for $g(t) - \lambda f(t+\xi)$ will be less than $2\pi/n$.

The inequality (3.14) in our notations has the form

$$||\Im(D) f|| \leq ||\Im(D) g||. \tag{3.16}$$

Suppose that instead of (3.16) the inequality of the opposite sence holds true, i.e. suppose that if

$$\lambda_0 = ||\Im(D)g||/||\Im(D)f||$$

then $|\lambda_0| < 1$. Choose $\xi = \xi_0$ such that the point where the function $\Im(D)(\lambda_0 f)(\cdot\xi_0)||$ reaches its norm coincides with the point where the function $||\Im(D)g||$ reaches its norm. Then for $\varepsilon = \pm 1$ the lenght of the maximals interval of the constant sign for the difference $g(t) - \varepsilon\lambda_0 f(t - \xi_0)$ will be less then $2\pi/n$, and consiquently, according to (3.15), the difference $\Im(D)g(t) - \varepsilon\lambda_0\Im(D)f(t + \xi_0)$ will have on the period no less then $2n$ sign changes. Besidrs that for $\varepsilon = +1$ or $\varepsilon = -1$ this difference will have at least one multiple zero. But then this difference being the trigonometric polynomial of order $n$ should have strictly more than $2n$ zeros on the period (if we take into account their multiplicities) Contradiction which we obtained proves the inequality (3.14) which becoms an equality for polynomials of the form $\tau(t) = a \cos n(t - t_0)$.

## 3.2 The inequalities of Zigmund type

If we will use the results of sections 1.8 and 3.1 we obtain more general inequalities without loss of its exactness.

**Theorem 3.2.1.** *Let $X$ be an arbitrary real linear space of periodic functions, which contains the space $C$ and $||\cdot||$ be a translation invariant norm in $X$, i.e. $||f(\cdot + \alpha)|| = ||\cdot||$. Then for any trigonometric polynomial $\tau \in T_{2n+1}$ the inequalities*

$$||\tau^{(k)}|| \le n^k ||\tau||, \ k = 1, 2, \dots,$$

*hold true. These inequalities become equalities for polynomials of the form $\tau(t) = a \cos n(t + \alpha)$.*

To prove the theorem 3.2.1 it suffices to take in the theorem 1.8 $T_{2n+1}$ as manifolds $H$ and $U$ and define $A_0$ as the operator of $k$-times differentiation, $A_0 = D^k$, $D = d/dt$, $A_1$ as identical operator, $\psi$ as function $\psi(u) = n_k u$, $x$ as polynomial $\tau$ and then notice that in view of inequality (3.2) all conditions of theorem 1.8.1 are sutisfied.

We recall that the notation write $f \prec g$ for functions $f, g \in L_1$, $f, g \ge 0$ denote that

$$\int_0^t r(f, u)du \le \int_0^t r(g, u)du, \ 0 \le t \le 2\pi,$$

where $r(f, u)$ is a decreasing rearrangement of the function $f(t) \geq 0$ on the $[0, 2\pi)$.

Since for any fixed $t \in [0, 2\pi]$ the value

$$\int_0^t r(|f, u)|du$$

is an translaten invariant norm in the linear space $L_1$ we obtain from the theorem 3.2.1 that the next theorem is valid.

**Theorem 3.2.2.** *For any polynomial $\tau \in T_{2n+1}$ and for $k = 1, 2, \dots$ the exact inequality*

$$|\tau^{(k)}| \prec n^k |\tau|$$

*holds true.*

¿From this in view of the theorem 1.3.11 we obtain next result.

**Theorem 3.2.3.** *For arbitrary $N$-function $\Phi$, for any trigonometric polynomial $\tau \in T_{2n+1}$ and for all $k = 1, 2, \dots$*

$$\int_0^{2\pi} \Phi(|\tau^{(k)}(t)|)dt \leq \int_0^{2\pi} \Phi(n^k |\tau(t)|)dt. \tag{3.17}$$

*In particular*

$$||\tau^{(k)}||_p \leq n^k ||\tau||_p, \quad 1 \leq p \leq \infty. \tag{3.18}$$

*For polynomial of the form $\tau(t) = a \cos n(t + \alpha)$ inequalities (3.17) and (3.18) become equalities.*

Analogously from the inequality (3.6) we obtain on the base of the theorem 1.8.1 that the next theorem is true.

**Theorem 3.2.4.** *Under the conditions of the theorem 3.2.1 for arbitrary polynomial $\tau \in T_{2n+1}$, $\alpha \in R$*

$$||\tau' \sin\alpha + n\tau \cos\alpha|| \leq n||\tau|| \tag{3.19}$$

*and for aebitrary $N$-function $\Phi$*

$$\int_0^{2\pi} \Phi(|\tau'(t) \sin\alpha + n\tau(t) \cos\alpha|)dt \leq \int_0^{2\pi} \Phi(n|\tau(t)|)dt. \tag{3.20}$$

*In particular*

$$||\tau' \sin\alpha + n\tau \cos\alpha||_p \leq n||\tau||_p,\ 1 \leq p \leq \infty. \tag{3.21}$$

With the help of theorems 3.1.4, 1.8.1 and 1.3.11 we establish that the next theorem is true.

**Theorem 3.2.5.** *Under the conditions of theorem 3.2.1 for any trigonometric polynomial $\tau \in T_{2n+1}$ and for $0 < h < \pi/n$ or $0 \le h \le \pi/(2n)$ respectively the next exact inequality*

$$\|\tau^{(k)}\| \le \left(\frac{n}{2\sin nh}\right)^k \|\Delta_h^k(\tau)\| \le n^k\|\tau\|, \; k = 1, 2, \ldots; \tag{3.22}$$

*holds true. Further for arbitraty $N$-function $\Phi$ and for precisely same $h$*

$$\int_0^{2\pi} \Phi(|\tau^{(k)}(t)|)dt \le \int_0^{2\pi} \Phi\left(\left(\frac{n}{2\sin nh}\right)^k |\Delta_h^k(\tau, t)|\right) dt \le$$

$$\int_0^{2\pi} \Phi(n^k|\tau(t)|)dt.$$

Finally note that from the theorems 3.1.8 and 1.8.1 it follows

**Theorem 3.2.6.** *Let $n, m \in N$, $\Im_m(x)$ be an algebraic polynomial of degree $m$ having real coefficients, $x_k$, $k = 1, 2, \ldots, n$ be its zeros and $\Im_m(D)$ be the corresponding to $\Im_m$ differential operator. Under the conditions of the theorem 3.2.1 for $n > 2\max\{|\Im x_k| : k\}$ for any trigonometric polynomial $\tau \in T_{2n+1}$*

$$\|\Im_m(D)\tau\| \le |\Im_m(in)|\,\|\tau\|$$

*and for arbitrary $N$-function $\Phi$*

$$\int_0^{2\pi} \Phi(|\Im_m(D)\tau(t)|)dt \le \int_0^{2\pi} \Phi(|\Im_m(in)||\tau(t)|)dt,$$

*in particular*

$$\|\Im_m(D)\tau(\cdot)\|_p \le |\Im_m(in)|\,\|\tau(\cdot)\|_p, \quad 1 \le p \le \infty.$$

*All inequalities are exact.*

Denote by $\Omega^+$ the class of functions $\Phi$ which are non-decreasing on $[0, \infty)$ and such that $u\Phi'(u)$ is non decreasing on $[0, \infty)$ also. Any $N$-function $\Phi$ is a function of the class $\Omega^+$. Such functions as $\ln u$, $\ln^+ u$, $\ln(1 + u^p)$, $u^p (p > 0)$ belong to the class $\Omega^+$.

In addition to the theorem 3.2.3 we present the next proposition.

**Theorem 3.2.7.** *If $\Phi \in \Omega^+$, then for any trigonometrical polynomial $\tau \in T_{2n+1}$ the exact inequality*

$$\int_0^{2\pi} \Phi(|\tau'(t)|)dt \leq \int_0^{2\pi} \Phi(n|\tau(t)|)dt,$$

*holds true. In particular for all $p > 0$*

$$||\tau'||_p \leq n||\tau||_p,$$

*where the value $||f||_p$ for $p < 1$ in the similar way as for $p \geq 1$ defined by the next formula*

$$||f||_p = \left(\int_0^{2\pi} |f(t)|^p dt\right)^{1/p}.$$

## 3.3 Operators of multiplicators type

One direction of generalization of Bernstein's inequality is connected with consideration more general than $d^k/dx_k$ i.e. operators $\Upsilon : T_{2n+1} \to T_{2n+1}$ of the form

$$\Upsilon\tau(t) = \sum_{k=-n}^{n} \lambda_k c_k e^{ikt}, \tag{3.23}$$

where

$$\tau(t) = \sum_{k=-n}^{n} c_k e^{ikt},\ c_{-k} = \overline{c}_k.$$

The investigation of the problem on estimates of the norm of $\Upsilon\tau(t)$ for $\tau \in T_{2n+1}$ we start in the space $L_2$. In view of Parseval's equality

$$||\Upsilon\tau||_2^2 = 2\pi \sum_{k=-n}^{n} |c_k|^2|\lambda_k|^2 \leq$$

$$\max_{-n\leq k\leq n} |\lambda_k|^2 \sum_{k=-n}^{n} |c_k|^2 = \max_{-n\leq k\leq n} |\lambda_k|^2||\tau||_2^2,$$

and consequently

$$||\Upsilon\tau||_2 \leq \max_{-n\leq k\leq n} |\lambda_k|||\tau||_2. \tag{3.24}$$

In order that function $\Upsilon\tau(t)$ be a polynomial from $T_{2n+1}$ it is necessary that for $k = 1, 2, \ldots, n$ the eqialities $\lambda_k = \overline{\lambda}_k$ holds true.Then (3.24) may be write in the form

$$||\Upsilon\tau||_2 \leq \max_{-n\leq k\leq n} |\lambda_k|||\tau||_2. \tag{3.25}$$

The inequality (3.25) is best possible. Really let $\max\{|\lambda_k| : -n \leq k \leq n\} = |\lambda_j|$ and

$$\tau(t) = e^{ijt} + e^{-ijt}. \tag{3.26}$$

Then

$$||\tau||_2 = 2\sqrt{\pi},\ ||\Upsilon\tau||_2 = 2|\lambda_j|\sqrt{\pi}$$

and therefore for polynomial (3.26) inequality (3.25) becomes an equality.

Thus we proved

**Theorem 3.3.1.** *If in (3.23) $\lambda_k = \overline{\lambda}_k$, $k = 1, 2, \ldots, n$ then for any polynomial $\tau \in T_{2n+1}$ the exact inequality (3.25) holds true. Furthermore if $|\lambda_1| \leq |\lambda_2| \leq \ldots \leq |\lambda_n|$, then*

$$||\Upsilon\tau||_2 \leq |\lambda_n|\, ||\tau||_2.$$

Consideration of the analogous questions in other spaces is more difficult problem and require of using of other methods, which begins from proof of Bernstein's inequality based on the interpolate formula of M.Riesz (see below (3.14)).

We present rather simple and highly general scheme of receiving the interpolate formula and exact estimations for $||\Upsilon\tau||$, there $||\cdot||$ is an arbitrary translate invariant norm.

Let operator $\Upsilon : T_{2n+1} \to T_{2n+1}$ is of the forn (3.23). Then $\Upsilon\tau(t)$ we can present in the form of convolution

$$\Upsilon\tau(t) = \frac{1}{2\pi}\int_0^{2\pi} U_n(u)(t-u)du \tag{3.27}$$

of polynomial $\tau(t)$ with the kernel

$$U_n(t) = \sum_{k=-n}^{n} \lambda_k e^{ikt}. \tag{3.28}$$

We shall say that operanor $\Upsilon$ belongs to the class $\Re_n$ if $\lambda_0 = 0$ and for some $\theta \in R$ for all $k = 1, \ldots, n$

$$\lambda_k = \mu_k e^{i\theta},\quad \mu_k \geq 0,\quad \mu_n \neq 0, \tag{3.29}$$

where $\mu_n \neq 0$.

We connect with the operator $\Upsilon \in \Re_n$ the special polynomial

$$K_n(t) = K_n(\lambda, t) = \mu_n + 2\sum_{k=1}^{n-1} \mu_{n-k} \cos kt. \tag{3.30}$$

**Proposition 3.3.2.** *Let $n \in N$, $\Upsilon \in \Re_n$ and $u_k = (k\pi - \theta)/n$, $k = 1, 2, \ldots, 2n$. Then for any polynomial $\tau \in T_{2n+1}$*

$$\Upsilon\tau(t) = \frac{1}{2n}\sum_{k=1}^{2n}(-1)^k K_n(\Upsilon, u_k)\tau(t - u_k). \tag{3.31}$$

**Proof.** Put

$$g(t) = e^{i(nt+\theta)}\sum_{k=1}^{n-1}\mu_{n-k}e^{ikt} + e^{-i(nt+\theta)}\sum_{k=1}^{n-1}\mu_{n-k}e^{-ikt}$$

Then as it is easy to see

$$U_n(t) + g(t) = e^{i(nt+\theta)}K_n(t) + e^{-i(nt+\theta)}K_n(t) = \\ = 2K_n(t)\cos(nt+\theta),$$

and since $g \perp T_{2n+1}$, we obtain from (3.27)

$$\Upsilon\tau(0) = \frac{1}{2\pi}\int_0^{2\pi}\tau(-t)[U_n(t) + g(t)]dt = \\ \frac{1}{2\pi}\int_0^{2\pi}2\tau(-t)K_n(t)\cos(nt+\theta)dt. \tag{3.32}$$

If

$$D_{2n}(t) = 1 + 2\sum_{k=1}^{2n}\cos kt$$

(Dirichlet's kernel), then

$$\frac{1}{2n}\sum_{k=1}^{2n}(-1)^k D_{2n}(t - k\pi/n) = 2\cos nt. \tag{3.33}$$

Put $t = u + \theta/n$ in equaluty (3.33) and substitude obtained expression for $2\cos(nu+\theta)$ in (3.32). We obtain

$$\Upsilon\tau(0) = \\ \frac{1}{2\pi}\int_0^{2\pi}\tau(-t)K_n(t)\frac{1}{2n}\sum_{k=1}^{2n}(-1)^k D_{2n}\left(t - \frac{k\pi-\theta}{n}\right)dt. \tag{3.34}$$

Taking into account that for any trigonometric polynomial $\tau(t)$ of order not more $2n$, $\tau \in T_{4n+1}$,

$$\frac{1}{2\pi}\int_0^{2\pi} \tau(t) D_{2n}\left(t - \frac{k\pi - \theta}{n}\right) dt = \tau\left(\frac{k\pi - \theta}{n}\right),$$

and also that $\tau(-\cdot)K_n(\cdot) \in T_{4n+1}$ the equality (3.34) we may write in the form

$$\Upsilon\tau(0) = \frac{1}{2n}\sum_{k=1}^{2n}(-1)^k K_n\left(t - \frac{k\pi - \theta}{n}\right)\tau\left(\frac{\theta - k\pi}{n}\right).$$

Applying now this formula to polynomial $\tau(u+t)$, we obtain (3.31).

Setting in (3.23) $\lambda_k = ik$ (i.e. $\lambda_k = k\, e^{i\pi/2}$) we see that the operator of differentiation

$$\frac{d}{dt}\tau(t) = \sum_{k=-n}^{n} ik\, c_k e^{ikt},\ \tau(t) = \sum_{k=-n}^{n} c_k e^{ikt},$$

belongs to the class $\Re_n$. By formula (3.30) we have

$$K_n\left(\frac{d}{dt}, t\right) = n + 2\sum_{k=1}^{n-1}(n-k)\cos kt = \\ = \sum_{k=1}^{n-1} D_k(t) = nF_n(t),$$

where

$$F_n(t) = \frac{1}{2n}\left(\frac{\sin(nt/2)}{\sin(t/2)}\right)^2 \tag{3.35}$$

is Fejer's kernel. Now from proposition 3.3.2 it follows the classical interpolational Riesz's formula

$$\tau'(t) = \frac{1}{n}\sum_{k=1}^{2n}\frac{(-1)^{k+1}}{(2\sin(u_k/2))^2}\tau(t - u_k). \tag{3.36}$$

where $u_k = (2k+1)\pi/(2n)$, $k = 1, 2, \ldots, 2n$.

Putting in (3.36) $t = 0$ and $\tau(t) = \sin nt$ we obtain

$$\frac{1}{n}\sum_{k=1}^{2n}\frac{1}{(2\sin(u_k/2))^2} = n.$$

Therefore for any of norm $||\cdot||$ which is translate invariant

$$||\tau|| \leq \frac{1}{n}\sum_{k=1}^{2n} \frac{1}{(2\sin(u_k/2))^2}||\tau(\cdot + u_k)|| = n||\tau||,$$

and we obtain the new proof of the theorems 4.1.1 and 3.2.1.

In general case from the proposition 3.3.2 we get

**Corollary 3.3.3.** *If operator $\Upsilon$ belongs to the class $\Re_n$ then for translate invariant norm $||\cdot||$ on the linear space $T_{2n+1}$ and for any polynomial $\tau \in T_{2n+1}$*

$$||\Upsilon\tau|| \leq \frac{1}{n}\sum_{k=1}^{2n}\left|K_n\left(\Upsilon, \frac{k\pi - \theta}{n}\right)\right| ||\tau||. \tag{3.37}$$

We note important cases when exactness of inequality (3.15) is guaranteed.

**Corollary 3.3.4.** *Let a operator $\Upsilon \in \Re_n$ is such that all values*

$$K_n(\Upsilon, (k\pi - \theta)/n), \quad k = 1, 2, \ldots, 2n,$$

*have the same sign. Then for translate invariant norm $||\cdot||$ for any trigonometric polynomial $\tau \in T_{2n+1}$ the exact inequality*

$$||\Upsilon\tau|| \leq |\mu_n|\,||\tau|| \tag{3.38}$$

*holds true.*

**Proof.** Really, in this case for $\varepsilon = 1$ and $\varepsilon = -1$ we will have

$$\frac{1}{2n}\sum_{k=1}^{2n}\left|K_n\left(\Upsilon, \frac{k\pi - \theta}{n}\right)\right| = \frac{\varepsilon}{2n}\sum_{k=1}^{2n} K_n\left(\Upsilon, \frac{k\pi - \theta}{n}\right) = \varepsilon\mu_n, \tag{3.39}$$

hence in view of (3.38) we obtain (3.38).

If $\tau(t) = \cos nt$ then $\Upsilon\,\tau(t) = \mu_n \cos(nt + \theta)$

$$||\tau|| = ||\cos n(\cdot)|| \tag{3.40}$$

and

$$||\Upsilon\tau|| = |\mu_n|\,||\cos n(\cdot)|| \tag{3.41}$$

Taking into account (3.37) and (3.39)-(3.41) we see that for $\tau(t) = \cos nt$ the inequality (3.38) becomes an equality.

Applying two times to the right part of (3.30) the Abel transformations, we obtain

$$K_n(\Upsilon, t) = 2\sum_{k=0}^{n-2}(k+1)\Delta_2\mu_{n-k}F_k(t) + 2nF_{n-1}(t),$$

where $\Delta\mu_k = \mu_k - \mu_{k+1}$, $\Delta^2 = \Delta\mu_k - \Delta\mu_{k+1}$ and $F_k$ is a Fejer's kernels.

Taking into account that Fejer's kernels are nonnegative we get the following statement.

**Corollary 3.3.5.** *If operator $\Upsilon \in \Re_n$ is such that $\mu_k \geq 0$ and for $k = 0, 1, \ldots, n-2$ $\Delta^2\mu_{n-k} \geq 0$. Then for any translate invariant norm $||\cdot||$ and for any polynomial $\tau \in T_{2n+1}$ the exact inequality*

$$||\Upsilon\tau|| \leq |\mu_n|\,||\tau|$$

*holds true.*

For polynomial $\tau(t) = \cos nt$ we we will have

$$||\Upsilon\tau|| = |\mu_n|\,||\cos n(\cdot)||$$

and

$$\sum_{k=1}^{2n}\left|K_n\left(\Upsilon, \frac{k\pi - \theta}{n}\right)\right| \cdot ||\tau|| = |\mu_n|\,||\cos n(\cdot)||.$$

The operator of differentiation of fractional order $r > 0$ we can deine on the set $T_{2n+1}$ puting in (3.23) $\lambda_0 = 0$, $\lambda_k = k^r e^{ir\pi/2}$ if $k > 0$ and $\lambda_k = |k|^r e^{-ir\pi/2}$ if $k < 0$. We will write in this case $\tau^{(r)}$ instead of $\Upsilon\tau$.It is obvious that for $r \geq 1$ for this operater satisfies the condition of the corollery 3.3.5.

So the next statement holds true.

**Theorem 3.3.6.** *For any trigonometrical polynomial $\tau \in T_{2n+1}$ and for all real $r \geq 1$ the exact inequality*

$$||\tau^{(r)}|| \leq n^r ||\tau|| \tag{3.42}$$

*holds true, where $||\cdot||$ is an arbitrary translate invariant norm in the linear space $T_{2n+1}$. In particular*

$$|\tau^{(r)}| \prec n^r |\tau|. \tag{3.43}$$

¿From (3.43) taking into account theorem 1.3.11 we deduce the next proposition.

**Theorem 3.3.7.** *For any N-function $\Phi$ and for any trigonometrical polynomial $\tau \in T_{2n+1}$ for all real $r \geq 1$*

$$\int_0^{2\pi} \Phi(|\tau^{(r)}(t)|)dt \leq \int_0^{2\pi} \Phi(n^r |\tau(t)|)dt. \tag{3.44}$$

*In particular for all $p \in [1, \infty]$*

$$||\tau^{(r)}||_p \prec n^r ||\tau||_p. \tag{3.45}$$

If polynomial $\tilde{\tau}(t)$ which is conjugate with the polynomial

$$\tau(t) = \sum_{k=-n}^{n} c_k e^{ikt}, \; c_k = \overline{c}_k,$$

we can define by means of the relation

$$\tilde{\tau}(t) = i \sum_{k=-n}^{n} (sgn\, k) c_k e^{ikt}, \; c_k = \overline{c}_k.$$

It followes from this that for any fixed $\alpha \in R$ the operator

$$\Upsilon\tau(t) := \tau'(t)\cos\alpha - \tilde{\tau}'(t)\sin\alpha \tag{3.46}$$

we can write in the form

$$\Upsilon\tau(t) = \sum_{k=-n}^{n} k c_k e^{ikt} e^{-i(\alpha+\pi/2)}. \tag{3.47}$$

¿From (3.47) we see that operator (3.46) belongs to the class $\Re_n$, for this operator in (3.28) and (3.29) we have $\mu_k = k$ and $\theta = -\alpha - \pi/2$. Thus we can use the proposition 3.3.2 which after simple transformations yields us to the next interpolate formula;

$$\tau'(t)\cos\alpha - \tilde{\tau}'(t)\sin\alpha =$$

$$\frac{1}{n} \sum_{k=1}^{2n} \frac{(-1)^{k+1} + \sin\alpha}{(2\sin(u_k/2))^2} \tau(t + u_k), \tag{3.48}$$

where

$$u_k = (2\alpha + (2k-1)\pi)/(2n).$$

¿From (3.48) it follows that the operator (3.47) satisfies the conditions of the corollary 3.3.4 and for this operatir $\nu_n = n$. Thus in view of (3.48) the next theorem is valid.

**Theorem 3.3.8.** *For any $\alpha \in R$ for any polynomial $\tau \in T_{2n+1}$ and any translate invariant norm $||\cdot||$ the exact inequality*

$$||\tau'(\cdot)\cos\alpha - \tilde{\tau}'(\cdot)\sin\alpha|| \leq n||\tau||. \tag{3.49}$$

*holds true.*

*In particular*

$$|\tau'(\cdot)\cos\alpha - \tilde{\tau}'(\cdot)\sin\alpha| \prec n|\tau(\cdot)|. \tag{3.50}$$

Taking into account (3.50) and the theorem 1.3.11 we obtain the next proposition.

**Theorem 3.3.9.** *For any $N$-function $\Phi$ and any trigonometrical polynomial $\tau \in T_{2n+1}$ and for any $\alpha \in R$*

$$\int_0^{2\pi} \Phi(|\tau'(t)\cos\alpha - \tilde{\tau}'(t)\sin\alpha|)dt \leq \int_0^{2\pi} \Phi(n|\tau(t)|)dt. \tag{3.51}$$

*In particular for any $p \in [1,\infty]$*

$$||\tau'\cos\alpha - \tilde{\tau}'\sin\alpha||_p \prec n||\tau||_p. \tag{3.52}$$

*Inequalities (3.51) and (3.52) are best possible.*

¿From the inequality (3.52) for $p = \infty$ it follows the next

**Corrolery 3.3.10.** *For any $\tau \in T_{2n+1}$*

$$\sqrt{(\tau'(t))^2 + (\tilde{\tau}'(t))^2} \leq n||\tau||_C. \tag{3.53}$$

For $\alpha = 0$ from theorems 3.3.8 and 3.3.9 we deduce the next

**Corrolary 3.3.11.** *For any $\tau \in T_{2n+1}$ for any $r \in N$ and for any translate invariant norm the exact inequality*

$$||\tilde{\tau}^{(r)}|| \leq n^r ||\tau||, \tag{3.54}$$

*holds true, in particular*

$$|\tilde{\tau}^{(r)}| \prec n^r |\tau|. \tag{3.55}$$

*For any N-function $\Phi$*

$$\int_0^{2\pi} \Phi(|\tilde{\tau}^{(r)}(t)|)dt \leq \int_0^{2\pi} \Phi(n^r |\tau(t)|)dt. \tag{3.56}$$

*In particular for all $p \in [1, \infty]$*

$$||\tilde{\tau}^{(r)}||_p \prec n^r ||\tau||_p. \tag{3.57}$$

¿From proposition 3.3.2 and corollaries 3.3.3 and 3.3.4 one can deduce other interesting inequalities. Some of them we present after the commentaries.

## 3.4 Bernstein's inequalities in various metrics

In this section we by means of the general comparison theorem for rearrangement representations (theorem 1.4.6) will obtain a series of inequalities, which estimate norms of derivatives of trigonometric polynomials by norms of polynomials in other space.

In view of lemma 3.1.1 a polynomial $\tau \in T_{2n+1}$ has the $\mu$-property relatively to $2\pi/n$-right function $A \cos nt$, if $A \geq ||\tau||_C$. Taking into account the Bernstein's inequality, we see that if

$$q_n(t) = ||\tau||_C \cos nt,$$

then firstly $\tau(t)$ has $\mu$-property relatively to $q_n(t)$, and secondly, $\tau'(t)$ has a $\mu$-property relatively to $q_n'(t)$. So in view of the theorem 1.4.6 the next theorem holds true.

**Theorem 3.4.1.** *If $\tau \in T_{2n+1}$ then for $k = 1, 2, \ldots$ and for any $\lambda \in R$*

$$(\tau^{(k)} - \lambda)_\pm \prec (q_n^{(k)} - \lambda)_\pm \tag{3.58}$$

*and, consequently,*

$$|\tau^{(k)}| \prec n^k \|\tau\|_C |\cos(\cdot)|. \tag{3.59}$$

*Inequalities (3.58) and (3.59) become equalities for polynomials of the form*

$$\tau(t) = a\cos n(t+\alpha),\ a, \alpha \in R.$$

In the sequel in the statements with the exact inequalities we will not indicate the exstremal function $a\cos n(t+\alpha)$.

¿From the theorem 3.4.1 in view of theorem 1.3.11 it follows that the next theorem holds true.

**Theorem 3.4.2.** *For any N-function $\Phi$ and for any polynomial $\tau \in T_{2n+1}$ for all $k = 1, 2, \ldots$ the exact inequality holds true*

$$\int_0^{2\pi} \Phi(|\tau^{(k)}(t)|)dt \le \int_0^{2\pi} \Phi(n^k \|\tau\|_C |\cos t|)dt. \tag{3.60}$$

*In particular, for all $p \in [1, \infty]$*

$$\|\tau^{(k)}\|_p \le n^k \|\tau\|_C \|\cos(\cdot)\|_p. \tag{3.61}$$

¿From the theorems 3.4.2 and 3.3.9 we obtain

**Corollary 3.4.3.** *For any natural number $k > 1$ and any $\alpha \in R$*

$$\int_0^{2\pi} \Phi(|\tau^{(k)}(t)\cos\alpha - \tilde{\tau}^{(k)}\sin\alpha|)dt \le$$

$$\int_0^{2\pi} \Phi(n^k \|\tau\|_C |\cos t|)dt. \tag{3.62}$$

*In particular the next exact inequalities hold true*

$$\int_0^{2\pi} \Phi(|\tilde{\tau}^{(k)}|)dt \le \int_0^{2\pi} \Phi(n^k \|\tau\|_C |\cos t|)dt \tag{3.63}$$

*and*

$$\|\tilde{\tau}^{(k)}\|_p \le n^k \|\tau\|_C \|\cos(\cdot)\|_p. \tag{3.64}$$

Really in view of inequality (3.51) for any $\alpha \in R$

$$\int_0^{2\pi} \Phi(|\tau^{(k)}(t)\cos\alpha - \tilde{\tau}^{(k)}\sin\alpha|)dt \leq$$

$$\int_0^{2\pi} \Phi(n^{k-1}|\tau'(t)|)dt.$$

But due to (3.60) the right part of this inequality is not more tham value

$$\int_0^{2\pi} \Phi(n^k\|\tau\|_C|\cos t|)dt.$$

Considering the functions $\tau(t)-\lambda$ (where $\lambda$ is a constant of best uniform approximation for $\tau(t)$) and $E_0(\tau)_C \cdot \cos nt$, we analogously to the theorems 3.4.1, 3.4.2 and corollary 3.4.3 obtain the next theorem.

**Theorem 3.4.4.** *For any $N$-function $\Phi$ and for any nonnegative polynomial $\tau \in T_{2n+1}$ for all $k = 1, 2, \ldots$ the exact inequality*

$$\int_0^{2\pi} \Phi(|\tau^{(k)}(t)|)dt \leq \int_0^{2\pi} \Phi\left(\frac{1}{2}n^k\|\tau\|_C|\cos t|\right)dt$$

*holds true.*

*If $k \geq 1$, then for any $\alpha \in R$*

$$\int_0^{2\pi} \Phi(|\tau^{(k)}(t)\cos\alpha - \tilde{\tau}^{(k)}\sin\alpha|)dt \leq$$

$$\int_0^{2\pi} \Phi\left(\frac{1}{2}n^k\|\tau\|_C|\cos t|\right)dt.$$

*In particular, for all $p \in [1, \infty]$*

$$\|\tilde{\tau}^{(k)}\|_p \leq \frac{1}{2}n^k\|\tau\|_C\|\cos(\cdot)\|_p, \; k = 2, 3, \ldots.$$

Let now $\Upsilon : T_{2n+1} \to T_{2n+1}$ be an operator of the form (3.23), i.e.

$$\Upsilon\tau(t) = \sum_{\nu==-n}^{n} \lambda_\nu c_\nu e^{i\nu t}, \tag{3.65}$$

if

$$\tau(t) = \sum_{\nu==-n}^{n} c_\nu e^{i\nu t}.$$

Let also an operator $\Upsilon$ belongs to the class $\Im_n$, i.e. $\lambda_0 = 0$ and for $\nu = 1, 2, \ldots, n$ we have $\lambda_\nu = \mu_\nu e^{i\theta}$, where $\mu_\nu \geq 0$, $\mu_n \neq 0$ and $\theta$ is some number. Assume ,finally that for $\nu = 1.2, \ldots, n-2$

$$\Delta^2 \mu_{n-\nu} \geq 0.$$

Then, taking into account, that $(\Upsilon\tau)' = \Upsilon\tau'$ and using the corollary 3.3.5, we obtain, that for any $N$-function $\Phi$

$$\int_0^{2\pi} \Phi(|(\Upsilon\tau)'(t)|)dt \leq \int_0^{2\pi} \Phi(\mu_n|\tau'(t)|)dt. \tag{3.66}$$

Estimating the right part of (3.66) by means of the theorem 3.4.2 we get the next statement.

**Theorem 3.4.5.** *If an operator $\Xi \in \Im_n$ is such that $\Delta^2\mu_{n-\nu} \geq 0$ for all $k = 1, 2, \ldots, n-2$ then for any $N$-function $\Phi$ and any trigonometric polynomial $\tau \in T_{2n+1}$ the exact inequality*

$$\int_0^{2\pi} \Phi(|(\Upsilon\tau)'(t)|)dt \leq \int_0^{2\pi} \Phi(\mu_n||\tau||_C|\cos t|)dt \tag{3.67}$$

*holds true.*

*In particular, for all real number $r \geq 2$*

$$\int_0^{2\pi} \Phi(|\tau^{(r)}(t)|)dt \leq \int_0^{2\pi} n^r||\tau||_C|\cos t|)dt \tag{3.68}$$

*and*

$$||\tau^{(k)}||_p \leq n^r||\tau||_C||\cos(\cdot)||_p. \tag{3.69}$$

The theorem 3.4.5 we can formulate in the next form.

**Theorem 3.4.5'.** *If operator $\Xi \in \Im_n$ is such that $\Delta^2\mu_{n-\nu}/(n-\nu) \geq 0$ for all $k = 1, 2, \ldots, n-2$ then for any $N$-function $\Phi$ and for any trigonometric polynomial $\tau \in T_{2n+1}$ the exact inequality*

$$\int_0^{2\pi} \Phi(|\Upsilon\tau(t)|)dt \leq \int_0^{2\pi} \Phi(\mu_n|\tau(t)|)dt$$

*holds true.*

Let now $\Im_m(D)$ be a differential operator of the form (3.10). By means of lemma 3.1.1 and theorem 3.1.8 we get that polynomials $\Im_m(D)\tau(t)$ and $D\Im_m(D)\tau(t)$ posses $\mu$-property relatively to function

$$||\tau||_C |\Im_m(in)| \cos nt$$

and

$$n||\tau||_C |\Im_m(in)| \sin nt$$

accordingly if

$$n > 2 \max\{|Im\, x_k| : k\}$$

($x_1, x_2, \ldots, x_m$ are zeros of $\Im_m(x)$.

Using the theorem 1.4.6 we see that the next theorem holds true.

**Theorem 3.4.6.** *For $n > 2\max\{|Im\, x_k| : k\}$ and for any polynomial $\tau \in T_{2n+1}$ the exact inequality*

$$|D\Im_n(D)\tau| \prec |n\Im_m(in)|||\tau||_C| \cos(\cdot)| \tag{3.70}$$

*holds true. Consequently for any N-function $\Phi$*

$$\int_0^{2\pi} \Phi(|D\Im_n(D)\tau(t)|)dt \leq$$

$$\int_0^{2\pi} \Phi(|n\Im_m(in)|||\tau||_C| \cos(t)|)dt. \tag{3.71}$$

*In particular for all $p \in [1, \infty]$*

$$||D\Im_n(D)\tau||_p \leq |n\Im_m(in)|||\tau||_C|| \cos(\cdot)||_p. \tag{3.72}$$

**Remark.** The fact that in inequalities (3.70)-(3.72) we write $D\Im_m(D)$, is essential. For arbitrary operator $\Im_m(D)$ of the form (3.10) the inequality

$$||\Im_n(D)\tau||_p \leq |\Im_m(in)|||\tau||_C|| \cos(\cdot)||_p \tag{3.73}$$

is generally speaking not true.

Really let for example, $p = 1$, $\Im_1(D) = D + n$ and $\tau(t) = 1 + \alpha \sin nt$ where $\alpha$ is sufficiently small positive number. Then $||\tau||_C = 1 + \alpha$ and $||\Im_1(D)\tau||_1 = 2\pi n$. If in this case (3.73) holds true we have

$$||\Im_1(D)\tau||_1 \leq |in + n|||\tau||_C|| \cos(\cdot)||_1,$$

or what is equivalent

$$2\pi n \le 4n\sqrt{2}(1+\alpha).$$

But the last inequality is not true if $\alpha$ is sufficiently small, since $2\pi > 4\sqrt{2}$.

Now, with the theorem 1.7.8 we will obtain the exact inequalities, which give us the estimates of $L_1$-norms of derivatives $\tau^{(k)}(t)$ of the polynomials $\tau \in T_{2n+1}$ by means of $L_p$-norms $||\tau||_p$, $1 \le p \le \infty$ of the polynomials $\tau$.

**Theorem 3.4.7.** *For all $k = 1, 2, \ldots$ and $1 \le p \le \infty$ for any polynomial $\tau \in T_{2n+1}$ the exact inequality*

$$||\tau^{(k)}||_1 \le n^k \frac{||\cos(\cdot)||_1}{||\cos(\cdot)||_p} ||\tau||_p \tag{3.74}$$

*holds true.*

**Proof.** If $\tau \in T_{2n+1}$ then $\mu(\tau') \le 2n$ and in view of (3.18) for $p = 1$

$$V_0^{2\pi}(\tau^{(j)}) \le n^{j-k} V_0^{2\pi}(\tau^{(k)}).$$

¿From this and from theorem 1.7.8 it follows that for all $j > k$

$$\frac{V_0^{2\pi}(\tau^{(k)})}{||\varphi_{1,j-k}||_C} \le n^{\frac{(k+1)(j-k)}{j+1/p}} \left(\frac{4||\tau||_p}{||\varphi_{1,j}||_p}\right)^{\frac{j-k}{j+1/p}} \left(V_0^{2\pi}(\tau^{(j)})\right)^{\frac{k+1/p}{j+1/p}}. \tag{3.75}$$

But since

$$\varphi_{1,j}(t) = \frac{4}{\pi} \sum_{\nu=1}^{\infty} \frac{\sin((2\nu-1)t + \frac{\pi j}{2})}{(2\nu-1)^{j+1}}$$

then for all $p \in [1, \infty]$

$$\lim_{j\to\infty} ||\varphi_{1,j}||_p = \frac{4}{\pi} ||\cos(\cdot)||_p.$$

So passing to the limit for $j \to \infty$ from (3.75) we obtain the inequality

$$V_0^{2\pi}(\tau^{(k)}) \le 4n^{k+1} \frac{||\tau||_p}{||\cos(\cdot)||_p}, \; k = 0, 1, \ldots,$$

that is equivalent to the inequality (3.74).

If a real number $r \ge 2$ is fixed, then using successively the inequalities (3.47) for $p = 1$ and (3.74) we obtain

$$||\tau^{(r)}||_1 \le n^{r-1} ||\tau'||_1 \le n^{r-1} n \frac{||\cos(\cdot)||_1}{||\cos(\cdot)||_p} ||\tau||_p.$$

**Corollary 3.4.8.** *For any real $r \geq 2$ and for any $\tau \in T_{2n+1}$ the exact inequality*

$$\|\tau^{(r)}\|_1 \leq n^r \frac{\|\cos(\cdot)\|_1}{\|\cos(\cdot)\|_p} \|\tau\|_p,\ 1 \leq p \leq \infty$$

*holds true.* Analogously from inequality (3.53) and from theorem 3.4.7 we deduce:

**Corollary 3.4.9.** *For any natural $k \geq 2$, real $\alpha \in R$ and for any trigonometrical polynomial $\tau \in T_{2n+1}$ the exact inequality*

$$\left\|\tau^{(k)}(\cdot)\cos\alpha - \tilde{\tau}^{(k)}(\cdot)\sin\alpha\right\|_1 \leq$$

$$n^k \frac{\|\cos(\cdot)\|_1}{\|\cos(\cdot)\|_p} \|\tau\|_p,\ 1 \leq p \leq \infty$$

*holds true. In particular for $\alpha = \pi/2$*

$$\|\tilde{\tau}^{(k)}\|_1 \leq n^k \frac{\|\cos(\cdot)\|_1}{\|\cos(\cdot)\|_p} \|\tau\|_p,\ 1 \leq p \leq \infty.$$

Note finally that the next theorem which is analogous to the theorem 3.4.5 holds true.

**Theorem 3.4.10.** *If an operator $\Upsilon \in \Im_m$ is such that*

$$\Delta^2 \left(\frac{\mu_{n-k}}{n-k}\right) \geq 0,\ k = 0, 1, \ldots, n-2,$$

*then for any polynomial $\tau \in T_{2n+1}$ and for all $1 \leq p \leq \infty$ the exact unequality*

$$\|\Upsilon\tau\|_1 \leq \mu_n \frac{\|\cos(\cdot)\|_1}{\|\cos(\cdot)\|_p} \|\tau\|_p$$

*holds true.*

## 3.5 Various metrics inequalities of Bernstein type

Let first $1 \leq p \leq q \leq \infty$. Then with the help of Holder's inequality for any polynomial $\tau \in T_{2n+1}$ we obtain

$$\|\tau\|_p^p = \int_0^{2\pi} |\tau(t)|^p \, 1dt \leq$$

$$(2\pi)^{(q-p)/q}\left(\int_0^{2\pi}|\tau(t)|^q dt\right)^{p/q}=(2\pi)^{1-p/q}||\tau||_q^p.$$

Consequently

$$||\tau||_p \le (2\pi)^{\frac{1}{p}-\frac{1}{q}}||\tau||_q. \tag{3.76}$$

The case $1 \le q \le p \le \infty$ is more difficult. In this case we need the next theorem which is of the independent interest also.

**Theorem 3.5.1.** *For any $n \in N$ and for arbitrary polynomial $\tau \in T_{2n+1}$ the next inequality*

$$\sqrt{\frac{2\pi}{2n+1}}||\tau||_C \le ||\tau||_2, \tag{3.77}$$

*holds true. This inequality becomes an equality for the Dirichlet kernel*

$$\tau(t) = D_n(t) = 1 + 2\sum_{\nu=1}^{n}\cos\nu t.$$

Really, using the identity

$$\tau(t) = \frac{1}{2\pi}\int_0^{2\pi}\tau(t+u)D_n(u)du,$$

and applying the inequality Caushy we obtain

$$|\tau(t)| \le \frac{1}{2\pi}\int_0^{2\pi}|\tau(t+u)||D_n(u)|du \le$$

$$\frac{1}{2\pi}||\tau||_2||D_n||_2. \tag{3.78}$$

Since

$$||D_n||_2 = \sqrt{2\pi\left(n+\frac{1}{2}\right)},$$

then from (3.72) we immediately obtain the inequality (3.71).

**Theorem 3.5.2.** *If $1 \le q \le p \le \infty$, then for any trigonometric polynomial $\tau \in T_{2n+1}$ the next inequality*

$$||\tau||_p \le \left(\frac{nq_0+1}{2\pi}\right)^{\frac{1}{q}-\frac{1}{p}}||\tau||_q, \tag{3.79}$$

*holds true, where $q_0$ is the minimal even number which is more or equal $q$.*

**Proof.** Applying the inequality (3.71) to polynomial $(\tau(t))^{\frac{q_0}{2}}$ of order no more $nq_0/2$ we will have

$$||\tau||_C^{q_0/2} = ||\tau^{q_0/2}||_C \leq \left(\frac{nq_0+1}{2\pi}\right)^{1/2} \left(\int_0^{2\pi} |\tau^{q_0/2}(t)|^2 dt\right)^{1/2} =$$

$$\left(\frac{nq_0+1}{2\pi}\right)^{1/2} \left(\int_0^{2\pi} |\tau(t)|^{q_0} dt\right)^{1/2} \leq$$

$$\left(\frac{nq_0+1}{2\pi}\right)^{1/2} ||\tau||_C^{(q_0-q)/2} \left(\int_0^{2\pi} |\tau(t)|^{q} dt\right)^{1/2} \leq$$

$$\left(\frac{nq_0+1}{2\pi}\right)^{1/2} ||\tau||_C^{(q_0-q)/2} ||\tau||_q^{q/2},$$

i.e.

$$||\tau||_C \leq \left(\frac{nq_0+1}{2\pi}\right)^{1/q} ||\tau||_q. \tag{3.80}$$

Now if $p \geq q$ then with the help of (3.74) we obtain

$$\int_0^{2\pi} |\tau(t)|^p dt = \int_0^{2\pi} |\tau(t)|^{p-q} |\tau(t)|^q dt \leq$$

$$||\tau||_C^{p-q} ||\tau||_q^q \leq \left(\frac{nq_0+1}{2\pi}\right)^{(p-q)/q} ||\tau||_q^p,$$

and the inequality (3.71) is proved.

**Remark.** Let $F_n(t)$, $n = 1, 2, \ldots$ be a sequence of Fejer's polynomials (3.36). Then there exists independent of $n$ constant $c > 0$ such that

$$||F_n||_p \geq cn^{1/q-1/p} ||F_n||_q. \tag{3.81}$$

Thus on the class of all trigonometrical polynomials the estimation (3.73) is exact on order when $n \to \infty$.

Now we will obtain one exact inequality of various metrics for polynomial $\tau \in T_{2n+1}$ having all real zeros.

**Theorem 3.5.3.** *If all zeros of polynomial $\tau \in T_{2n+1}$ are real, then for any $\lambda > 0$*

$$mes\{t : t \in [0,2\pi], |\tau(t)| > \lambda\} \leq$$

$$mes\{t : t \in [0,2\pi], ||\tau||_C |\cos nt| > \lambda\}, \tag{3.82}$$

*or, that is equivalent, for all $t \in [0,2\pi]$*

$$r(|\tau|,t) \leq ||\tau||_C\, r(|\cos(\cdot)|,t). \tag{3.83}$$

*Inequalities (3.76) and (3.77) become the equalities for $\tau(t) = \cos nt$.*

**Proof.** If a polynomial $\tau \in T_{2n+1}$ has only real zeros $x_1 \leq x_2 \leq \ldots x_{2n} < x_1 + 2\pi$ then we can write $\tau(t)$ in the form

$$\tau(t) = a \prod_{k=1}^{2n} \sin \frac{t - x_k}{2}.$$

But then

$$\tau'(t) = \frac{a}{2} \sum_{\nu=1}^{2n} \cos \frac{t - x_\nu}{2} \prod_{k \neq \nu} \sin \frac{t - x_k}{2}$$

and, consequently,

$$\frac{\tau'(t)}{\tau(t)} = \frac{1}{2} \sum_{\nu=1}^{2n} \cot \frac{t - x_\nu}{2}. \tag{3.84}$$

In the proof of inequality (3.76) we may assume that $||\tau||_C = 1$. Instead of $mes\{t : t \in [0,2\pi], |f(t)| > y\}$ we will write $mes\{|f(t)| > y\}$. Taking into account (3.78) in view of theorem 1.6.2 for any $y > 0$ we obtain

$$mes\left(\frac{|\tau'(t)|}{n|\tau(t)|} > y\right) = arc\tan \frac{1}{y}. \tag{3.85}$$

Furthermore, using the inequality (3.5) we will have

$$mes\left(\frac{|\tau'(t)|}{n|\tau(t)|} > y\right) = mes\{|\tau'(t)|^2 > n^2 y^2 |\tau(t)|^2\} \leq$$

$$mes\{n^2 - n^2\tau^2(t) > n^2 y^2 \tau^2(t)\} \leq$$

$$mes\left(\tau^2(t) < \frac{1}{1+y^2}\right) = 2\pi - mes\left(\tau^2(t) > \frac{1}{1+y^2}\right).$$

Puting $\lambda = (1+y^2)^{-1/2}$ and taking into account (3.79) we obtain

$$mes\{|\tau(t)| > \lambda\} \leq 2\pi - mes\left(\frac{|\tau'(t)|}{n|\tau(t)|} > \frac{\sqrt{1-\lambda^2}}{\lambda}\right) =$$

$$2\pi - 4arc\tan\frac{\lambda}{\sqrt{1-\lambda^2}} = 2\pi - 4arc\sin\lambda = mes\{|\cos nt| > \lambda\}.$$

The theorem 3.5.3 is proved.

¿From (3.77) we deduce

**Corollary 3.5.4.** *If all zeros of polynomial $\tau \in T_{2n+1}$ are real then for any function $\Phi(t)$ which is nondecreasing on $(0,\infty)$ we have*

$$\int_0^{2\pi} \Phi(|\tau(t)|)dt \leq \int_0^{2\pi} \Phi(||\tau||_C|\cos t|)dt.$$

*In particular*

$$||\tau||_p \leq ||\tau||_C||\cos(\cdot)||_p,\ 0 < p < \infty,$$

*where*

$$||\tau||_p := \left(\int_0^{2\pi} |\tau(t)|^p dt\right)^{1/p},\ p > 0.$$

A slight change of the proof of theorem 3.5.3 and in particular using of the inequality (3.6) instead of (3.5) allows us to prove the next theorem.

**Theorem 3.5.5.** *If all zeros of polynomial $\tau \in T_{2n+1}$ are real, then for any $\lambda > 0$*

$$mes\{t : t \in [0,2\pi],\ \tau(t) > \lambda\} \leq$$

$$mes\{t : t \in [0,2\pi],\ ||\tau||_C \cos^2(t/2) > \lambda\},$$

*or, that is equivalent, for any $t \in [0,2\pi]$*

$$r(\tau,t) \leq ||\tau||_C\, r(\cos^2(\cdot),t).$$

*These inequalities are exact.*

## 3.6 Inequalities for complex-valued polynomials

Denote by $T_{2n+1}^C$, $n = 0, 1, \dots$, the set of all complex-valued trigonometric polynomials of order no more $n$, i.e. the set of functions of the form

$$\tau(t) = \sum_{k=-n}^{n} c_k e^{ikt}, \; t \in R,$$

where $c_k$, $k = 0, \pm 1, \pm 2, \dots, \pm n$ are abitrary complex numbers. It is clear that $T_{2n+1}^C$ is complex linear space of dimension $2n + 1$.

In this section we will show that if we have the inequalities of Bernstein type for the real trigonometric polynomials we can obtain in some cases their analogs for polynomials from $T_{2n+1}^C$.

As in a real case let for $f : R \to C$

$$||f||_C = vrai \sup\{|f(t)| : t\}$$

and if $0 < p < \infty$

$$||f||_p = \left( \int_0^{2\pi} |f(t)|^p \, dt \right)^{1/p}.$$

**Theorem 3.6.1.** *Let $n = 0, 1, \dots$. For any $\tau \in T_{2n+1}^C$*

$$||\tau'||_\infty \leq n ||\tau||_\infty. \tag{3.86}$$

*This inequality becomes an equality if and only if*

$$\tau(t) = a e^{\pm int}, \; a \in C.$$

**Proof.** For any $t \in R$ we have

$$|\tau'(t)| = \sqrt{(Re\, \tau'(t))^2 + (Im\, \tau'(t))^2} =$$

$$\sup\{\alpha Re\, \tau'(t) + \beta Im\, \tau'(t) : \alpha^2 + \beta^2 = 1, \; \alpha, \beta \in R\} =$$

$$\sup\{(\alpha Re\, \tau(t) + \beta Im\, \tau(t))' : \alpha^2 + \beta^2 = 1, \; \alpha, \beta \in R\}.$$

If $\tau \in T_{2n+1}^C$, then as it easy to verify $Re\, \tau \in T_{2n+1}$ and $Im\, \tau \in T_{2n+1}$ and consequently $\alpha Re\, \tau + \beta Im\, \tau \in T_{2n+1}$. Applying theorem 3.1.1 we get

$$|\tau'(t)| = \sup\{(\alpha Re\, \tau(t) + \beta Im\, \tau(t))' : \alpha^2 + \beta^2 = 1, \; \alpha, \beta \in R\} \leq$$

$$n \sup\{||\alpha Re\, \tau(\cdot) + \beta Im\, \tau(\cdot)||_\infty : \alpha^2 + \beta^2 = 1,\, \alpha, \beta \in R\} =$$

$$n \left\|\sup\{\alpha Re\, \tau(\cdot) + \beta Im\, \tau(\cdot) : \alpha^2 + \beta^2 = 1,\, \alpha, \beta \in R\}\right\|_\infty =$$

$$n \left\| \sqrt{(Re\, \tau(\cdot))^2 + (Im\, \tau(\cdot))^2} \right\|_\infty = n||\tau||_\infty .$$

The theorem is proved.

**Theorem 3.6.2.** *Let $n = 0, 1, \ldots$ and $1 \le p < \infty$. For any polynomial $\tau \in T^C_{2n+1}$*

$$||\tau'||_p \le n||\tau||_p . \tag{3.87}$$

*This inequality becomes an equality if and only if*

$$\tau(t) = ae^{\pm int},\ a \in C.$$

**Proof.** We will use as it was in the proof of the theorem 3.2.1 the Stein method. For any $2\pi$-periodic function $g : R \to C$ such that $|g| \in L_{p'}$, $(1/p + 1/p' = 1)$ we obtain $\tau * g \in T^C_{2n+1}$. In view of theorem 3.6.1 we will have

$$||(\tau * g)'||_\infty \le n||\tau * g||_\infty$$

and therefore

$$||\tau'||_p = \sup \left( \left| \int_0^{2\pi} \tau'(-t) g(t) dt \right| : ||g||_{p'} \le 1 \right) \le$$

$$\max_x \sup \left( \left| \int_0^{2\pi} \tau'(x - t) g(t) dt \right| : ||g||_{p'} \le 1 \right) =$$

$$\sup\{||(\tau * g)'||_\infty : ||g||_{p'} \le 1\} \le n \sup\{||\tau * g||_\infty : ||g||_{p'} \le 1\} = n||\tau||_p .$$

The theorem is proved.

**Theorem 3.6.3.** *Let $n = 0, 1, \ldots$ and $0 \le p \le q \le \infty$. For any polynomial $\tau \in T^C_{2n+1}$ the exact inequality*

$$||\tau||_p \le (2\pi)^{1/p - 1/q} n||\tau||_q \tag{3.88}$$

*holds true.*

**Proof.** Applying the Holder's inequality we will have

$$\|\tau'\|_p^p \leq (2\pi)^{1-p/q} \left( \int_0^{2\pi} |\tau'(t)|^q dt \right)^{p/q} = (2\pi)^{1-p/q} \|\tau'\|_q^p.$$

Therefore

$$\|\tau'\|_p \leq (2\pi)^{1/p-1/q} \|\tau'\|_q,$$

where in view of theorem 3.6.2 we deduce

$$\|\tau'\|_p \leq (2\pi)^{1/p-1/q} n \|\tau\|_q,$$

and inequality (3.83) is proved. Exactness of this inequality follows from the fact that for polynomials of the form $\tau(t) = ae^{\pm int}$, $a \in C$ this inequality becomes an equality.

## 3.7 Commentaries

Sec.3.1. Theorem 3.1.2 was proved by S.M. Bernstein with using of the method different from the one given in this book. The reasonings connected with the comparison theorems for derivatives apparently have been employed by S.B. Stechkin [1] under the proving of Bernstein's type inequalities.

The numerous investigations directed to the intensification, generalisation and making more precise the Bernstein's type inequalities and also its different applications one can find in the monographs of A.F. Timan [1], N.I. Ahiezer [6], and also in the articles of E.A. Gorin [2], V.S. Videnski [8] and etc. Bernstein's type inequalities in their different variants have been proved for integral functions of exponential type in the articles of S.N. Bernstein [13], B.M. Levitan [1], R. Boas [2], S.M. Nikolski [5] and many other authors.

The theorem 3.1.3 was proved by G. Szego [2]. Theorem 3.1.4 for $h = \pi/n$ was proved by S.M. Nikolski [3], and in generqal case it was proved by S.B. Stechkin [1].

Sec.3.2. Theorem 3.2.1 was proved by A. Zigmund [2] with the using of another method. Theorems 3.2.2 and 3.2.3 were marked by G.G. Lorentz [4]. Theorem 3.2.7 was obtained by V.V. Arestov.

Sec.3.3. Given in this section proofs were suggested by S.A. Pichugov. The proofs of Bernstein's type inequalities received with the help of interpolation formulas of different type one can find in the books of A. Zigmund [2], R. Boas [2], A.F. Timan [1] and N.I. Ahiezer [6].

Sec.3.4. Theorems 3.4.2 - 3.4.4 were proved by another way by L.V. Tajkov [2]. Theorem 3.4.7 belongs to A.A. Ligun [10].

Sec.3.5. At first unusual inequalities of different metrics for trigonometric polynomials were proved by S.M. Nikolski [5], who established in particular theorem 3.5.2. Theorem 3.5.3 was proved by Pichorides [2].

Sec.3.6. Theorems 3.6.1 - 3.6.3 may be easy obtained from theorem 3.1.2.

## 3.8 Supplementary results

**1.** If

$$\tau(t) = \sum_{k=-n}^{n} c_k e^{ikt} \tag{3.89}$$

then

$$\left| \sum_{k=-n}^{-1} k c_k e^{ikt} \right| + \left| \sum_{k=1}^{n} k c_k e^{ikt} \right| \leq n \|\tau\|_\infty .$$

Moreover

$$\|\tau'\|_\infty \leq \left(n - \frac{1}{2}\right) \|\tau\|_\infty + \frac{1}{2}(|c_{-n}| + |c_n|)$$

and

$$|c_{-n}| + |c_n| \leq \|\tau\|_\infty ,$$

Frappier K. [3].

**2.** For any trigonometric polynomial of the form (3.89) the next inequality

$$\left| \sum_{k=-n}^{-1} (b^{-k} - 1) c_k e^{ikt} \right| + \left| \sum_{k=1}^{n} (b^k - 1) c_k e^{ikt} \right| \leq (b^n - 1)\|\tau\|_\infty ,\ b \geq 1$$

is valid.

Frappier K. [3].

If $0 < \delta < h < \pi/n$ then for arbitrary polynomial $\tau \in T_{2n+1}$ and for any natural $r$

$$\left(\frac{n}{2 \sin n\delta}\right)^r \|\Delta_\delta^r \tau\|_\infty \leq \left(\frac{n}{2 \sin nh}\right)^r \|\Delta_h^r \tau\|_\infty$$

Boas R. [2].

**3.** For aby polynomial $\tau \in T_{2n+1}$

$$\|\tau\|_{C[\pi,\pi]} \leq \left(\tan^{2n} \frac{a}{4} + \cot^{2n} \frac{a}{4}\right) \|\tau\|_{C[-a,a]} .$$

Goncharov V.P.[1].

**4.** For arbitrary integral function of exponential type $\sigma$ the inequality (S.N.Bernstein [10])

$$\sup_x |f^{(k)}(x)| \leq \sigma \sup_x |f(x)|$$

holds true. This inequality becomes an equality if and only if

$$f(z) = a\cos(\sigma z - \alpha).$$

Moreover (R. Boas [2])

$$\int_{-\infty}^{\infty} |f^{(k)}(t)|^p\, dt \leq \sigma^{pk} \int_{-\infty}^{\infty} |f(t)|^p\, dt,\ p \geq 1,$$

and (N.I. Ahiezer [6])

$$\int_{-\infty}^{\infty} |\sin\alpha f'(t) + \sigma\cos\alpha f(t)|^p\, dt \leq \sigma^{pk} \int_{-\infty}^{\infty} |f(t)|^p\, dt,\ p \geq 1,\ \alpha \in R.$$

**5.** If $f(t)$ is the integral function of exponential type $\sigma$, then for $p \geq 2$ and $n \geq 0$

$$\left(\int_{-\infty}^{\infty} |f^{(k)}(t)|^p\, dt\right)^{1/p} \leq \left(\frac{\sigma}{\pi(2n+1)}\right)^{(q-2)/(2q)} \sigma^n \left(\int_{-\infty}^{\infty} |f(t)|^2\, dt\right)^{1/2}.$$

Ibragimov I.I. [2].

# Chapter 4

# Inequalities of Markov type

Methods of obtaining strict inequalities for derivatives of trigonometric polynomials were based on the theorem of comparison of functions and their rearrangements. In the theorems of comparison of functions invariability with respect to shift of argument has been used essentialy. Functions defined on the finite segment including algebraic polynomials do not possess such invariability and it significantly narrows possibilities of study of their extremal properties. Some strict inequalities for polynomials are obtained in this chapter, in particular, with the help of reduction to the periodical case.

## 4.1 Markov's inequalities

**Theorem 4.1.1.** *Let $n \in N$, $-1 < x < 1$. Then for any algebraic polynomial $p \in P_n$ the inequality*

$$|p'(x)| \leq \frac{n\|p\|_{C[-1,1]}}{\sqrt{1-x^2}} \tag{4.1}$$

*holds true. An equality in (3.58) for $x = \cos[(2k-1)\pi/(2n)]$, $k = 1, 2, \ldots, n$ holds true for polynomials of the form $p(x) = aT_n(x)$.*

**Proof.** A function $\tau(t) = p(\cos t)$ is a trigonometric polynomial of order at most $n$. Consequently in view (3.1)

$$|\tau'(t)| \leq n \ \max\{|\tau(t)| : \ t\}.$$

But $\max\{|\tau(t)| : t\} = ||p||_{C-1,1]}$ and $|\tau'(t) = -p'(\cos t)\sin t$. Thus (below in this paragraph $||\cdot|| = ||\cdot||_C$),

$$|p'(\cos t)| \leq n||p||_C / \sin t,$$

and change of variables $\cos t = x$ gives us to inequanity (4.1).

**Proposition 4.1.2.** *Let $n \in N$ and $x \in (-1,1)$. Then for any algebraic polynomial $p \in P_n$ the inequality*

$$|p'(x)| \leq n\sqrt{\frac{||p||_C^2 - p^2(x)}{1-x^2}} \leq \frac{n||p||_C}{\sqrt{1-x^2}} \tag{4.2}$$

*holds true.*

**Proof.** Estimating $\tau'(t)$ with the help of inequality (3.7), and taking into account that $|\cos(t+\pi/n) - \cos t| \leq 2\sin(\pi/(2n))$, we get

$$|p'(\cos t)| = \frac{|\tau'(t)|}{|\sin t|} \leq \frac{n}{2|\sin t|}\omega\left(\tau;\ \frac{\pi}{n}\right) \leq$$

$$\frac{n}{2|\sin t|}\omega\left(p;\ 2\sin\frac{\pi}{2n}\right)_{C[-1,1]},$$

($\omega(f;t)_{C[-1,1]}$ is the modulus of continuity of the function $f \in C[-1,1]$) and now the change of variables $\cos t = x$ immediately gives us the next inequality which is more precise than (3.1).

**Proposition 4.1.3.** *Let $n \in N$ and $x \in (-1,1)$, then for any algebraic polynomial $p \in P_n$ the inequality*

$$|p'(x)| \leq \frac{n}{2\sqrt{1-z^2}}\omega\left(p;\ 2\sin\frac{\pi}{2n}\right)_{C[-1,1]} \tag{4.3}$$

*holds true.*

By means of reduction to periodic case one can obtain some other generalization of Bernstein's inequality (4.2). Thus for instance inequalities (3.18), (3.61) and (3.74) immediately brings us to inequalities

$$\left(\int_{-1}^{1}|p'(t)|^q(1-x^2)^{(q-1)/2}dx\right)^{1/q} \leq n\left(\int_{-1}^{1}|p(t)|^q(1-x^2)^{-1/2}dx\right)^{1/q}, \tag{4.4}$$

$$\left(\frac{1}{2}\int_{-1}^{1}|p'(t)|^q(1-x^2)^{(q-1)/2}dx\right)^{1/q} \leq n||\cos(\cdot)||_q||p||_C, \tag{4.5}$$

$$\int_{-1}^{1} |p'(t)|dx \le \frac{2^{1-1/q}n}{\|\cos(\cdot)\|_q} \left( \int_{-1}^{1} |p(t)|^q (1-x^2)^{-1/2} dx \right)^{1/q}, \tag{4.6}$$

are carrying out for all $n \in N$, $q \in [1, \infty]$ and $p \in P_n$.

It follows immediately from (4.1) that for all $n \in N$, $k = 1, 2, \ldots, n$ and for any algibraic polynomial $p \in P_n$ the inequality

$$|p^{(k)}| \le \frac{n!}{(n-k)!}(1-x^2)^{-k/2}\|p\|_C$$

holds true.

But this inequality is not exact for $k \ge 2$.

The inequalities (4.1)-(4.3) become crude neir the ends of the segment $[-1, 1]$. Uniform estimation which is exact on the set of polynomials of order at most $n$ is given us by the inequality of A.A.Markov.

**Theorem 4.1.4.** *For all $n \in N$ and for any polynomial $p \in P_n$ the inequality*

$$\|p'\|_C \le n^2 \|p\|_C, \tag{4.7}$$

*holds true and becomes an equality for polynomial of form $p(x) = aT_n(x)$.*

This theorem we deduce from the next lemma.

**Lemma 4.1.5.** *For any algebraic polynomial $p \in P_{n-1}$ the inequality*

$$\|p\|_C \le n\|p(\cdot)\sqrt{1-(\cdot)^2}\|_C \tag{4.8}$$

*holds true.*

**Proof.** Let

$$x_k = x_k^n = \cos[(2k-1)\pi/(2n)], \ k = 1, 2, \ldots, n,$$

are zeros of Chebyshev polynomial $N_n(x)$ and

$$p_*(x) = \sum_{k=1}^{n} p(x_k) \frac{T_n(x)}{T_n'(x_k)(x-x_k)}.$$

Then $p_*(x_k) = p(x_k)$, $k = 1, 2, \ldots, m$ and consequently by theorem 2.1.1 $p_*(x) \equiv p(x)$.

Thus

$$p(x) = \sum_{k=1}^{n} p(x_k) \frac{T_n(x)}{T_n'(x_k)(x-x_k)}.$$

But

$$T_n'(x_k) = n(-1)^{k-1}/\sqrt{1-x_k^2},$$

and therefore

$$p(x) = \frac{T_n(x)}{n}\sum_{k=1}^{n}\frac{(-1)^{k-1}p(x_k)\sqrt{1-x_k^2}}{(x-x_k)}.$$

Let $|x| \in (\cos(\pi/(2n)), 1]$. Then all differences $x - x_k$ have the same sign and therefore

$$|p(x)| \leq \frac{|T_n(x)|}{n}\sum_{k=1}^{n}\frac{|p(x_k)|\sqrt{1-x_k^2}}{|x-x_k|} \leq$$

$$\frac{|T_n(x)|}{n}||p(\cdot)\sqrt{1-(\cdot)^2}||_C\sum_{k=1}^{n}\frac{1}{|x-x_k|} =$$

$$\frac{|T_n(x)|}{n}||p(\cdot)\sqrt{1-(\cdot)^2}||_C|\sum_{k=1}^{n}\frac{1}{x-x_k}|.$$

But

$$T_n(x) = \prod_{k=1}^{n}(x-x_k),$$

and consequently

$$T_n'(x) = T_n(x)\sum_{k=1}^{n}\frac{1}{x-x_k}.$$

Thus for $|x| \in (\cos(\pi/(2n)), 1]$

$$\frac{|p(x)|}{||p(\cdot)\sqrt{1-(\cdot)^2}||_C} \leq |\frac{T_n(x)}{n}||\frac{T_n'(x)}{T_n(x)}|$$

$$|T_n'(x)|/n \leq T_n'(1)/n = n,$$

or

$$|p(x)| \leq n||p(\cdot)\sqrt{1-(\cdot)^2}||_C.$$

For $|x| \in [0, \cos(\pi/(2n))]$ the inequalities

$$\frac{1}{n} \leq \sin\frac{\pi}{2n} \leq \sqrt{1-z^2},$$

hold true and consequently for such $x$

$$|p(x)| \leq n|p(x)\sqrt{1-x^2}| \leq n||p(\cdot)\sqrt{1-(\cdot)^2}||_C. \tag{4.9}$$

Really,if $p \in P_n$, then $p' \in P_{n-1}$ and in view of lemma 4.1.5

$$\|p'\|_C \le n \max\{|p'(x)|\sqrt{1-x^2} : x\}$$

and to complete the proof of (4.7) it remains to use the inequality (4.1).

Using inequality (4.3) instead of (4.1) we obtain the following making more precise of inequality (4.7)

$$\|p'\|_C \le \frac{n^2}{2}\omega(p; 2\sin\pi/(2n))_{C[-1,1]}.$$

It is known many different generalizations and making more precise of Markov's inequality. Before all note that in the process of proving inequality (4.7) we in fact proved somewhat more exact inequality, namely: we shown that if

$$\Phi_n(x) = \begin{cases} n/\sqrt{1-x^2}, & |x| \le \cos\pi/(2n); \\ |T_n'(x)|, & \cos\pi/(2n) \le |x| \le 1, \end{cases}$$

then for all $x \in [-1,1]$ and $p \in P_n$

$$|p'(x)| \le \Phi_n(x) \le \Phi_n(1) = T_n'(1) = n^2.$$

By analogy we may prove that for $p \in P_n$ and $k = 1, 2, \ldots, n$

$$|p^{(k)}(x)| \le |T_n^{(k)}(x) + i\, s_n(x)|, \ \ |x| \le 1, \tag{4.10}$$

where

$$s_n(x) = \sin n\, arc\cos x = \frac{1}{n}\sqrt{1-x^2}T_n'(x).$$

Inequality (4.10) for $k = 1$ is identity of (4.1). We may proved that near the the hends of the segmemt $[-1,1]$ the inequality

$$|p^{(k)}(x)| \le |T^{(k)(x)}|\,\|p\|_C, \ \ -1 < x \le y^*_{n,k}, \ y_{n,k} \le x < 1, \tag{4.11}$$

where $y^*_{n,k}$ is left xero of $s^{(k)}(x)$ and $y_{n,k}$ is right zero of its from the interval $(-1,1)$. With the help of induction we may proved that the function

$$|T_n^{(k)}(x) + i\, s_n(x)|^2$$

may be reprezented in the interval $(-1,1)$ in the form of even power series having positive coefficients. From thi follows that this function is nondicreasing on $[-1,1]$. From this and from (4.10) and (4.11) it follows that for any polynomial $p \in P_n$ and for all $k = 1, 2, \ldots$ the exact inequality

$$|p^{(k)}(x)| \le |T^{(k)}(1)|\,\|p\|_C \le 2^k(2k)!\,\frac{n^2(n^2-1)\ldots(n^2-(k-1)^2)}{k!}\,\|p\|_C. \tag{4.12}$$

## 4.2 Inequalities for polynomials of special form

Denote by $P_N^*$ the set of all polynomials $p \in P_n$ having all zeros outside the unit disk $\{|z| \leq 1\}$ and by $P_n^{**}$ the set of all polynomials $p \in P_n$ of the form

$$p(x) = \sum_{k=0}^{b} a_{n,k}(1+x)^k(1-x)^{n-k},$$

where $a_{n,k} \geq 0$, $k = 0, 1, \ldots, n$. It is clear that if $p_1 \in P_n^{**}$ and $p_2 \in P_m^{**}$ then $p_1 p_2 \in P_{n+m}^{**}$. Besides that since

$$a - x = \frac{a-1}{2}(1-x) + \frac{a+1}{2}(1+x)$$

and

$$4(x-z)(x-\overline{z}) = |1+z|^2(1-x)^2 + 2(|z|^2-1)(1-x)(1+x) + |z-1|^2(1+x)^2,$$

then next lemma takes place.

**Lemma 4.2.1.** *If $p \in P_n^*$, then $p(\cdot) sgn\, p(0) \in P_n^{**}$.*

**Theorem 4.2.2.** *For all $n = 3, 4, \ldots$, $|x| \leq 1$ and $p \in P_n^{**}$ and consequently in view of lemma 4.2.1 for all $p \in P_n^*$ the inequalities*

$$-np(x) \leq p'(x) \leq \frac{ne}{2} p\left(x\left(1-\frac{2}{n}\right)\right) \tag{4.13}$$

*and*

$$|p''(x)| \leq \frac{en(n-1)}{2} |p\left(x\left(1-\frac{2}{n}\right)\right)|. \tag{4.14}$$

*hold true. Consequently ($\|\cdot\|_C = \|\cdot\|_{C[-1,1]}$)*

$$\|p'\|_C \leq \frac{en}{2}\|p\|_C \tag{4.15}$$

*and*

$$\|p''\|_C \leq \frac{en(n-1)}{2}\|p\|_C. \tag{4.16}$$

*Besides that*

$$\lim_{n\to\infty} \frac{\|q'_{n,1}\|_C}{n\|q_{n,1}\|_C} = \frac{e}{2} \tag{4.17}$$

*and*

$$\lim_{n\to\infty} \frac{\|q''_{n,1}\|_C}{n(n-1)\|q_{n,1}\|_C} = \frac{e}{2} \tag{4.18}$$

*where*

$$q_{n,k}(x) = (1+x)^k(1-x)^{n-k}. \tag{4.19}$$

Note that relations (4.17) and (4.18) show us that in inequalities (4.15) and (4.16) constant $e/2$ simultaneoulsly for all $n$ we can not diminish.

**Proof.** Since

$$q'_{n.k}(x) = n\frac{v-x}{1-x^2}q_{n,k}(x), \tag{4.20}$$

where $v = 2k/n - 1$, puting $t = x(1-2/n)$ we obtain

$$r_{n,k}(x) := \frac{q'_{n,k}(x)}{nq_{n,k}(t)} = \frac{v-x}{1-t^2}\left(\frac{1+x}{1-t}\right)^{k-1}\left(\frac{1-x}{1-t}\right)^{n-k-1} =$$

$$\frac{v-x}{1-t^2}\left(1+\frac{x-t}{1+t}\right)^{k-1}\left(1-\frac{x-t}{1-t}\right)^{n-k-1}.$$

¿From this and from inequality $1+u \le e^u$ it follows that for $k>0$ and $q'_{n,k} > 0$

$$r_{n,k} \le \frac{v-x}{1-t^2}\exp(x-t)\left(\frac{k-1}{1+t} - \frac{n-k-1}{1-t}\right) := \psi(y)$$

where

$$\psi(y) = ye^{2xy-a}, \tag{4.21}$$

and

$$y = \frac{v-x}{1-t^2}, \quad a = \frac{8x^2(n-1)}{n^2(1-t^2)}.$$

Let at first $-1 \le x \le -1/3$.Taking into account the relation

$$\max\{\psi(y) : \ y>0\} = -\frac{1}{2x}e^{a-1},$$

we obtain

$$r_{n,k} \le -\frac{1}{2x}e^{a-1} = -\frac{1}{2x}\exp\left(\frac{8x^2(n-1)}{n^2(1-t^2)} - 1\right) := \Psi(x).$$

The function $\Psi(x)$ decreases for $x \in [-1, -1/3]$. Therefore

$$\Psi(x) \le \Psi(-1) = e/2.$$

Thus if $x \in [-1, -1/3]$ and $q'_{n,k} > 0$ then

$$r_{n,k} \leq e/2. \tag{4.22}$$

Let now $x \in [-1/3, 0]$. Then

$$2x^2 - 2x \leq 1 - x^2 \leq 1 \leq t^2$$

and consequently

$$\frac{v - x}{1 - t^2} \leq \frac{1 - x}{1 - t^2} \leq -\frac{1}{2x}$$

Besides that the function $\psi(y)$ defined by (4.21) increases on the interval $[0, -1/(2x)]$. Therefore

$$r_{n,k} \leq \psi\left(\frac{1 - x}{1 - t^2}\right) = \frac{1 - x}{1 - t^2} \exp\left(2x\frac{1 - x}{1 - t^2} + \frac{8x^2(n - 1)}{n^2(1 - t^2)}\right) < \frac{e}{2}.$$

Thus for all $k = 1, 2, \ldots, n$ and $x \in [-1, 1]$, such that $q'_{n,k} > 0$ , the inequality (4.22) holds true.

If $x \in [-1, 0]$ and $q'_{n,k} < 0$, then due to (4.19)

$$\frac{-q''_{n,k}(x)}{nq_{n,k}(x)} = \frac{x - v}{1 - x^2} < 1$$

or

$$q'_{n,k}(x) \geq -nq_{n,k}. \tag{4.23}$$

Let $p$ be an arbitrary polynomial from the set $P_n^{**}$ Then there exist numbers $a_{n,k} \geq 0,\ k = 0, 1, \ldots, n$ such that

$$p(x) = \sum_{k=1}^{n} a_{n,k} q_{n,k}(x).$$

In view of (4.22) for $x \in [-1, 0]$ the relations

$$p'(x) = a_{n,0}q'_{n,0}(x) + \sum_{k=1}^{n} a_{n,k}q'_{n,k}(x) \leq$$

$$a_{n,0}q'_{n,0}(x) + \frac{en}{2}\sum_{k=1}^{n} a_{n,k}q_{n,k}(t)$$

hold true and since

$$q'_{n,k}(x) \leq 0 \leq \frac{en}{2}q_{n,0}(t)$$

then

$$p'(x) \leq \frac{en}{2} \sum_{k=1}^{n} a_{n,k} q_{n,k}(t) = \frac{en}{2} p(t).$$

Analogously from (4.23) deduce the left inequality (4.13) for $x \in [-1, 0]$. Applying proved inequalities to polynomial $p(-x)$ we verify that inequalities (4.13) are valid and for $x \in [0, 1]$.

It is easy to deduce from (4.19) that

$$q'_{n,k}(x) = n(n-1)\theta_{n,k}(x)(1+x)^{k-2}(1-x)^{n-k-2},$$

where

$$\theta_{n,k} = (v-x)^2 - \frac{4k(n-k)}{n^2(n-1)}.$$

Therefore for $t = x(1 - 2/n)$

$$r^*_{n,k}(x) := \frac{q''_{n,k}(x)}{n(n-1)q_{n,k}(x)} =$$

$$\frac{\theta_{n,k}(x)}{(1-t^2)^2} \left(1 + \frac{x-t}{1+t}\right)^{k-2} \left(1 - \frac{x-t}{1-t}\right)^{n-k-2}, \tag{4.24}$$

and for $k = 2, 3, \ldots, n-2$

$$|r^*_{n,k}(x)| \leq \frac{|\theta_{n,k}(x)|}{(1-t^2)^2} \exp\left(\frac{2x}{n}\left(\frac{k-2}{1+t} + \frac{n-k-2}{1-t}\right)\right) =$$

$$\frac{|\theta_{n,k}(x)|}{(1-t^2)^2} \exp 2x \frac{v - t + 4t/n}{1-t^2} := \Psi_1(x).$$

Reasoning as in the proof of the first inequality (4.13) we can show that for $|x| < 1$ and $3, 4, \ldots, n-2$

$$|r^*_{n,k}(x)| \leq \Psi_1(x) \leq e/2. \tag{4.25}$$

Considering the functions $|r^*_{n,k}(x)|$ for $k = 0, 1, 2$ and $n-1, n$ we see that for these functions also the inequalities $|r^*_{n,k}(x)| \leq e/2$ hold true. Thus for all $|x| < 1$ and $k = 0, 1, \ldots, n$

$$|q''_{n,k}(x)| \leq \frac{e}{2} n(n-1) q_{n,k}(t)$$

and consequently

$$|p,'(x)| \leq \sum_{k=1}^{n} a_{n,k} |q''_{n,k}(x)| \leq$$

$$= \frac{en(n-1)}{2} \sum_{k=1}^{n} a_{n,k} q_{n,k}(x) = \frac{en(n-1)}{2} p(t).$$

The inequalities (4.15) and (4.16) directly follow from the inequalities (4.13) and (4.14). To prove the relations (4.17) and (4.18) we may use direct calculations.

The analog of the theorem 4.2.2 in the space $L_2[-1,1]$ is given to us by next theorem

**Theorem 4.2.3.** *Let all zeros of apolymomial $p \in P_n$ are real and lie outside the interval $[-1,1]$. Then*

$$\int_{-1}^{1} (1-t^2)(p'(t))^2 dt \le \frac{n}{2} \int_{-1}^{1} p^2(t) dt \tag{4.26}$$

*and*

$$\int_{-1}^{1} (1-t^2)(p'(t))^2 dt \le \frac{n(n+1)(2n+3)}{2(2n+1)} \int_{-1}^{1} (1-t^2) p^2(t) dt. \tag{4.27}$$

*Besides that for any polynomial of the form*

$$p(t) = a(1+t)^k (1-t)^{n-k}, \quad k = 0, 1, \ldots, n$$

*inequality (4.26) becomes an equality.*

To prove theorem 4.2.3 we need two lemmas which are useful for solution of other extremal problems for algebraic polynomials.

**Lemma 4.2.4.** *If all zeros $t_1, t_2, \ldots, t_n$ of polynomial $p \in P_n$ are real, then*

$$\int_{-1}^{1} (1-t^2)(p'(t))^2 dt = \frac{n}{2} \int_{-1}^{1} p^2(t) dt +$$

$$\frac{1}{2} \int_{-1}^{1} q^2(t) \sum_{k=1}^{n} \frac{1-t_k^2}{(t-t_k)^2} dt. \tag{4.28}$$

**Proof.** Since

$$\int_{-1}^{1} (1-t^2)(p'(t))^2 dt = \int_{-1}^{1} (1-t^2) p'(t) dp(t) =$$

$$2 \int_{-1}^{1} t p'(t) p(t) dt - \int_{-1}^{1} (1-t^2) p''(t) p(t) dt;$$

$$p'(t) = p(t)\Sigma(t);$$

$$p''(t) = p'(t)\Sigma(t) + p(t)\Sigma'(t);$$

$$\Sigma(t) := \sum_{k=1}^{n} \frac{1}{t - t_k},$$

then

$$\int_{-1}^{1} (1-t^2)(p'(t))^2 dt = 2\int_{-1}^{1} tp^2(t)\Sigma(t)dt -$$

$$\int_{-1}^{1} (1-t^2)p(t)[p'(t)\Sigma(t) + p(t)\Sigma'(t)]dt =$$

$$2\int_{-1}^{1} tp^2(t)\Sigma(t)dt - \int_{-1}^{1} (1-t^2)p^2(t)\Sigma'(t)dt - \int_{-1}^{1} (1-t^2)(p'(t))^2 dt.$$

Therefore

$$2\int_{-1}^{1} (1-t^2)p'(t)dp(t) = \int_{-1}^{1} p^2(t) \sum_{k=1}^{n} \left( \frac{2t}{t-t_k} + \frac{1-t^2}{(t-t_k)^2} \right) dt =$$

$$\int_{-1}^{1} p^2(t) \left( n + \sum_{k=1}^{n} \frac{1-t_k^2}{(t-t_k)^2} \right) dt,$$

that is equivalent to the statement of the lemma.

**Lemma 4.2.5.** *Under the conditions of theorem 4.2.3*

$$\int_{-1}^{1} p^2(t)dt \le \frac{(n+1)(2n+3)}{4n+2} \int_{-1}^{1} (1-t^2)p^2(t)dt$$

**Proof.** It follows from lemma 2.4.1 that there exist numbers $a_{n,k} \ge 0,\ k = 1, 2, \ldots, 2n$ such that

$$p^2(t) = \sum_{k=0}^{2n} a_{n,k}(1+t)^k(1-t)^{2n-k}.$$

Therefore

$$\int_{-1}^{1}(1-t^2)p^2(t)dt = \sum_{k=0}^{2n} a_{n,k}\int_{-1}^{1}(1+t)^{k+1}(1-t)^{2n-k+1}dt =$$

$$\sum_{k=0}^{2n} a_{n,k}\frac{2k(2n-k)+4n+2}{(2n+3)(n+1)}\int_{-1}^{1}(1+t)^{k}(1-t)^{2n-k}dt \geq$$

$$\frac{4n+2}{(2n+3)(n+1)}\sum_{k=0}^{2n} a_{n,k}\int_{-1}^{1}(1+t)^{k}(1-t)^{2n-k}dt =$$

$$\frac{4n+2}{(2n+3)(n+1)}\int_{-1}^{1}p^2(t)dt.$$

**Proof of theorem 4.2.3.** The inequality (4.26) immediately follows from lemma 4.2.4, since by condition $1-t_k^2 \leq 0$, $k=1,2,\ldots,n$.

Comparing the inequality (4.26) and the statement of lemma 2.4.5, we immediataly obtain the inequality (4.27).

## 4.3 Estimates of the norms of polynomials

In this section we shall present analogues of the results of section 3.5 for algebraic polynomials. Here we shall consider in the main various inequalities for norms of the form

$$||f||_{L_p^{\alpha,\beta}} = \left(\int_{-1}^{1}|f(x)|^q\omega(x)dx\right)^{1/q}, \qquad (4.29)$$

where

$$\omega(x) = (1-x)^{\alpha}(1+x)^{\beta}. \qquad (4.30)$$

Instead of $C[-1,1]$ and $L_q^{0,0}$ we'll write $C$ and $L_q$ respectively.

We shall first present some exact inequalities.

Note that if $p$ is an arbitrary algebraic polynomial $P_n$ then $\tau(t) = p(\cos t)$ is an even trigonometrical polynomial from $T_{2n+1}$ and so in view of theorem 3.5.1

$$\max\{|\tau(t|:\ t\in[0,2\pi]\} \leq \sqrt{\frac{2n+1}{\pi}}\left(\int_0^{\pi}|\tau(t)|^2dt\right)^{1/2}.$$

¿From this with the help of the change of variable $t = arc\cos x$ we obtain

$$\|p\|_C \leq \sqrt{\frac{2n+1}{\pi}} \left( \int_{-1}^{1} \frac{p^2(x)}{\sqrt{1-x^2}} dx \right)^{1/2} =$$

$$\sqrt{\frac{2n+1}{\pi}} \|p\|_{L_2^{-1/2,-1/2}} \tag{4.31}$$

This inequality becomes an equality if and only if

$$p(x) = a \left( \frac{1}{2} + \sum_{k=1}^{n} T_n(x) \right) = a \left( \frac{1}{2} + \sum_{k=1}^{n} \cos n \; arc\cos x \right) =$$

$$= a \frac{\sin(n+1) arc\cos x + \sin n \; arc\cos x}{\sin x} = a((n+1)U_{n+1}(x) + nU_n(x)),$$

i.e. if and only if

$$p(x) = a(U_{n+1}(x) + \frac{n}{n+1} U_n(x)).$$

Replacing in (4.31) $p(x)$ by $p(x)(1-x^2)$ we obtain

$$|p(x)|(1-x^2) \leq \sqrt{\frac{2n+5}{\pi}} \int_{-1}^{1} p^2(x)\sqrt{1-x^2}dx, \tag{4.32}$$

$$p \in P_n, \quad |x| \leq 1.$$

Analogously using the proposition 2.9.6 and theorems 2.9.7 and 2.9.8 instead of the theorem 3.5.1, we obtain the inequalities

$$\|p\|_C \leq \frac{n+1}{\pi} \int_{-1}^{1} \frac{p(x)}{\sqrt{1-x^2}} dx; \tag{4.33}$$

$$\int_{-1}^{1} \frac{p(x)}{\sqrt{1-x^2}} dx \leq \frac{1}{1-\cos \pi/(n+2)} \int_{-1}^{1} \sqrt{\frac{1-x}{1+x}} p(x) dx; \tag{4.34}$$

$$\int_{-1}^{1} \frac{p(x)}{\sqrt{1-x^2}} dx \leq$$

$$\frac{2}{1-\cos \pi/([n/2]+2)} \int_{-1}^{1} p(x)\sqrt{1-x^2}dx. \tag{4.35}$$

Here $p \in P_n^+$ and $P_n^+$ is the set of all nonnegative on the interval $[-1, 1]$ polynomials from $P_n$.

These inequalities are also exact on the set $P_n^+$ and, for example, inequality (4.33) becomes an equality (see proof of the proposition 2.9.5) if and only if then

$$p(x) = a\frac{1-\cos(n+1)arc\cos x}{1-x} a\frac{1-T_(n+1)(x)}{1-x}.$$

Note also that if at inequalities (4.33), (4.34) and (4.35) we replace $p(x)$ by $p(x)(1+x)(1-x)$ or $p(x)(1-x^2)$, we shall obtain some new inequalities for $p \in P_n^+$:

$$|p(x)| \leq \frac{n+3}{\pi\sqrt{1-x^2}} \int_{-1}^{1} p(x)\sqrt{1-x^2}dx;$$

$$\int_{-1}^{1} \sqrt{\frac{1+x}{1-x}} p(x)dx \leq \frac{1}{1-\cos\pi/(n+3)} \int_{-1}^{1} p(x)\sqrt{1-x^2}dx;$$

$$\int_{-1}^{1} p(x)\sqrt{1-x^2}dx \leq$$

$$\frac{2}{1-\cos\pi/([(n+2)/2]+2)} \int_{-1}^{1} p(x)(\sqrt{1-x^2})^3 dx.$$

But these inequalities are not exact on the set $P_n^+$.

**Theorem 4.3.1.** *If $\alpha,\beta > -1$, $\max\{\alpha,\beta\} = \delta_+ \geq -1/2$, $\delta_- = \min\{\alpha,\beta\}$ and $\gamma = \alpha+\beta+1$, then for any polynomial $p \in P_n$ and any $x \in [-1,1]$ the inequality*

$$p^2(x) + \frac{1-x^2}{n(n+\gamma}(p'(x))^2 \leq \theta_{n,0}(\alpha,\beta)||p||^2_{L_2^{\alpha,\beta}}, \tag{4.36}$$

*where*

$$\theta_{n,0}(\alpha,\beta) = \frac{\Gamma(n+\gamma+1)\Gamma(n+\delta_++2)}{2^\gamma} n!\Gamma(n+\delta_-+1)\Gamma(1+\delta_+)\Gamma(2+\delta_+). \tag{4.37}$$

*holds true.*

*If $\alpha \geq \beta$ then inequality (4.36) becomes an equality for $x = 1$ and*

$$p(x) = a\sum_{k=1}^{n} \tilde{J}_k^{(\alpha,\beta)}(1)\tilde{J}_k^{(\alpha,\beta)}(x) \tag{4.38}$$

*and if $\beta \geq \alpha$ then for $x = -1$*

$$p(x) = a \sum_{k=1}^{n} \tilde{J}_k^{(\alpha,\beta)}(-1) \tilde{J}_k^{(\alpha,\beta)}(x) \tag{4.39}$$

*Here and below $\tilde{J}_k^{(\alpha,\beta)}(x)$ are orthonormed (in metric of space $L_2^{\alpha,\beta}$) Jacoby's polynomials.*

**Proof.** If $p \in P_n$, then in view of theorem 2.5.1

$$p(x) = a \sum_{k=1}^{n} c_k \tilde{J}_k^{(\alpha,\beta)}(x),$$

where

$$c_k = \int_{-1}^{1} p(x) \tilde{J}_k^{(\alpha,\beta)}(x)(1-x)^{\alpha}(1+x)^{\beta} dx, \; k = 1, \ldots, n.$$

Therefore in view of theorem 2.5.1 and inequality of Caushy-Bunyakowski we have

$$||p||^2_{L_2^{\alpha,\beta}} = \sum_{k=1}^{n} c_k^2,$$

and

$$p^2(x) \leq \sum_{k=1}^{n} c_k^2 \sum_{k=1}^{n} (\tilde{J}_k^{(\alpha,\beta)}(x))^2 =$$

$$||p||^2_{L_2^{\alpha,\beta}} \sum_{k=1}^{n} (\tilde{J}_k^{(\alpha,\beta)}(x))^2$$

and

$$(p'(x))^2 \leq ||p||^2_{L_2^{\alpha,\beta}} \sum_{k=1}^{n} ((\tilde{J}_k^{(\alpha,\beta)}(x))')^2$$

i.e.

$$p^2(x) + \frac{1-x^2}{n(n+\gamma)} (p'(x))^2 \leq \theta_n(\alpha, \beta, x) ||p||^2_{L_2^{\alpha,\beta}},$$

where

$$\theta_n(\alpha, \beta, x) = \sum_{k=1}^{n} (\tilde{J}_k^{(\alpha,\beta)}(x))^2 + \frac{1-x^2}{n(n+\gamma)} \sum_{k=1}^{n} ((\tilde{J}_k^{(\alpha,\beta)}(x))')^2$$

If $|x_0| = 1$, then this inequality becomes an equality if and only if

$$c_k = a\tilde{J}_k^{(\alpha,\beta)}(x_0), k = 1, 2, \ldots, n.$$

To complete the proof of theorem 4.3.1 it remains make use the statement of corollary 2.5.11.

Now we present some corollaries of theorem 4.3.1.

We first note that from (4.36) it follows immediately that for any polynomial $p \in P_n$ the inequality

$$||p||_C \leq \sqrt{\theta_{n,0}(\alpha, \beta)}||p||_{L_2^{\alpha,\beta}}, \tag{4.40}$$

$$\alpha, \beta > -1, \quad \max\{\alpha, \beta\} \geq -1/2,$$

holds true and this inequality becomes an equality for polynomials of the form (4.38) if $\alpha \geq \beta$ and for polynomials of the form (4.39) if $\alpha < \beta$.

In particular for $p \in P_n$ and $|x| \leq 1$ we have

$$p^2(x) + \frac{1 - x^2}{n^2}(p'(x))^2 \leq \frac{2n+1}{\pi}\int_{-1}^{1}\frac{p^2(x)}{\sqrt{1 - x^2}}dx; \tag{4.41}$$

$$p^2(x) + \frac{1 - x^2}{n(n+1)}(p'(x))^2 \leq \frac{(n+1)^2}{2}\int_{-1}^{1}p^2(x)dx; \tag{4.42}$$

$$p^2(x) + \frac{1 - x^2}{n(n+2)}(p'(x))^2 \leq$$

$$\frac{(n+1)(2n+3)(n+2)}{3\pi}\int_{-1}^{1}p^2(x)\sqrt{1 - x^2}dx. \tag{4.43}$$

Consequently

$$||p||_C \leq \sqrt{\frac{2n+1}{\pi}}||p||_{L_2^{-1/2,-1/2}}; \tag{4.44}$$

$$||p||_C \leq \frac{n+1}{\sqrt{2}}||p||_{L_2}; \tag{4.45}$$

$$||p||_C \leq \sqrt{\frac{(n+1)(2n+3)(n+2)}{3\pi}}||p||_{L_2^{1/2,1/2}}. \tag{4.46}$$

Besides it follows from (4.40) and lemma 2.5.8 if $\delta_+ = \max\{\alpha, \beta\} > -1/2$, then when $n \to \infty$

$$||p||_C \leq \theta(\alpha, \beta)(n^{\delta_+ + 1} + O(n^{\delta_+})||p||_{L_2^{\alpha,\beta}},$$

where

$$\theta(\alpha,\beta) = (\Gamma(\delta_+ + 1)\Gamma(\delta_+ + 2)2^{\alpha+\beta+1})^{-1/2}.$$

Note that as it follows from Markov's inequality if $p \in P_n$ and $||p||_C = |p(x_0)|$, $x_0 \in [1,1]$ then

$$|p(x_0)| - |p(x)| \le |p(x_0) - p(x)| \le ||p||_C|x - x_0| \le n^2|x - x_0|||p||_C,$$

i.e.

$$||p||_C(1 - n^2|x - x_0|) \le |p(x)|.$$

But then (we consider for definitness that $x_0 \le 0$ )

$$||p||^q_{L_q} = \int_{-1}^{1} |p(x)|^q dx \ge \int_{x_0}^{x_0+1/n^2} |p(x)|^q dx \ge$$

$$||p||^q_{L_q} \int_{x_0}^{x_0+1/n^2} (1 - n^2(x - x_0))^q dx = \frac{n^2}{q+1}||p||^2_C,$$

i.e.

$$||p||_C \le n^{2/q}(q+1)^{1/q}||p||_{L_q};$$

Consequently, for $0 < q < s < \infty$

$$||p||^s_{L_s} = \int_{-1}^{1} |p(x)|^q|p(x)|^{s-q}dx \le ||p||_C^{s-q} \int_{-1}^{1} |p(x)|^q dx \le$$

$$n^{2(s-q)/q}(1+q)^{(s-q)/q}||p||^{s-q}_{L_q}||p||^q_{L_q} = ((n^2(q+1))^{1/q-1/s}||p||_{L_q})^s$$

or

$$||p||_{L_s}(n^2(q+1))^{1/q-1/s}||p||_{L_q},\ 0 < q < s < \infty. \tag{4.47}$$

If $0 < s < q < \infty$, then by Holder's inequality

$$||p||^s_{L_s} = \int_{-1}^{1} |p(x)|^s dx \le \left(\int_{-1}^{1} |p(x)|^q|p(x)|^{s-q}dx\right)^{s/q} 2^{1-s/q},$$

i.e.

$$||p||_{L_s} \le 2^{1/q-1/s}||p||_{L_q},\ 0 < s < q < \infty. \tag{4.48}$$

Thus, for all $q, s \in (0,\infty)$

$$||p||_{L_s} \le C_{q,s}n^{2/q-2/s}||p||_{L_q}, \tag{4.49}$$

where

$$C_{q,s} = (q+1)^{1/q-1/s},$$

if $0 < q \leq s < \infty$ and

$$C_{q,s} = 2^{1/q-1/s}$$

if $0 < q < s < \infty$.

The inequality (4.49) is exact in the sense of order. To see this it is sufficiently to consider the polynomial $p(x) = 1$, if $0 < s < q < \infty$, and sequence of polynomials

$$p_{4n}(x) = (U_n(x))^4, \ n = 1, 2, \ldots.$$

We present now the analalogue of the theorem 4.3.1 for nonnegative on interval $[-1, 1]$ polynomials in the metric of the space $L_1^{\alpha,\alpha}$.

First we present two general lemmas which are of independent interest.

Let $q \geq 1$, $x_0 \in [-1, 1]$, $\omega(x)$- is an integrable, nonnegative function such that $\omega(x) \neq 0$ almost everywhere on the interval $[-1, 1]$

$$A_{n,q}^{+}(x_0) = \min_{p \in P_n,\, p(x_0)=1} \int_{-1}^{1} |p(x)|^q \omega(x) dx \tag{4.50}$$

and let $p_q^+ \in P_n^+$ is polynimial such that $p_q^+(x_0) = 1$ and

$$\int_{-1}^{1} |p_q^+(x)|^q \omega(x) dx = A_{n,q}^{+}(x_0). \tag{4.51}$$

**Lemma 4.3.2.** *All zeros of polynomial $p_q^+(x)$ are real.*

**Proof.** If the statement of this lemma is not true, then there exist numbers $a, b \in R$ such that

$$a(x - x_0)^2 + b(x - x_0) + 1 > 0, \ x \in R,$$

and polynomial $p_0 \in P_{n-2}^+$ such that $p_0(x_0) = 1$ and

$$p_q^+(x) = p_0(x)(a(x - x_0)^2 + b(x - x_0) + 1)$$

But then, if

$$f(\delta) = \int_{-1}^{1} |p_0(x)|^q |(a + \delta)(x - x_0)^2 + b(x - x_0) + 1|^q \omega(x) dx$$

then $f'(0) = 0$, i.e.

$$q \int_{-1}^{1} |p_0(x)|^q |a(x - x_0)^2 + b(x - x_0) + 1|^{q-1} (x - x_0)^2 dx = 0,$$

but this is impossible.

In the sequel the $p_q^+(x)$ denote that polynomial $p_q^+(x)$ (if $p_q^+(x)$ is not unique) which have the least degree.

**Lemma 4.3.3.** *All zeros of the polynomial $p_q^+(x)$ lie inside the interval $[-1,1]$.*

**Proof.** If this statement is not true then in view of lemma 4.3.2 there exist number $b$ and polynomial $p_1 \in P_{n-1}^+$ such that $p_1(x_0) = 1,\ b(x - x_0) + 1 > 0$ $(x \in [-1,1])$ and

$$p_q^+(x) = p_1(x)(b(x - x_0) + 1).$$

But then, if

$$f_1(\delta) = \int_{-1}^{1} |p_1(x)|^q |(b+\delta)(x - x_0) + 1|^q \omega(x) dx$$

then $f_1'(0) = 0$, i.e.

$$q \int_{-1}^{1} |p_1|^q |b(x - x_0) + 1|^{q-1} (x - x_0) dx = 0,$$

¿From this and from the criterion of best approximation element in the space $L_{q,\omega}[-1,1]$ it follows that

$$\min_b \int_{-1}^{1} (p_1(x) + b(x - x_0)p_1(x))^q \omega(x) dx = \int_{-1}^{1} p_1^q(x)\omega(x) dx$$

and consequently

$$\int_{-1}^{1} (p_q^+(x))^q \omega(x) dx \geq \int_{-1}^{1} p_1^q(x)\omega(x) dx.$$

But it is impossible as degree of the polynomial $p_1(x)$ is less then degree of the polynomial $p_q^+(x)$.

Next lemma follows immediately from lemma 4.3.3.

**Lemma 4.3.3.** *If polynomial $p_q^+(x)$ has degree $2n$, then there exists a polynomial $g \in P_n$ having all zeros inside $[-1,1]$ and such that $p(x) = g^2(x)$, and if polynomial $g \in P_m$ has degree $2m + 1$, then there exists a polynomial $g \in P_m$ having all zeros inside $[-1,1]$ and such that $p(x) = (1 + x)g^2(x)$ or $p(x) = (1 - x)g^2(x)$.*

Let

$$\omega_q^+(x) = \omega(x)(1 + x)^q, \quad \omega_q^-(x) = \omega(x)(1 - x)^q$$

Denone by $c_{n,q}(\omega, x)$ ($|x| \leq 1$) the least constant $c$ in the inequality

$$|p(x)| \leq c\|p\|_{L_{q,\omega}[-1,1]} \ (p \in P_n).$$

i.e. let

$$c_{n,q} = \min_{p \in P_n} \frac{\|p\|_{L_{q,\omega}[-1,1]}}{p(x)} \tag{4.52}$$

Let now

$$c^0_{2n,q}(\omega, x) = (c_{n,2q}(\omega, x))^2 \tag{4.53}$$

and

$$c^0_{2n+1,q} = \tag{4.54}$$

$$\max\{(1-x)(c_{n,2q}(\omega_q^-, x))^2, \ (1+x)(c_{n,2q}(\omega_q^+, x))^2\},$$

$$c^+_{n,q}(\omega, x) = \max\{c^0_{m,q}(\omega, x) : \ m = 1, 2, \ldots, n\}. \tag{4.55}$$

**Theorem 4.3.5.** *If $n \in N$, $q \geq 1; \alpha, \beta \geq -1$ and $\omega(x)$ is summable nonnegative function such that $\omega \neq o$ almost everywhere inside $[-1,1]$, then for $[-1,1]$ and any polynomial $p \in P_n^+$ the exact inequality*

$$|p(x)| \leq c^+_{n,q}(\omega, x)\|p\|_{L_{q,\omega}[-1,1]} \tag{4.56}$$

*is valid.*

**Proof.** Let $p_q^+ \in P_n^+$, $p_q^+(x_0) = 1$ and let condition (4.51) is fulfilled. If $n = 2m$ and polynomial $p_q^+$ has degree $2m_0$ ($m_0 \leq m$) , then by lemma 4.3.4 there exists a polynomial $g \in P_{m_0}$ such that $p_q^+(x) = g^2(x)$.

In view of (4.52)

$$|g(x)| \leq c_{m_0,2q}(\omega, x) \leq \|g\|_{L_{2q,\omega}[-1,1]} \leq c_{m,2q}(\omega, x)\|g\|_{L_{2q,\omega}[-1,1]}$$

or

$$p_q^+(x_0) = g^2(x) \leq (c_{m_0,2q}(\omega, x))^2\|g\|^2_{L_{2q,\omega}[-1,1]} \leq c^+_{n,q}(\omega, x)\|P_q^+\|_{L_{q,\omega}[-1,1]}.$$

If $n = 2m$ and polynomial has odd degree $2m_0 + 1$ ($m_0 < m$), then by lemma 4.3.4 there exists a polynomial $g \in P_{m_0}$ such that $p_q^+(x) = (1+x)g^2(x)$ or $p_q^+(x) = (1-x)g^2(x)$. Besides that by definition (4.52)

$$g^2(x) \leq (c_{m_0,2q}(\omega_q^+, x))^2\|g\|^2_{L_{2q,\omega_q^+}[-1,1]}$$

and

$$g^2(x) \leq (c_{m_0,2q}(\omega_q^-, x))^2\|g\|^2_{L_{2q,\omega_q^-}[-1,1]}$$

Thus in the first case

$$p_q^+(x) = (1+x)g^2(x) \le (1+x)(c_{m_0,2q}(\omega_q^+,x))^2\|g\|^2_{L_{2q,\omega_q^+}[-1,1]} =$$

$$= (1+x)(c_{m_0,2q}(\omega_q^+,x))^2\|p_q^+\|^2_{L_{q,\omega}[-1,1]} \le$$

$$c_{n,q}^+(\omega,x)\|p_q^+\|_{L_{q,\omega}[-1,1]}$$

and in the second case

$$p_q^+(x) = (1-x)g^2(x) \le (1-x)(c_{m_0,2q}(\omega_q^-,x))^2\|g\|^2_{L_{2q,\omega_q^-}[-1,1]} =$$

$$(1-x)(c_{m_0,2q}(\omega_q^-,x))^2\|p_q^+\|^2_{L_{q,\omega}[-1,1]} \le$$

$$= c_{n,q}^+(\omega,x)\|p_q^+\|_{L_{q,\omega}[-1,1]}.$$

## 4.4 Various metrics inequalities for polynomials

In this section we will present a series of inequalities which give the estimations for derivatives of algebraic polynomials by norm of polynomials in other space. As in the case of trigonometric polynomials (see sec. 3.4) we pay great attention to the exact results.

Here as in the previous section we will write $\|f\|_C$ and $\|f\|_{L_q^{\alpha,\beta}}$ instead of $\|f\|_{C[-1,1]}$ and $\|f\|_{L_q^{\alpha,\beta}[-1,1]}$ and also $\|f\|_{L_q^\alpha}$ and $\|f\|_{L_q}$ instead of $\|f\|_{L_q^{\alpha,\alpha}}$ and $\|f\|_{L_q^{0,0}}$ relatively.

First of all note the inequality which directly follows from the inequalities (3.45), (3.61) and (3.47).

**Proposition 4.4.1.** *If $q \in [1,\infty]$ $n \in N$ and*

$$c_q = \left(\int_0^\pi |\cos t|^q dt\right)^{1/q},$$

*then for all $p \in P_n$ the inequalities*

$$\|p'\|_{L_q^{(q-1)/2}} \le n\|p\|_{L_q^{-1/2}}; \tag{4.57}$$

$$\|p'\|_{L_q^{(q-1)/2}} \le c_q n\|p\|_C \tag{4.58}$$

*and*

$$\|p'\|_{L_1} \le 2c_q^{-1}\|p\|_{L_q^{-1/2}} \tag{4.59}$$

*hold true. These inequalities become the equalities if and only if* $p(x) = aT_n(x)$*, where* $T_n(x)$ *is a Chebyshev's polynomial of first kind.*

We will prove, for example, the inequality (4.57). If $p \in P_n$, then $\tau(t) = p(\cos t)$ is an even trigonometric polynomial from $T_{2n+1}$. Therefore in view of (3.45)

$$\int_0^\pi |\tau'(t)|^q dt \leq n \int_0^\pi |\tau(t)|^q dt \tag{4.60}$$

and inequality becomes an equality if and only if $\tau(t) = a\cos nt$. If we write (4.60) in the form

$$\int_0^\pi |p'(\cos t)|^q (\sin t)^{q-1} d(-\cos t) \leq n \int_0^\pi |p(\cos t)|^q (\sin t)^{-1} d(-\cos t)$$

and making the change of variable $x = \cos t$ we immediately obtain the inequality (4.57).

**Proposition 4.4.2.** *If* $\alpha, \beta > -1$, $k \in N$, $k \leq n$ *then for any polynomial* $p \in P_n$ *we have*

$$\|p_{(k)}\|_{L_2^{\alpha+k,\beta+k}} \leq \eta_{n,k}(\alpha,\beta)\|p\|_{L_2^{\alpha,\beta}}, \tag{4.61}$$

*where*

$$\eta_{n,k}(\alpha,\beta) = \|(\tilde{J}_n^{\alpha,\beta})(k)\|_{L_2^{\alpha+k,\beta+k}} = \left(\frac{\Gamma(n+1)\Gamma(n+k+\gamma)}{\Gamma(n+1-k)\Gamma(n+\gamma)}\right)^{1/2}, \tag{4.62}$$

$$\gamma = \alpha + \beta + 1$$

*and* $\tilde{J}_n^{\alpha,\beta}$ *is orthonormed in* $L_2^{\alpha,\beta}$ *Jackoby's polynomials.*

*In particular*

$$\int_{-1}^1 (p^{(k)}(x))^2 (1-x^2) dx \leq \frac{(n+k)!}{(n-k)!} \int_{-1}^1 p^2(x) dx$$

*and*

$$\int_{-1}^1 (p^{(k)}(x))^2 \sqrt{1-x^2} dx \leq \frac{n(n+1-k)!}{n-k} \int_{-1}^1 \frac{p^2(x)}{\sqrt{1-x^2}} dx.$$

*Inequality becomes an equality if and only if* $p(x) = a\tilde{J}_n^{\alpha,\beta}(x)$.

**Proof.** If $p \in P_n$, then

$$p(x) = \sum_{\nu=0}^{n} c_\nu \tilde{J}_n^{\alpha,\beta}(x)$$

and

$$p^{(k)}(x) = \sum_{\nu=k}^{n} c_\nu (\tilde{J}_n^{\alpha,\beta}(x))^{(k)}.$$

It follows from these inequalities, theorem 2.5.1 that

$$||p^{(k)}||^2_{L_2^{\alpha+1,\beta+1}} = \sum_{\nu=k}^{n} c_\nu^2 ||(\tilde{J}_n^{\alpha+1,\beta+1}(x))^{(k)}||^2_{L_2^{\alpha+k,\beta+k}} =$$

$$\sum_{\nu=k}^{n} c_\nu^2 \frac{\Gamma(n+1)\Gamma(n+k+\gamma)}{\Gamma(n+1-k)\Gamma(n+\gamma)} \leq \sum_{\nu=k}^{n} c_\nu^2 \eta_{\nu,k}^2 \leq$$

$$\eta_{n,k}^2 \sum_{\nu=k}^{n} c_\nu^2 \leq \eta_{n,k}^2 ||p||^2_{L_2^{\alpha,\beta}},$$

since $\eta_{\nu,k} < \eta_{n,k}$, $\nu = k, \ldots, n-1$ and this inequality becomes an equality if and only if $c_k = 0$, $k = 0, 1, \ldots, n-1$.

**Proposition 4.4.3.** *Let $\alpha, \beta > -1$, $k \in N$, $k \leq n$. Then for any polynomial $p \in P_n$ holds true the next exact on $P_n$ inequality*

$$||p(k)||_C \leq \theta_{n,k}(\alpha,\beta) ||p||_{L_2^{\alpha,\beta}}, \tag{4.63}$$

*where*

$$\theta_{n,k}(\alpha,\beta) = \max\{\theta^+_{n,k}(\alpha,\beta); \theta^-_{n,k}(\alpha,\beta)\}$$

*and $\theta^+_{n,k}(\alpha,\beta)$ and $\theta^-_{n,k}(\alpha,\beta)$ are defined by means of equality (2.93).*

( As to constants $\theta_{n,1}(\alpha,\beta)$ and $\theta_{n,k}(\alpha,\beta)$ they were calculated in lemma 2.5.8 and corollary 2.5.11.)

**Proof.** If $p \in P_n$ then

$$p(x) = \sum_{\nu=0}^{n} c_\nu \tilde{J}_n^{\alpha,\beta}(x)$$

and, consequently,

$$|p^{(k)}(x)| = |\sum_{\nu=k}^{n} c_\nu (\tilde{J}_n^{\alpha,\beta}(x))^{(k)}| \leq \left(\sum_{\nu=k}^{n} ((\tilde{J}_n^{\alpha,\beta}(x))^{(k)})^2\right)^{1/2} ||p||_{L_2^{\alpha,\beta}}.$$

To complete the proof of proposition we must use the corollary 2.5.7, lemma 2.5.8 and corollary 2.5.11.

We present some partial cases of the inequality (4.63):

$$||p||_C \leq \sqrt{\frac{2\pi}{2n+1}}||p||_{L_2},$$

$$||p'||_C \leq \sqrt{\frac{n^2+2n+6}{6}}||p||_{L_2}.$$

The main result of this section gives us the next statement.

**Theorem 4.4.4.** *Let $\Phi(t)$ be an arbitrary $N$-function. Then for any polynomial $p \in P_n$ holds true the exact on $P_n$ inequality*

$$\int_{-1}^{1} \Phi(|p'(x)|)dx \leq \int_{-1}^{1} \Phi(||p||_C|T_n'(x)|)dx, \tag{4.64}$$

*and for any polynomial $p \in P_n$ such that $p(\pm 1) = 0$ the exact inequality holds true:*

$$\int_{-1}^{1} \Phi(|p'(x)|)dx \leq \int_{-1}^{1} \Phi\left(\cos\frac{\pi}{2n}||p||_C\left|T_n'\left(x\cos\frac{\pi}{2n}\right)\right|\right)dx. \tag{4.65}$$

*The inequality (4.66) becomes an equality if and only if*

$$p(x) = aT_n(x), \tag{4.66}$$

*and inequality (4.66) becomes an equality if only if*

$$p(x) = aT_n\left(x\cos\frac{\pi}{2n}\right). \tag{4.67}$$

*In particular for all $q \in [1, \infty]$ and for $p \in P_n$*

$$||p'||_{L_q} \leq ||T_n'||_{L_q}||p||_C \tag{4.68}$$

*and for $p \in P_n$ such that $p(\pm 1) = 0$*

$$||p'||_{L_q} \leq \left\|\cos\frac{\pi}{2n}T_n'\left(\cdot\cos\frac{\pi}{2n}\right)\right\|_{L_q}||p||_C. \tag{4.69}$$

*Besides that, it follows from (4.66), (4.67) and from theorem 1.3.11 that for $p \in P_n$ on the interval $[-1, 1]$*

$$p' \prec ||p||_C T_n' \tag{4.70}$$

*and if $p(\pm 1) = 0$ then*

$$p'(\cdot) \prec ||p||_C \cos\frac{\pi}{2n}T_n'\left(\cdot\cos\frac{\pi}{2n}\right). \tag{4.71}$$

**Proof.** Denote by $P_n^*$ the set of all algebraic polynomials $p \in P_n$ such that

$$p(1) = |p(-1)| = ||p||_C$$

and at every point $\theta \in [-1,1]$ such that $p'(\theta) = 0$ the equality $|P(\theta)| = 1$ is valid.

By $P_n^{*,0}$ we denote the set of all algebraic polynomials $p \in P_n$ such that $p(1) = p(-1) = 0$, $p(1+0) > 0$, $||p||_C = 1$ and at every point $\theta$ such that $p'(\theta) = 0$ the equality $|p(\theta)| = 1$ is valid.

It is clear that if $p \in O_n^*$ or $p \in P_n^{*,0}$ then all zeros of the derivative $p'(x)$ belonging to interval $(-1,1)$ are the points of alternans of the polynomial $p(x)$. It is obvious also that $T_n \in P_n^*$ and

$$T_{n,*} = T_n\left(\cdot \cos\frac{\pi}{2n}\right) \in P_n^{*,0}.$$

Denote by $\theta_k$, $k = 1,2,\ldots,n$ the points of alternans of Chebyshev's polynomial $T_n(x)$. As we already noticed in section 4.1

$$T_n(\theta_k) = (-1)^{n-k}, \; k = 0,1,\ldots,n.$$

By $x_k$, $k = 0,1,\ldots,m$ we denote the points of alternance of the polynomial $p \in P_n^*$. Since $p(1) = p(x_m) = 1$, then

$$p(x_k) = (-1)^{m-k} \quad k = 0,1,\ldots,m.$$

Note that if $m = n$, then $p(x) \equiv T_n(x)$.

Consider the partition of segment [-1, 1]]

$$I_0 = [x_0,x_1],\ldots,I_i = [x_i,0], \; I_{i+1} = [0,x_{i+1}],\ldots,I_m = [x_{m-1},x_m]$$

and define the points $t_1$ and $t_2$ by means of conditions

$$t_1 \in [\theta_i,\theta_{i+1}], \; T_n(t_1) = p(0),$$

$$t_2 \in [\theta_{i+n-m},\theta_{i+n+1-n}], \; T_n(t_2) = p(0)$$

and let

$$J_0 = [\theta_0,\theta_1],\ldots,J_i = [\theta_i,t_1],$$
$$J_{i+1} = [t_2,\theta_{i+n+1-m}],\ldots,J_m = [\theta_{n-1},\theta_n].$$

Further we'll distinguish next supplementary statement.

**Lemma 4.4.5.** *Let $p \in P_n^*$ and points $\xi$ and $\eta$ are defined by the conditions $\xi \in J_k$, $\eta \in I_k$, $k \geq i+1$, $|p(\eta)| \leq 1$ and $P(\eta) = T_n(\xi)$. Then for any $\alpha \leq 1/2$ the inequality*

$$(1-\eta^2)|p'(\eta)| \leq (1-\xi^2)^\alpha |T_n'(\xi)| \tag{4.72}$$

*holds true and this inequality becomes an equality if and only if $p(x) = T_n(x)$.*

**Proof.** It had been noted that if $m = n$,then $p(x) \equiv T_n(x)$. So it is sufficiently to consider only the case $m < n$ and show that in this case the inequality (4.72) is strict.

First we proved that $x_k < \theta_{n+k-m}$, $k = i+1, \ldots, m-1$. Let first $k = m-1$. If $x_{m-1} \geq \theta_{n-1}$ then on the segment $[\theta_n - 1, \theta_n]$ the difference $p(x) - T_n(x)$ has no less three zeros (if we take into account their muliplicities). But on the segment $I = [-1, 1]$ this difference has at most $n+1$ zeros (with account of their multiplicities) but that is impossible. So $x_{m-1} < \theta_{n-1}$. Further we'll use the method of induction. Let $x_k < \theta_{n+k-m}$ for $k = \nu, \ldots, m-1$ and $x_{\nu-1} \geq \theta_{n+\nu-1-m}$. Then on the segment $[\theta_{n+\nu-1-m}, \theta_{n+\nu-m}]$ the difference $p(x) - T_m(x)$ has at least two sign changes or at least four zeros with account of their multiplicities (if $x_{\nu-1} = \theta_{n+\nu-1-m}$). But then on all interval $I$ this difference has at least than $n+1$ zeros (calculated with their multiplicities) that is impossible.

So we proved that if $p(x) \neq T_n(x)$, $p \in P_n^*$ then

$$x_k < \theta_{n+k-m}, \quad k = i+1, \ldots, m-1. \tag{4.73}$$

Prove now that $\eta < \xi$. Really, if it is not so then in view of (4.74) the difference $p(x) - T_n(x)$ has at most two zeros inside the interval $J_k$ and then on all interval $I$ it has at least $n+1$ zeros (calculated with their multiplicities) what is impossible.

Thus $0 < \eta < \xi < 1$ and consequently, for any $\alpha < 1/2$

$$(1-\eta^2)^{\alpha-1/2} < (1-\xi^2)^{\alpha-1/2}. \tag{4.74}$$

Consider the trigonometrical polynomial $\tau(t) = p(\cos t)$. From obvious equalities

$$T_n(x) = \cos(n \, arc\cos x),$$

and

$$p(x) = \tau(arc\cos x)$$

it follows that

$$(1-\xi^2)^\alpha T_n'(\xi) = n(1-\xi^2)^{\alpha-1/2} \sin(n \, arc\cos \xi)$$

and

$$(1-\eta^2)^\alpha p'(\eta) = -(1-\eta^2)^{\alpha-1/2}\tau'(arc\cos\eta),$$

and since

$$\cos(n\ arc\cos\xi) = T_n(\xi) = p(\eta) = \tau(arc\cos\eta),$$

then, in view of lemma 3.1.1 and proposition 1.4.2

$$|n\sin(n\ arc\cos\xi)| > |\tau'(arc\cos\eta)|$$

and consequently, in view of (4.74) for $\alpha \le 1/2$

$$(1-\eta^2)^\alpha|p'(\eta)| < (1-\xi^2)^\alpha|n\sin(n\ arc\cos\xi)| = (1-\xi^2)\alpha|T_n'(\xi)|.$$

Denote by $s_k,\ k=1,\ldots,n-1$ the points of alternans of the polynomial $T_{n,*}$ and let $s_0=-1,\ s_n=1$. By $y_k,\ k=1,\ldots,m-1$ we denote the points of alternans of the polynomial $p\in P_n^{*,0}$ and let $y_0=-1,\ y_m=1$.

Consider the partition of the interval $I=[-1,1]$

$$Q_0=[y_0,y_1],\ldots,Q_\nu=[y_\nu,0],$$
$$Q_{\nu+1}=[0,y_{\nu+1}],\ldots,Q_m=[y_{m-1},y_m],$$

Define the points $t_3$ and $t_4$ by the equalities

$$t_3\in[y_\nu,y_{\nu+1}],\ T_{n,*}(t_3)=p(0),$$
$$t_4\in[s_{\nu+n-m},s_{\nu+n+1-m}],\ T_{n,*}(t_4)=p(0)$$

and set

$$G_0=[s_0,s_1],\ldots,G_\nu=[s_\nu,t_3],$$
$$G_{\nu+1}=[t_4,s_\nu+1+n-m],\ldots,G_m=[s_{n-1},s_n].$$

Almost word for word repeating the proof of lemma 4.4.5 we establish the justice of the next statement.

**Lemma 4.4.6.** *Let $p\in P_n^{*,0}$ and points $\xi$ and $\eta$ are defined by the conditions $\xi\in G_k,\ \eta\in Q_k,\ k\ge\nu+1,\ |p(\eta)|<1$ and $p(\eta)=T_{n,*}$. Then for any $\alpha\le 1/2$*

$$(1-\eta^2)^\alpha|p'(\eta)| \le (1-\xi^2)^\alpha|T_{n,*}'(\xi)|.$$

*This inequality becomes an equality if and only if $p(x)=T_{n,*}(x)$.*

Prove now that on the interval $[0,1]$ for any polynomial $p\in P_n^*$ the inequality

$$(p')_\pm \prec (T_n')_\pm \tag{4.75}$$

holds true, i.e.

$$\int_0^x r_0((p')_+,t)dt \le \int_0^x r_0((T_n')_+,t)dt \tag{4.76}$$

and

$$\int_0^x r_0((p')_-,t)dt \le \int_0^x r_0((T_n')_-,t)dt \tag{4.77}$$

where $r_0(f,t)$ is an equimeasurable decreasing rearrangement (see sec. 1.3) of the narrowing of the function $f$ on $[0,1]$.

First of all we note that if

$$\delta(x) = \int_0^x (r_0((T_n')_+,t) - r_0((p')_+,t))dt$$

in view of (1.36) for all polynomial $p \in P_n^*,\ \ p \ne T_n$

$$\delta(1) = \int_0^1 (r_0((T_n')_+,t) - r_0((p')_+,t))dt =$$

$$\beta(n) - \int_0^1 (p')_+(t)dt \ge \beta(m+1) - \int_0^1 (p')_+(t)dt \ge 0,$$

where $\beta(1) = 1,\ \beta(2) = \beta(3) = \beta(4) = 2,\ \beta(5) = 3,\ \beta(6) = \beta(7) = \beta(8) = 4,\ \beta(9) = 5,\ \beta(11) = \beta(12) = 6,\ \beta(13) = 7, \ldots$. Therefore if the inequality (4.76) is not true then there exists a point $y \in (0,1)$ such that $\delta'(y) = 0$ and $\delta(y) < 0$, i.e. such that

$$A + r_0((p')_+,y) = r_0((T_n')_+,y) \tag{4.78}$$

and

$$\int_0^y r_0((p')_+,t)dt \ge \int_0^y r_0((T_n')_+,t)dt. \tag{4.79}$$

Put now

$$E = \{x :\ x \in (0,1),\ p'(x) > A\}$$

and denote by $(\alpha_\nu, \beta_\nu)$, $\nu = 1, \ldots, l$ the composing intervals of the set $E$. Then

$$\int_0^y r_0((p')_+,t)dt = \sum_{\nu=1}^{l} \int_{\alpha_\nu}^{\alpha_\nu} p'(t)dt = \sum_{\nu=1}^{l} (p(\beta_\nu) - p(\alpha_\nu))$$

and

$$mes\ E = \sum_{\nu=1}^{l} (\beta_\nu - \alpha_\nu) = y.$$

It is clear that every interval $(\alpha_\nu, \beta_\nu)$, $\nu = 1, \dots, l$ is the subset of one of the intervals $I_k$, $k = i+1, \dots, m$.

Choose points $\alpha_\nu^*$ and $\beta_\nu^*$ from corresponding intervals $J_k$ as follows

$$p(\alpha_\nu) = T_n(\alpha_\nu^*),\ p(\beta_\nu) = T_n(\beta_\nu^*).$$

Then in view of lemma 4.4.5 for $\alpha = 0$

$$\beta_\nu - \alpha_\nu \geq \beta_\nu^* - \alpha_\nu^*$$

and equality is possible if and only if $p(x) = T_n(x)$. Taking into account this fact we using the proposition 1.3.1 obtain

$$\int_0^y r_0((p')_+, t)dt = \sum_{\nu=1}^{l} (T_n(\beta_\nu^*) - T_n(\alpha_\nu^*) =$$

$$\sum_{\nu=1}^{l} \int_{\alpha_\nu^*}^{\beta_\nu^*} T_n'(t)dt = \sum_{\nu=1}^{l} \int_{\alpha_\nu^*}^{\beta_\nu^*} (T_n')_+(t)dt \leq \int_0^{mes E^*} r_0((T_n')_+, t)dt,$$

where

$$mes\ E^* = mes \cup_{\nu=1}^{l} (\alpha_\nu^*, \beta_\nu^*) = \sum_{\nu=1}^{l} (\beta_\nu^* - \alpha_\nu^*) \leq mes\ E = y.$$

Thus

$$\int_0^y r_0((p')_+, t)dt \leq \int_0^y r_0((T_n')_+, t)dt,$$

that contradicts to (4.79).

So inequalities (4.75) are valid. But then in wiev of proposition 1.3.6 for any polynomial $p \in P_n^*$ on the segment $[0,1]$ the following inequality is valid

$$p' \prec T_n'. \tag{4.80}$$

Consider the polynomial

$$q(x) = p(-x)\ sgn\ p(-1).$$

Then $q \in P_n^*$ and if $r_{00}(f, x)$ is the rearrangement of the narrowing on $[-1, 0]$ of the function $f(x)$, then

$$r_0(q', x) \equiv r_{00}(p', x).$$

It follows from this and from (4.80) that on the segment $[0,1]$ inequality (4.80) holds true. From these facts and from the corollary 1.3.12 it follows that on the segment $[-1,1]$

$$p' \prec T_n', \tag{4.81}$$

i.e for any polynomial $p \in P_n^*$ for all $x \in [0,2]$ the inequality

$$\int_0^x r(p',t)dt \le \int_0^x r(T_n',t)dt \tag{4.82}$$

is valid and for any $x > 0$ the inequality becomes an equality if and only if $p(x) = T_n(x)$.

Alalogously it may be proved that if $p \in P_m^{*0}$, then for all $x \in [0,2]$

$$\int_0^x r(p',t)dt \le \cos\frac{\pi}{2n}\int_0^x r\left(T_n'\left(\cdot\cos\frac{\pi}{2n}\right),t\right)dt \tag{4.83}$$

or

$$p' \prec T_{n,*}'. \tag{4.84}$$

Denote by $P_{n,a,b}$ the set of all algebraic polynomials $p \in P_n$ such that $p(-1) = a,\ p(1) = b,\ ||p||_C = 1$ and consider the extremal problem:

$$K_{n,\alpha}(a,b) = \max_{p\in P_{n,a,b}} \int_{-1}^1 \Phi(|p'(x)|(1-x^2)^\alpha\frac{dx}{(1-x^2)^\alpha}. \tag{4.85}$$

The next lemma holds true.

**Lemma 4.4.7.** *Let $\Phi(t)$ be a strictly, continuously differentiable and convex from up on $[0,\infty)$ $N$-function, $\alpha \in (0,1)$ and $P_*$ be a solution of the extremal problem (4.85). Then if $p_*'(\theta) = 0$ and $\theta \in (-1,1)$ then $|p_*(\theta)| < 1$.*

**Proof.** Suppose that the assertion of the lemma is not true, i.e. suppose that there exists a point $y \in (-1,1)$ such that $p_*'(y) = 0$ and $|p_*(y)| < 1$.

Consider the function

$$g(x) = \frac{x^2-1}{x-y}p_*'(x).$$

It is clear that $g \in P_n$. Taking into account that $g(\pm 1) = 0$ and for all points $x \in (-1,1)$ such that $|p_*(x)| = 1$ the function $g(x)$ vanishes, we write

$$||p_* + \varepsilon g||_C = ||p_*||_C + \varepsilon\delta(\varepsilon),$$

where $\delta(\varepsilon) \to 0$ for $\varepsilon \to 0$. So if

$$\sigma_0(\varepsilon) = ||p_* + \varepsilon g||_C,$$

then $\sigma_0'(0) = 0$. Besides that $\sigma_0(0) = 1$. So if

$$p_\varepsilon(x) = \frac{p_*'(x) + \varepsilon g(x)}{\sigma_0(\varepsilon)},$$

then $p_\varepsilon \in P_{n,a,b}$ and

$$\frac{d}{d\varepsilon} p'_\varepsilon(x)|_{\varepsilon=0} = g'(x), \quad p'_\varepsilon(x) = \frac{d}{dx} p_\varepsilon(x). \tag{4.86}$$

Let

$$\sigma(\varepsilon) = \int_{-1}^{1} \Phi(|p'_\varepsilon(x)|(1-x^2)^\alpha) \frac{dx}{(1-x^2)^\alpha}$$

and

$$\psi(t) = \Phi'(t)t - \Phi(t).$$

Then

$$\sigma'(\varepsilon) = \int_{-1}^{1} \Phi'(|p'_\varepsilon(x)|(1-x^2)^\alpha) sgn\ p'_\varepsilon(x) \frac{d}{d\varepsilon} p'_\varepsilon\, dx$$

and in view of (4.86)

$$\sigma'(0) = \int_{-1}^{1} \Phi'(|f(x)|) sgn\, f(x) \left( \frac{f(x)(1-x^2)^{1-\alpha}}{y-x} \right)' dx,$$

where

$$f(x) = p'_*(x)(1-x^2)^\alpha$$

Further

$$\sigma'(0) = \int_{-1}^{1} \Phi'(|f(x)|)|f(x)| \left( \frac{(1-x^2)^{1-\alpha}}{y-x} \right)' dx +$$

$$\int_{-1}^{1} \frac{(1-x^2)^{1-\alpha}}{y-x} d\Phi(|f(x)|) = \int_{-1}^{1} \psi(|f(x)|) \left( \frac{(1-x^2)^{1-\alpha}}{y-x} \right)' dx. \tag{4.87}$$

Since $\Phi(x)$ is continuously differentiable, convex from up and increasing on $[0, \infty)$ function, then

$$\Phi(t) = (0) + \Phi'(\theta)t = \Phi'(\theta)t < \Phi'(t)t, \ \theta \in (0,1),$$

i.e. for all $t \in (0, \infty)$ inequality $\psi(t) > 0$ holds true. Consequently almost everywhere on $(-1, 1)$

$$\psi(|f(x)|) > 0. \tag{4.88}$$

Besides that

$$\left( \frac{(1-x^2)^{1-\alpha}}{y-x} \right)' = \frac{l(y,x)}{(y-x)^2(1-x^2)^\alpha},$$

where

$$l(y,x) = 2(1-\alpha)(x^2 - xy) + 1 - x^2.$$

Since $l(y,x)$ is linear on $y$ and for $x \in [-1,1]$

$$l(1,x) = (1-x)(2x\alpha + 1 - x)$$

and

$$l(-1,x) = (1+x)(1+x-2x\alpha),$$

then $l(y,x) \geq 0$ for all $x,y \in [-1,1]$. But then for $x \neq y,\ x,y \in [-1,1]$

$$\left(\frac{(1-x^2)^{1-\alpha}}{y-x}\right)' > 0.$$

¿From this and from relations (4.88) and (4.89) we obtain $\sigma'(0) > 0$. It means that there exists $\varepsilon_0 > 0$ such that for a polynomial $p_{\varepsilon_0} \in P_{n,a,b}$ the next inequality will hold true

$$\int_{-1}^{1} \Phi(|p'_{\varepsilon_0}(x)|(1-x^2)^\alpha)\frac{dx}{(1-x^2)^\alpha} > \int_{-1}^{1} \Phi(|p'_*(x)|(1-x^2)^\alpha)\frac{dx}{(1-x^2)^\alpha}$$

But this inequality contradicts the fact that polynomial $p_*(x)$ is a solution of the extremal problem (4.85).

Lemma 4.4.7 for $a = b = 0$ can be written in the form as follows:

**Lemma 4.4.8.** *Let $\Phi(t)$ be a continuously differentiable, strictly convex from up $N$-function, $\alpha \in (0,1)$ and $p_*(x)$ is a solution of the extremal problem (4.85) for $a = b = 0$. Then $p_* \in P_n^{*0}$ or $-p_* \in P_n^{*0}$.*

Consider the next extremal problem: to find

$$\max_{p\in P_n,\ \|p\|_C=1} \int_{-1}^{1} \Phi(|p'(x)|(1-x^2)^\alpha)\frac{dx}{(1-x^2)^\alpha}. \tag{4.89}$$

**Lemma 4.4.9.** *Let $\Phi(t)$ be a strictly convex from up, continuously differentiable $N$-function, $\alpha \in (0,1)$ and $p_*(x)$ be a solution of the extremal problem (4.90). Then $|p_*(\pm 1)| = 1$.*

**Proof.** Let the assertion of lemma is not true and, for example, $|p_*(1)| < 1$. Then if

$$g_1(x) = (x+1)p'_*(x)$$

then (see the proof of lemma 4.4.6)

$$\frac{d}{d\varepsilon}\|p_* + \varepsilon g_2\|_C|_{\varepsilon=0} = 0$$

and therefore for the function

$$\sigma_1(x) = \int_{-1}^{1} \Phi\left(\frac{|p'_*(x) + \varepsilon g_1(x)|}{||p'_* + \varepsilon g_1||_C}(1 - x^2)^\alpha\right) \frac{dx}{(1 - x^2)^\alpha}$$

will be valid the equality

$$\sigma'_1(0) = \int_{-1}^{1} \Phi'(|f(x)|)|f(x)| \left(\frac{1 + x}{(1 - x^2)^\alpha}\right)' dx + \int_{-1}^{1} \frac{1 + x}{(1 - x^2)^\alpha} d\Phi(|f(x)|),$$

where the function $f(x)$ is defined by equality (4.87). So

$$\sigma'_1(0) = \int_{-1}^{1} \psi(|f(x)|) \left(\frac{1 + x}{(1 - x^2)^\alpha}\right)' dx =$$

$$\int_{-1}^{1} \psi(|f(x)|) \frac{1 - x + 2x\alpha}{1 - x} \frac{dx}{(1 - x^2)^\alpha} > 0, \tag{4.90}$$

and now to prove the lemma 4.4.9 it remains to repeat the end of the proof of lemma 4.4.6.

¿From lemmas 4.4.7 and 4.4.9 it immediately follows

**Lemma 4.4.10.** *Let the conditions of lemma 4.4.9 are satisfied. Then* $p \in P_n^*$ *or* $-p \in P_n^*$.

**Proof of theorem 4.4.3.** Let $\Phi(t)$ be the strictly convex from up, continuously differentiable $N$-function and $p_*(x)$ be a solution of extremal problem (4.90). Then in view of lemma 4.4.10 $\alpha = 0$ $p_* \in P_n^*$ or $-p \in P_n^*$. From this fact and from the inequality (4.81) (or (4.82)) and from the theotem 1.3.11 it follows that

$$\int_{-1}^{1} \Phi(|p'_*(x)|)dx \leq \int_{-1}^{1} \Phi(|T'_n(x)|)dx.$$

But then for any polynomial $p \in P_n$

$$\int_{-1}^{1} \Phi\left(\frac{|p'(x)|}{||p||_C}\right) dx \leq \int_{-1}^{1} \Phi(|T'_n(x)|)dx.$$

Besides that if $\Phi(t)$ is $N$-function, then for any $a > 0$ $\Phi(at)$ also is $N$-function. From this fact and from the last inequality (if we change $\Phi(t)$ by $\Phi(||p||_C t)$) it follows that for any continuously differentiable, strictly convex from up $N$-function the inequality (4.64) is valid.

The statement of the theorem in general case we obtain by means of the limited passage.

Using the lemma 4.4.8 instead of 4.4.10 and the inequality (4.83) instead of (4.82) we in analogous way obtain the inequality (4.65).

## 4.5 Estimates of derivatives in complex domain

Denote by $P_n^C$ the set of algebraic polynomials of degree $n$ with complex coefficient, i.e. the set of functions of the form

$$p(z) = a_0 z^n + a_1 z^{n-1} + \ldots + a_n, \quad z \in C$$

In this sections we present a series of exact estimations of derivatives of polynomials $p \in P_n^C$ on unit disk $|z| \leq 1$ or on unit circle in the complex plane $C$. First of all we note that the next theorem is valid.

**Theorem 4.5.1.** *Let $n = 1, 2, \ldots$. For any algebraic polynomial $p \in P_n^C$*

$$\max_{|z|\leq 1} |p'(z)| \leq n \max_{|z|\leq 1} |p(z)|. \tag{4.91}$$

*Inequality (4.91) becomes an equality for polynomials of the form $P(z) = az^n$, $a \in C$.*

**Proof.** Inequality (4.91) we can write in the next equivalent form

$$\max_{|z|=1} |p'(z)| \leq n \max_{|z|=1} |p(z)|. \tag{4.92}$$

If $\tau(\theta) = p(e^{i\theta})$, $\theta \in R$ then $\tau \in T_{2n+1}^C$. Besides that, as it is easy to verify,

$$|p'(e^{i\theta})| = |\tau'(\theta)|. \tag{4.93}$$

Using the theorem 3.6.1, we obtain the inequality

$$||p'(e^{i\theta})||_\infty \leq n||p(e^{i\theta})||_\infty.$$

Taking into account (4.93) and applying to polynomial $\tau$ the theorem 3.6.1 we get

$$\max_\theta |p'(e^{i\theta})| = ||\tau'||_\infty \leq n||\tau||_\infty = n \max_\theta |p(e^{i\theta})|.$$

It is equivalent to inequality (4.92).

The fact that for $P(z) = az^n$, $a \in C$ inequality (4.91) becomes an equality is obvious.

Theorem is proved.

¿From the theorem 3.6.2 it immediately follows

**Theorem 4.5.2.** *For any polynomial $p \in P_n^C$, $n = 1, 2, \ldots$ and for all $p \in [1, \infty)$ the inequality*

$$\left(\int_0^{2\pi} |p'(e^{i\theta})|^p d\theta\right)^{1/p} \leq n \left(\int_0^{2\pi} |p(e^{i\theta})|^p d\theta\right)^{1/p} \tag{4.94}$$

*takes place. This one becomes an equality if $P(z) = az^n$, $a \in C$.*

Finally from the theorem 3.6.3.immedietly follows

**Theorem 4.5.3.** *If $n = 1, 2, \ldots$, $0 < p \leq q \leq \infty$ and $q \geq 1$ then*

$$\left(\int_0^{2\pi} |p'(e^{i\theta})|^p d\theta\right)^{1/p} \leq (2\pi)^{1/p - 1/q} n \left(\int_0^{2\pi} |p(e^{i\theta})|^q d\theta\right)^{1/q}. \tag{4.95}$$

*In particular for any $p \in (0, \infty]$*

$$\left(\int_0^{2\pi} |p'(e^{i\theta})|^p d\theta\right)^{1/p} \leq (2\pi)^{1/p} \max_{|z| \leq 1} |p(z)|. \tag{4.96}$$

*Inequalities (4.95) and (4.96) become equalities for polynomials of the form $P(z) = az^n$, $a \in C$.*

Inequalities (4.91) and (4.94)-(4.96) we can make more precise if we assume that a polynomial $p \in P_n^C$ has no zeros inside the disk $|z| \leq 1$.

**Theorem 4.5.4.** *If an algebraic polynomial $p \in P_n^C$ of degree $n$ has all zeros outside the unit disk $|z| < 1$, then*

$$\max_{|z| \leq 1} |p'(z)| \leq \frac{n}{2} \max_{|z| \leq 1} |p(z)|. \tag{4.97}$$

*This inequality becomes an equality for polynomials of the form $p(z) = (\alpha + \beta z^n)/2$ where $|\alpha| = |\beta| = 1$.*

To prove that we need two auxiliary statements.

**Lemma 4.5.5.** *If $p(z)$ is a polynomial of degree $n$, $z_1, z_2, \ldots, z_n$ are zeros of the polynomial $z^n + a$ where $a$ is a complex number different from 0 and from $-1$ then for any complex number $t$*

$$tp'(t) = \frac{n}{1+a} p(t) + \frac{1+a}{na} \sum_{k=1}^{n} p(tz_k) \frac{z_k}{(z_k - 1)^2}. \tag{4.98}$$

**Proof.** Let $t$ be an arbitrary complex number. Consider the function

$$F_t(z) = \frac{p(tz) - p(t)}{z - 1}.$$

Sinse $p(t \cdot 1) - p(t) = 0$ (i.e. $z = 1$ is also a zero of the numerator) then $F_t(z)$ is polynomial of degree $n - 1$. So using the interpolational Lagrange formula we can present $F_t(z)$ in the form

$$F_t(z) = \sum_{k=1}^{n} F_t(z_k) \frac{z^n + a}{n z_k^{n-1}(z - z_k)}.$$

Since $z_k^n + a = 0$ and $z_k \neq 0$, $k = 1, 2, \ldots, n$ then $z_k^{n-1} = -a/z_k$ and consequently,

$$F_t(z) = \sum_{k=1}^{n} F_t(z_k) \frac{z_k(z^n + a)}{na(z_k - z)}. \tag{4.99}$$

Using the fact that $F_t(1) = tp'(t)$ we obtain from (4.99) next identity by $t$

$$tp'(t) = \frac{1+a}{an} \sum_{k=1}^{n} F_t(z_k) \frac{z_k}{z_k - 1} =$$

$$\frac{1+a}{an} \sum_{k=1}^{n} (p(tz_k) - p(t)) \frac{z_k}{(z_k - 1)^2} =$$

$$\frac{1+a}{an} \sum_{k=1}^{n} p(tz_k) \frac{z_k}{(z_k - 1)^2} - p(t) \frac{1+a}{an} \sum_{k=1}^{n} \frac{z_k}{(z_k - 1)^2}. \tag{4.100}$$

Puting in (4.100) $p(z) = t^n$ we obtain

$$\sum_{k=1}^{n} \frac{z_k}{(z_k - 1)^2} = -\frac{an^2}{(a+1)^2}. \tag{4.101}$$

Combining (4.100) and (4.101), we obtain

$$tp'(t) = \frac{1+a}{an} \sum_{k=1}^{n} p(tz_k) \frac{z_k}{(z_k - 1)^2} + p(t) \frac{1+a}{an} \frac{an^2}{(a+1)^2} =$$

$$\frac{1+a}{an} \sum_{k=1}^{n} p(tz_k) \frac{z_k}{(z_k - 1)^2} + p(t) \frac{n}{a+1}$$

and lemma is proved.

**Lemma 4.5.6.** *If $p(z)$ is a polynomial of degree $n$ then for any $z$, , $|z| = 1$*

$$|p'(z)| + |np(z) - zp'(z)| \leq n \max_{|z|=1} |p(z)|. \tag{4.102}$$

**Proof.** Let in the lemma 4.5.5 $a$ is an arbitrary complex number such that $|a| = 1$, $a \neq -1$. Then zeros $z_k$ of the polynomial $z^n + a$ are equal to 1 in absolute magnitude and $z_k \neq 1$, $n = 1, 2, \ldots, n$. So from (4.101) for $|t| = 1$ we obtain

$$|atp'(z) + tp'(t) - np(t)| = |\frac{(1+a)^2}{an} \sum_{k=1}^{n} p(tz_k) \frac{z_k}{(z_k - 1)^2}| \leq$$

$$\left|\frac{(1+a)^2}{an}\right| \sum_{k=1}^{n} \left|\frac{z_k}{(z_k - 1)^2}\right| \max_{|z|=1} |p(z)|. \tag{4.103}$$

Now, if $|z| = 1$ and $z \neq 1$ then $z/(z-1)^2$ is nonnegative real number. Really, it is easy to see that

$$e^{i\theta}/(e^{i\theta} - 1)^2 = -\frac{1}{4} \sin^2 \frac{\theta}{2}, \quad \theta \neq 0 (mod\, 2\pi)$$

and moreover if $|a| = 1$ and $a \neq -1$ then $(1+a)^2/a$ is a positive real number. So (taking into account (4.101)) we obtain

$$\left|\frac{(1+a)^2}{an}\right| \sum_{k=1}^{n} \left|\frac{z_k}{(z_k - 1)^2}\right| = -\frac{(1+a)^2}{an} \sum_{k=1}^{n} \frac{z_k}{(z_k - 1)^2} = n.$$

Consequently, from (4.103) we obtain that for $|t| = 1$, $|a| = 1$ and $a \neq -1$

$$|atp'(z) + tp'(t) - np(t)| \leq n \max_{|z|=1} |p(z)| \tag{4.104}$$

(for $a = -1$ this inequality is trivial).

Choose the argument of $a$ such that

$$|atp'(z) + tp(t) - np(t)| = |p'(z)| + |tp'(t) - np(t)|.$$

Then in view of (4.104) we obtain for $|z| = 1$

$$|p'(z)| + |tp'(t) - np(t)| \leq n \max_{|z|=1} |p(z)|.$$

Lemma is proved.

**Proof oftheorem 4.5.4..** If a polynomial $p(z)$ has no zeros inside $|z| < 1$ then

$$p(z) = c \prod_{j=1}^{n} (z - w_j)$$

where $|w_j| \geq 1$, $j = 1, 2, \ldots$. For points $e^{i\theta}$, $0 \leq \theta < 2\pi$ which are different from zeros of $p(z)$ we have

$$Re \left( \frac{e^{i\theta} p'(e^{i\theta})}{np(e^{i\theta})} \right) = \frac{1}{n} \sum_{j=1}^{n} Re \left( \frac{e^{i\theta}}{e^{i\theta} - W_j} \right) \leq$$

$$\frac{1}{n} \sum_{j=1}^{n} \left( \frac{1}{2} \right) = \frac{1}{2}.$$

It follows from this that

$$\left| Re \left( \frac{e^{i\theta} P'(e^{i\theta})}{P(e^{i\theta})} \right) \right| \leq \left| Re \left( 1 - \frac{e^{i\theta} P'(e^{i\theta})}{P(e^{i\theta})} \right) \right|$$

and since

$$Im \left( \frac{e^{i\theta} P'(e^{i\theta})}{P(e^{i\theta})} \right) = Im \left( 1 - \frac{e^{i\theta} P'(e^{i\theta})}{P(e^{i\theta})} \right)$$

for points $e^{i\theta}$ which are different from zeros of the polynomial $P(z)$ the next inequality takes place

$$\left| \frac{e^{i\theta} P'(e^{i\theta})}{nP(e^{i\theta})} \right| \leq \left| 1 - \frac{e^{i\theta} P'(e^{i\theta})}{nP(e^{i\theta})} \right|.$$

The last inequality is equivalent to the next inequality

$$|P'(e^{i\theta})| \leq |nP(e^{i\theta}) - e^{i\theta} P'(e^{i\theta})| \tag{4.105}$$

for points $e^{i\theta}$ which are different from zeros of $P(z)$. For points $e^{i\theta}$ which are zeros of $P(z)$ the inequality holds true in a trivial way.

Consequently for $|z| = 1$

$$|P'(z)| \leq |nP(z) - zP'(z)|. \tag{4.106}$$

Comparing the inequality (4.102) with (4.106) we obtain (if $|z| = 1$)

$$z|P'(z)| \leq |P'(z)| + |nP(z) - zP'(z)| \leq n \max_{|z|=1} |P(z)|.$$

Theorem is proved.

Next theorem makes more precise the theorem 4.5.2.

**Theorem 4.5.7.** *If a polynomial $p \in P_n^C$ has no zeros inside $|z| < 1$ then for any $p \geq 1$*

$$\left(\int_0^{2\pi} |P'(e^{i\theta})|^p \, d\theta\right)^{1/p} \leq C_p \left(\int_0^{2\pi} |P(e^{i\theta})|^p \, d\theta\right)^{1/p} \tag{4.107}$$

*where the constant $C_p$ is such that*

$$C_p^p = \frac{2\pi}{\int_0^{2\pi} |1 + e^{i\eta}| \, d\eta} = \frac{\sqrt{\pi}}{2^p} \frac{\Gamma(1 + p/2)}{\Gamma((1 + p)/2)}.$$

*The inequality (4.107) becomes an equality if $p(z) = \alpha + \beta z^n$, $|\alpha| = |\beta| = 1$.*

**Proof.** Let $p \in P_n^C$. The inequality (4.101) we can write in the form

$$\max_\theta \left| np(e^{i\theta}) - e^{i\theta} p'(e^{i\theta}) - e^{i(\theta+\eta)} p'(e^{i\theta}) \right| \leq n \max_\theta \left| p(e^{i\theta}) \right|.$$

¿From this as in the proof of the theorem 3.6.2 we deduce that for all $p \geq 1$

$$\int_0^{2\pi} \left| np(e^{i\theta}) - e^{i\theta} p'(e^{i\theta}) - e^{i(\theta+\eta)} p'(e^{i\theta}) \right|^p d\theta \leq$$

$$n^p \int_0^{2\pi} \left| p(e^{i\theta}) \right|^p d\theta. \tag{4.108}$$

Puting

$$A(\theta) = np(e^{i\theta}) - p'(e^{i\theta}) e^{i\theta}$$

and

$$B(\theta) = -e^{i\theta} P'(e^{i\theta})$$

and integrating the both part of (4.108) on variable $\eta$ we obtain

$$\int_0^\pi d\theta \int_0^{2\pi} |A(\theta) + B(\theta) e^{i\eta}|^p \, d\eta \leq 2\pi n^p \int_0^{2\pi} \left| p((e^{i\theta}) \right|^p d\theta. \tag{4.109}$$

If the polynomial $p(z)$ has no zeros inside the disk $|z| < 1$, then in view of the inequality (4.105) we will have

$$|A(\theta)| \leq |B(\theta)| \tag{4.110}$$

for all $\theta \in [0, 2\pi]$. But for arbitrary complex numbers $a$ and $b$ such that $|a| \geq |b|$ the next inequality

$$\int_0^{2\pi} |a + be^{i\eta}|^p \, d\eta \geq |b|^p \int_0^{2\pi} |1 + e^{i\eta}|^p \, d\eta \tag{4.111}$$

holds true (it is sufficiently to prove this inequality for $b = 1$, but in this case it follows from the obvious inequality $\left|a + e^{i\eta}\right| \geq \left|1 + e^{i\eta}\right|$). Comparing the inequalities (4.109)-(4.111), we obtain

$$\int_0^{2\pi} |B(\theta)|^p \, d\theta \int_0^{2\pi} \left|1 + e^{i\eta}\right|^p d\eta \leq 2\pi n^p \int_0^{2\pi} \left|p(e^{i\theta})\right|^p d\theta,$$

that is equivalent to the inequality (4.107).

Theorem is proved.

## 4.6 Commentaries

Sec.4.1. Theorem 4.1.4 has been proved by V.A. Markov [1]. This result excited a great number of investigations directed to the generalization, intensification and making more precise the inequality (1.7).The inequality (1.12) has been proved by A.A. Markov [1], theorem 4.1.1 has been proved by S.N. Bernstein [2]. The row of closed results is given in the section Supplementary results.

Sec.4.2. The possibility of making more precise Markov's inequality for polynomials which have only real zeros being outside of interval $(-1, 1)$, was noticed by P.Erdesh [1].Theorem 4.2.2 belongs to D.T. Scheik [1]. Theorem 4.2.3 was proved by A.K. Varma [1].

Sec. 4.3. The exact in order for $n \to \infty$ estimates of the norm for algebraic polynomials by their norms in another spaces belong to S.M.Nikolski [5].

Sec. 4.4. The main result of this section was proved by B.D.Bojanov [9]. We adduced the proof that is different from proof of Bojanov in order to show the technique of application the rearrangement comparison for functions that are given on finite segment.

## 4.7 Supplementary results

**1.** If $p \in P_n$ and $\cos(\pi/(2n)) \leq |x| \leq 1$, then

$$|p'(x)| \leq \frac{n \cdot \sin(n \, arc \cos x)}{\sqrt{1 - x^2} ||p||_{C[-1,1]}}.$$

Pashkovski S [1, P. 69].

**2.** Let $n \geq 4$, $1 \leq k \leq n$ and

$$B_{n,k}(x) = \sup_{p \in P_n} \frac{|p^{(k)}(x)|}{||p||_{C[-1,1]}}.$$

Then there exist the numbers $c_k$ such that for all $x \in (-1,1)$

$$\left| B_{n,k}(y) - \left(\frac{n}{\sqrt{1-x^2}}\right)^k \right| \leq c_k \frac{n^{k-1/2}}{(\sqrt{1-x^2})^{r_k}},$$

where $r_1 = 3$, $r_2 = 4$, $r_k = 3k - 3$ for $k \geq 3$.

Witly R [1].

**3.** For any polynomial $p \in P_n$ for which $p(1) = 0$ the next inequality

$$||p^{(k)}||_{C[-1,1]} \leq \left(\cos \frac{\pi}{4n}\right)^{2k} \left|T_n^{(k)}(1)\right| \; ||p||_{C[-1,1]}, \quad k = 1, 2, \ldots, n$$

is valid and if $p \in P_n$ and $p(\pm 1) = 0$ then

$$||p^{(k)}||_{C[-1,1]} \leq \left(\cos \frac{\pi}{2n}\right)^k \left|T_n^{(k)}\left(\cos \frac{\pi}{2n}\right)\right| ||p||_{C[-1,1]}, \quad k = 1, 2, \ldots, n.$$

Inequalities are exact.

Pier P. [1].

**4.** If $n \geq 2$

$$x_\nu = \left(\cos \frac{\nu\pi}{n}\right) / \cos\left(\frac{\pi}{2n}\right), \quad \nu = 1, 2, \ldots, n$$

$p \in P_n$ and $p(\pm 1) = 0$ then $|x| \leq (\cos(\pi/(2n)))^{-1}$ and $y \in R$ the exact inequality

$$|p^{(k)}(x + iy)| \leq \left(\cos \frac{\pi}{2n}\right)^k \left|T_n^{(k)}\left(1 + iy \cos \frac{\pi}{2n}\right)\right| \max_{1 \leq \nu \leq n} |p(x_\nu)|$$

holds true.

Frappier K. [1].

**5.** If $n \geq 2$ and $k = 1, 2, \ldots, n$ then for any polynomial $p \in P_n$ for which $p(\pm 1) = 1$ the exact inequality is valid

$$||p^{(k)}||_{C[-1,1]} \leq c_k \left\| \frac{p(\cdot)}{\sqrt{1 - (\cdot)^2}} \right\|_{C[-1,1]},$$

where

$$c_k = \left| \frac{d^k}{dx^k}((1-x^2)U_{n-2}(x))|_{x=1} \right|.$$

Rahman K.I. [4].

**6.** If $n = 1, 2, \ldots, \nu = 0, 1, \ldots, n$ and $p \in P_{2n+1}$ then

$$|p^{(2\nu)}(0)| \leq \frac{2^{2\nu} n(n+\nu-1)!}{(n-\nu)!} \|p\|_{C[-1,1]}$$

and

$$\left| p^{(2\nu+1)}(0) \right| \leq \frac{2^{2\nu}(2n+1)(n+\nu)!}{(n-\nu)!} \|p\|_{C[-1,1]}.$$

Inequalities turn into equalities for $p(x) = aT_{2n+1}(x)$.
Markov V.A. [1].

**7.** For all $p \in P_n$ and $t \in (-1, 1)$

$$|p'(t)| \leq \left( \frac{t^2}{1-t^2} + (n-1)^2 \right)^{1/2} \left\| \frac{p(\cdot)}{\sqrt{1-(\cdot)^2}} \right\|_{C[-1,1]}.$$

The inequality turns into equality for polynomial

$$p(t) = (1-t^2)U_{n-2}(t)$$

at such points from $t \in (-1, 1)$ that

$$(n-1)\sqrt{1-t^2} \tan((n-1)arc\cos t) = t.$$

Moreover

$$|(1-t^2)p''(t) - 2t/, p'(t)| \leq n\sqrt{\frac{t^2}{1-t^2} + n^2} \, \|p\|_{C[-1,1]}.$$

Rahman K.I. [4].

**8.** If $p \in P_n$ and $p(0) = 0$ then

$$\|p'\|_{C[-1,1]} \leq ((n-1)^2 + 1)\|p(t)/t\|_{C[-1,1]}.$$

The example $p(t) = t\,T_{n-1}(t)$ shows us that this inequality is exact.
Rahman K.I. [4].

**9.** If all zeros of a polynomial $p \in P_n$ don't lie in the disk $\{z : |z| < 1\}$ then for any $\alpha > 0$ the next exact inequality is valid

$$||p||_{C[-1,1]} \leq K_{n,\alpha} ||p(\cdot)(1-(\cdot)^2)^\alpha||_{C[-1,1]},$$

where

$$K_{n,\alpha} = (n+2\alpha)^{n+2\alpha}(4\alpha)^{-\alpha}(n+\alpha)^{-n-\alpha}.$$

Erdelei T., Sabadosh J. [1].

**10.** There exists a constant $c_r$ such that for any polynomial $p \in P_n$ all zeros of which don't lie in the disk $\{z : |z| < 1\}$ the next inequality

$$||p^{(r)}||_{C[-1,1]} \leq c_r n^r ||p||_{C[-1,1]}$$

holds true.

Lorentz G.G. [3].

**11.** There exist absolute constants $0 < c_1 < c_2$ such that for all $n \in N$, $k = 1, 2, \ldots, n$ and any polynomial $p \in P_n$

$$c_1 \frac{n\,T_n^{(k)}(1)}{(k+1)(n+1-k)} \leq \sup_{p \in P_n,\, p \not\equiv 0} \frac{||p^{(k)}||_{L_1[-1,1]}}{||p||_{L_1[-1,1]}} \leq c_2 \frac{n\,T_n^{(k)}(1)}{(k+1)(n+1-k)}.$$

Konyagin S.V. [2].

**12.** If $t \in [-1, 1]$ and $p \in P_n$ then exact inequality

$$||p||_{C[-1,1]} \leq T_n\left(\frac{2+t}{2-t}\right) r(p,t)$$

holds true.

Pashkovcki S. [1. P. 68].

**13.** If a polynomial $p \in P_n$ has not zeros in the interval $(-1, 1)$ then

$$\int_{-1}^{1} (1-x^2)(p'(x))^2 dx \leq \frac{n(n+1)(2n+3)}{4(2n+1)} \int_{-1}^{1} (1-x^2)(p(x))^2 dx$$

and if also $p(\pm 1) = 0$ then

$$\int_{-1}^{1} (p'(x))^2 dx \leq \frac{n(2n+1)(n-1)}{4(2n-3)} \int_{-1}^{1} (p(x))^2 dx.$$

The first inequality turns into equality for

$$p(x) = a(1 \pm x)^n$$

and second inequality turns into equality for

$$p(x) = a(1-x)(1+x)^{n-1}.$$

Erdesh P., Varma A.K. [1].

**14.** If $\alpha > 1$, $p \in P_n$, $p^{(k)}(0) \geq 0$, $k = 1, 2, \ldots, n$ then for $m = 1, 2, \ldots, n$

$$\int_0^\infty x^\alpha e^{-x} (p^{(m)}(x))^2 dx \leq c_{n,m}(\alpha) \int_0^\infty x^\alpha e^{-x} (p(x))^2 dx,$$

where

$$c_{n,m}(\alpha) = \begin{cases} m!(2m+\alpha)^{-2m}, & -1 < \alpha \leq \alpha_{n,m}, \\ n^2(n-1)^2 \ldots (n-m+1)^2 (2n+\alpha)^{-2m}, & \alpha \geq \alpha_{n,m}, \end{cases}$$

and $\alpha_{n,m}$ is unique positive root of the next equation

$$(2n+\alpha)^{2m} = \left(\frac{n}{m}\right)^2 (2m+\alpha)^{2m}.$$

Inequality is exact.

Milovanovich G., Milovanovich I. [1].

**15.** Let $V$ is a convex set in a space $R^m$, $\omega$ is arbitrary measurable subset $V$ and $\lambda = mes\, \omega / mes\, V$. Then for any algebraic polynomial $p(x)$ of degree $n$ of $m$ variables with real coefficients the next exact inequality is valid

$$\max_{x \in V} |p(x)| \leq b_n(\lambda) \max_{x \in \omega} |p(x)|,$$

where

$$b_n(\lambda) = T_n \left( \frac{1 + (\sqrt{1-\lambda})^{1/n}}{1 - (\sqrt{1-\lambda})^{1/n}} \right).$$

Moreover for all $0 < s < q \leq \infty$ the next inequality holds true

$$\left( \int_V |p(x)|^q dx \right)^{1/q} \leq$$

$$c \frac{(mes\, V)^{1/q}}{(mes\, \omega)^{1/s}} (n+1)^{2m(1/s-1/q)} b_n(\lambda) \left( \int_\omega |p(x)|^s dx \right)^{1/s},$$

where $c \leq 8^{1-s/q}$.

Ganzburg M.I. [3].

**16.** Let $\Delta$ is triangle domain in $R^2$,

$$p_k(x) = \sum_{|s| \le k} c_s x^s,$$

where $x^s = x_1^{s_1} x_2^{s_2}$, $|s| = s_1 + s_2$ and $D_\xi p_k(x)$ is derivative on way $\xi$ of the polynomial $p_k(x)$. Then

$$\frac{2k^2}{h} \le \sup_{\xi} \sup_{p_k \not\equiv 0} \frac{||D_\xi p_k||_{C(\Delta)}}{||p_k||_{C(\Delta)}} \le \frac{4k^2}{h},$$

where $h$ ??.

Madzhmiddinov D., Subbotin Y.N. [1].

**17.** For any polynonial $p \in P_n^C$ the exact inequality

$$\max_{|z| \le 1} |p'(z)| \le n \max_{|z| \le 1} |Re\, p(z)|$$

holds true.

Szego G. [2].

**18.** If a polynomial $p \in P_n^C$ has not zeros in the disk $\{z : |z| le 1\}$ then for all $0 < q < 1$ the exact inequality

$$\int_0^{2\pi} |p'(e^{it})|^q dt \le \int_0^{2\pi} |1 + e^{it}|^q dt \int_0^{2\pi} |p(e^{it})|^q dt$$

is valid.

Rahman K.I., Schmeisser G.O. [2].

**19.** If a polynomial $p \in P_n^C$ has not zeros in the disk $\{z : |z| le k\}$ , $k > 1$ and

$$p(z) = c_0 + \sum_{\nu=\mu}^{n} c_\nu z^\nu$$

then

$$\max_{|z|=1} |p'(z)| \le \frac{n}{k+1} \max_{|z|=1} |p(z)|.$$

If $n$ is multiply to $\mu$ then this inequality turns into equality for $p(z) = (z^\mu + k^\mu)^{n/\mu}$.

Chan T.N., Malik M.A. [1].

**20.** If $E_{a,b}$ is ellips having the semiaxis $a$ and $b$, $a > b$ then for any polynomial $p \in P_n^C$

$$\max_{z \in E_{a,b}} |p'(z)| \leq \frac{n}{b} \max_{z \in E_{a,b}} |p(z)|.$$

Sewell W.E. [1].

**21.** If $E_{a,b}$ is ellips with the focuses $(-1,0)$ and $(0,1)$ and having the sum of semiaxises $R > 1$ then for any polynomial $p \in P_n^C$ the exact inequality

$$\max_{z \in E_{a,b}} |p'(z)| \leq \frac{1}{2}(R^n + R^{-n})\|p\|_{C[-1,1]}.$$

is valid.

Bernstein S.N., Daffin R.J., Scheffer A.C. [2].

**22.** If $z_k$, $k = 1, 2, \ldots, n$ are zeros of the polynomial $z^n + 1$, then for any polynomial $p \in P_n^C$ for which $p(\beta) = 0$ the next exact inequality is valid

$$|p'(\beta)| \leq \frac{n}{2\beta} \max_{1 \leq k \leq n} |p(\beta z_k)|.$$

In particular, if $\beta = 1$, then

$$|p'(1)| \leq \frac{n}{2} \max_{1 \leq k \leq n} |p(z_k)|,$$

and inequality becomes an equality for $p(z) = z^n - 1$.

Aziz A., Mahammad Q.G. [1], Rahman K.I., Mahammad Q.G. [1].

**23.** If $z_k$, $k = 1, 2, \ldots, n$ are zeros of the polynomial $z^n + 1$, then for any polynomial $p \in P_n^C$ for which $p(\beta) = 0$, $\beta > 0$ the next inequality is valid

$$|p'(\beta)| \leq \frac{1 - beta^n}{1 - \beta^2} \max_{1 \leq k \leq n} |p(z_k)|.$$

For $\beta = 0$ and $\beta = 1$ this inequality is exact.

Azis A. [3].

**24.** If $z_k$, $k = 1, 2, \ldots, n$ are zeros of the polynomial $z^n + 1$, then for any polynomial $p \in P_n^C$ for which $p(\beta) = 0$, $\beta > 0$ the next inequality is valid

$$\frac{1}{2\pi} \int_0^{2\pi} \left| \frac{p(e^{i\theta})}{e^{i\theta - \beta}} \right|^2 d\theta \leq \frac{1 - beta^n}{1 - \beta^2} \max_{1 \leq k \leq n} |p(z_k)|^2$$

and inequality becomes an equality for $p(z) = z^n - \beta^n$.

Azis A. [3].

**25.** For any polynomial $p \in P_n^C$ such that $p(b) = 0$ the next inequalities are valid

$$\left(1 + b^2 - 2b\cos\frac{\pi}{n+1}\right)\int_0^{2\pi}\left|\frac{p(e^{it})}{e^{it}-b}\right|^2 dt \le \int^{2\pi}\left|p(e^{it})\right|^2 dt \le$$

$$\left(1 + b^2 + 2b\cos\frac{\pi}{n+1}\right)\int_0^{2\pi}\left|\frac{p(e^{it})}{e^{it}-b}\right|^2 dt.$$

The left inequality becomes an equality for

$$p(x) = c(x-b)\sum_{k=1}^{n} x^{k-1}\sin\frac{k\pi}{n+1}$$

and right does - for

$$p(x) = c(x-b)\sum_{k=1}^{n}(-x)^{k-1}\sin\frac{k\pi}{n+1}.$$

Milovanovich I. [1].

**26.** If all zeros $z_1, z_2, \ldots, z_n$ of a polynomial $p \in P_n^C$ are in domain $D = \{z \in C : Re\, z \ge 0,\ |z| \ge 1\}$, then

$$\max_{|z|=1}|p'(z)| \le \sum_{\nu=1}^{n}\frac{1}{1+|z_\nu|}\max_{|z|=1}|p(z)|.$$

Giroux A., Rahmam Q.I. [1].

**27.** If all zeros $z_1, z_2, \ldots, z_n$ of a polynomial $p \in P_n^C$ are in domain $D = \{z \in C :\ |z| \ge 1\}$, then in condition that $|p(z)|$ and $|p'(z)|$ attain maximum at the same point of the circle $|z| = 1$

$$\max_{|z|=1}|p'(z)| \le \sum_{\nu=1}^{n}\frac{1}{1+|z_\nu|}\max_{|z|=1}|p(z)|.$$

Aziz A. [2].

# Chapter 5

# Turan's type inequalities

In the chapters 3 and 4 were proved inequalities that estimate from above norms of derivatives of trigonometric and algebraic polynomials. Nontrivial estimations from below of norm of derivatives by norms of polynomials themselves are in general impossible. For instance, on the class $P_n$ of all algebraic polynomials $p$ of $n$-th degree, $n = 1, 2, \ldots$, inequality

$$||p'||_{C[-1,1]} \geq A \, ||p||_{C[-1,1]}$$

is impossible for all constants $A > 0$.

On the other hand, for algebraic polynomials that vanish only on the segment $[-1, 1]$ and for trigonometric polynomials that have only real zeros nontrivial estimations of this type are possible. They are the matter of this chapter.

## 5.1 The exact Turan's inequality

P.Turan proved that for algebraic polynomials $p(x)$ of degree $n$, having all zeros inside $[-1, 1]$, the inequality

$$||p'||_{C[-1,1]} \geq \frac{\sqrt{n}}{6} \, ||p||_{C[-1,1]} \tag{5.1}$$

holds true and the order of growht of the constants $\sqrt{n}\ /6$ in this inequality is exact when $n \to \infty$ .

The next theorem presents the exact inequality of type (5.1) for any fixed $n$ .

**Theorem 5.1.1** *If any algebraic polynomial p of degree n, $n = 1, 2, \ldots$, has all zeros inside* $[-1, 1]$ *then*

$$\|p'\|_{C[-1,1]} \geq A_n \|p\|_{C[-1,1]} \tag{5.2}$$

*where*

$$A_n = \begin{cases} \frac{n}{2}, \text{ if } n = 2, 3; \\ \frac{n}{\sqrt{n-1}}(1 - \frac{1}{n-1})^{(n-2)/2}, \text{ if } n = 4, 6, 8, \ldots; \\ \frac{n^2}{(n-1)\sqrt{n+1}}(1 - \frac{\sqrt{n+1}}{n-1})^{(n-3)/2}(1 + \frac{1}{\sqrt{n+1}})^{(n-1)/2}, \\ \text{if } n = 5, 7, 9, \ldots, \end{cases} \tag{5.3}$$

*and*

$$A_n = \sqrt{\frac{n}{e}} + O\left(\frac{1}{\sqrt{n}}\right) \tag{5.4}$$

*if* $n \to \infty$ .

*The inequality (5.1) becomes an equality for*

$$p(x) = a(1 \pm x)^n$$

*if* $n = 2, 3$ ; *for polynomials*

$$p(x) = a(1 - x)^{(n-1)/2}(1 + x)^{(n+1)/2}$$

*if* $n = 5, 7, 9, \ldots$ ; *and for polynomials*

$$p(x) = a(1 - x^2)^{n/2}$$

*if* $n = 4, 6, 8 \ldots$ .

**Proof.** The proof of the theorem may be obtained in elementary manner for $n = 2$ . For this reason we will assume that $n \geq 3$ . For polynomials of the form

$$p(t) = a(t - t_1)^n, \ a \in R, \ t_1 \in [-1, 1],$$

the inequality

$$\|p'\|_{C[-1,1]} \geq \frac{n}{2}\|p\|_{C[-1,1]} \geq A_n \|p\|_{C[-1,1]} \tag{5.5}$$

holds true and so it is sufficiently to prove the inequality (5.2) for polynomials $p(t)$ having at least two different zeros.

Any polynomial $p \in P_n$ having only real zeros can be written in the form

$$p(t) = a \prod_{k=1}^{n} (t - t_k),$$

where $a \in R$ and

$$t_1 \leq t_2 \leq \ldots \leq t_n. \tag{5.6}$$

Conversely, every set of the points which possesses the conditions (5.6) defines the polynomial

$$p(t) = p(t_1, \ldots, t_n) = \prod_{k=1}^{n} (t - t_k) \tag{5.7}$$

having zeros in these point.

If for the fixed $i$, $2 \leq i \leq n-1$ it will be $t_i < t_{i+1}$, then in the interval $(t_i, t_{i+1})$ there exists a unique point $\xi = \xi_i$ of the local extremum for polynomial $p(t)$ of the form (5.7). Let us denote by $\eta = \eta_i$ the nearest from the left to $\xi_i$ point of the local extremum of the polynomial $p'(t)$ . It is clear that such point exists and is uniquely determined by the set (5.6) with $t_i < t_{i+1}$ . Thus one can define on the set

$$M = \{(t_1, \ldots, t_n)^n) \in R : \ t_1 \leq \ldots \leq t_i \leq t_{i+1} \leq \ldots \leq t_n\} \tag{5.8}$$

the functions $\xi = \xi(t_1, \ldots, t_n)$ and $\eta = \eta(t_1, \ldots, t_n)$. It is not difficult to verify that the functions

$$p(\xi) = p(t_1, \ldots, t_n;\ \xi(t_1, \ldots, t_n))$$

and

$$p'(\eta) = p'(t_1, \ldots, t_n; \eta(t_1, \ldots, t_n))$$

are also continuous on $M$ and the function $p'(\eta)$ does not vanish on $M$ .

For $(t_1, \ldots, t_n) \in M$ let

$$F(t_1, \ldots, t_n) = \frac{|p(\xi)|}{|p'(\eta)|}. \tag{5.9}$$

It is follows from the facts presented above that the function $F$ is continuous on $M$ and as it is easily seen that this function is positive on $M$ .

Let $M^* = \{(t_1, \ldots, t_n) \in M : \ -1 \leq t_1, t_n \leq 1\}$ . Consider the extremal problem

$$F(t_1, \ldots, t_n) \rightarrow sup, \quad (t_1, \ldots, t_n) \in M^* \tag{5.10}$$

and the value of lower upper bound in (5.10) denote by $B_{n,i}$ .

Since

$$|p(\xi)| = \int_{t_i}^{\xi} p'(t)dt \leq \max_{\{} |p'(t)|(\xi - t^i) : t \in [t_i, \xi]\} \leq$$

$$|p'(\eta)|(\xi - t_i) \leq |p'(\eta)|(t_{i+1} - t_i),$$

we have

$$F(t_1, \ldots, t_n) \leq t_{i+1} - t_i.$$

Thus, $0 < B_{n,i} \leq 2$ and wittingly

$$B_{n.i} = sup\{F(t_1, \ldots, t_n) : (t_1, \ldots, t_n) \in M_n^*\},$$

where $M_n^* = \{(t_1, \ldots, t_n) \in M^* : |t_{i+1} - t_i| \geq B_{n,i}/2\}$ .

The set $M_n^*$ is a compactum, and since the function $F$ is continious on $M_n^*$, the supremum in (5.10) is attained in some point $(t_1, \ldots, t_n) \in M_n^* \subset M^*$ .

Let $\theta_1^* < \theta_2^* < \ldots < \theta_m^*$ - are all different zeros of the polynomial $p(t) = p(t_1^*, \ldots, t_n^*; t)$ of the multiplicities $l_1, \ldots, l_m$ respectively (i.e. $\sum_{k=1}^m l_k = n$), and let $(t_i^*, t_{i+1})^*) = (\theta_j^*, \theta_{j+1}^*)$ . It is clear that $-1 \leq \theta_1^*, \theta_m^* \leq 1$ .

Any point from the set

$$N = \{(\theta_1, \ldots, \theta_m) \in R^m : \theta_1 < \theta_2 < \ldots < \theta_m\}$$

uniquely determines the polynomial

$$p(t) = p(\theta_1, \ldots, \theta_m) = \prod_{k=1}^{m} (t - \theta_k)^{l_k}. \tag{5.11}$$

Let $\alpha$ be a unique point of the local extremum of $p(t)$ belonging to the interval $(\theta j, \theta j + 1)$ and let $\beta$ be the nearest to $\alpha$ from the left point of the local extremum for $p'(t)$ . Then

$$p'(\alpha) = 0, \ p''(\alpha) \neq 0; \tag{5.12}$$

$$p(\beta) = 0, \ p'''(\beta) \neq 0, \ p'(\beta) \neq 0. \tag{5.13}$$

Thus on the set $N$ the functions $\alpha = \alpha(\theta_1, \ldots, \theta_m)$ and $\beta = \beta(\theta_1, \ldots, \theta_m)$ are defined and in view of the implicit function theorem these functions have in $N$ the continuous partial derivatives $\frac{\partial \alpha}{\partial \theta_k}$ and $\frac{\partial \beta}{\partial \theta_k}$, $k = 1, \ldots, m$ . For this reason functions

$$\Psi(\theta_1, \ldots, \theta_m) = \frac{p(\alpha)}{p'(\beta)}$$

is continuously differentiable in $N$ .

It is obviously that if

$$N^* = \{(\theta_1, \ldots, \theta_m) \in N : \ -1 \le \theta_1, \ \theta_m \le 1\},$$

then

$$\max\{\Psi(\theta_1, \ldots, \theta_m) : \ (\theta_1, \ldots, \theta_m) \in N^*\} = \Psi(\theta_1^*, \ldots, \theta_m^*) = B_{n,i} =$$

$$\max\{F(t_1, \ldots, t_n) : \ (t_1, \ldots, t_n) \in N^*\} \tag{5.14}$$

and

$$\alpha^* = \alpha(\theta_1^*, \ldots, \theta_m^*) = \xi(t_1^*, \ldots, t_n^*)$$

$$\beta^* = \beta(\theta_1^*, \ldots, \theta_m^*) = \eta(t_1^*, \ldots, t_n^*).$$

The direct calculation shows that

$$p'(t) = p(t) \sum_{k=1}^{m} \frac{l_k}{t - \theta_k}$$

and for $k = 1, 2, \ldots, m$

$$\frac{\partial}{\partial \theta_k} p(t) = -\frac{p(t) l_k}{t - \theta_k},$$

$$\frac{\partial}{\partial \theta_k} p'(t) = -\frac{p'(t) l_k}{t - \theta_k} + \frac{p(t) l_k}{(t - \theta_k)^2} .$$

Therefore

$$\frac{\partial}{\partial \theta_k} \Psi(\theta_1, \ldots, \theta_m) = \frac{1}{p'(\beta)^2} \Big( p'(\beta) \frac{\partial}{\partial \theta_k} p(\alpha) - p(\alpha) \frac{\partial}{\partial \theta_k} p'(\beta) \Big) =$$

$$\frac{l_k}{p'(\beta)^2} \Big( p'(\beta) \Big( -\frac{p(\alpha)}{\alpha - \theta_k} + p'(\alpha) \frac{\partial \alpha}{\partial \theta_k} \Big)$$

$$p(\alpha) \Big( -\frac{p'(\beta)}{\beta - \theta_k} + \frac{p(\beta)}{(t - \theta_k)^2} \Big) - p(\alpha) p''(\beta) \frac{\partial \beta}{\partial \theta_k} \Big).$$

Taking into account the relations (5.12) and (5.13) we obtain after simple transformations

$$\frac{\partial}{\partial \theta_k} \Psi(\theta_1, \ldots, \theta_m) =$$

$$\frac{p(\alpha) l_k}{p'(\beta)(\beta - \theta_k)^2} \Big( -\frac{p(\beta)}{p'(\beta)} + (\alpha - \beta) \frac{\beta - \theta_k}{\alpha - \theta_k} \Big). \tag{5.15}$$

Since $(\theta_1^*, \ldots, \theta_m^*)$ - is a point of maximum for the function $\Psi$ on the set $N^*$ the next relations

$$\frac{\partial \Psi}{\partial \theta_1^*}(\theta_1^*, \ldots, \theta_m^*) \leq 0, \tag{5.16}$$

$$\frac{\partial \Psi}{\partial \theta_k}(\theta_1^*, \ldots, \theta_m^*) = 0 \ (k = 2, \ldots, m-1), \tag{5.17}$$

$$\frac{\partial \Psi}{\partial \theta_m}(\theta_1^*, \ldots, \theta_m^*) \geq 0 \tag{5.18}$$

hold true.

Taking into account the fact that the first factor in the right part of (5.15) is positiv and putting

$$\gamma = \frac{p(\beta^*)}{(\alpha^* - \beta^*)p'(\beta^*)},$$

the conditions (5.16)–(5.18) we can write in the form

$$\gamma \geq \frac{\beta^* - \theta_1^*}{\alpha^* - \theta_1^*}; \tag{5.19}$$

$$\gamma = \frac{\beta^* - \theta_k^*}{\alpha^* - \theta_k^*}, \ k = 2, \ldots, m-1; \tag{5.20}$$

$$\gamma \leq \frac{\beta^* - \theta_k^*}{\alpha^* - \theta_m^*}. \tag{5.21}$$

The function $\varphi(t) = (\beta^* - t)/(\alpha^* - t)$ is strictly decreasing from $(\beta^* + 1)/(\alpha^* + 1)$ to $-\infty$ in the interval $(-1, \alpha^*)$, and is strictly decreasing from $+\infty$ to $(\beta^* - 1)/(\alpha^* - 1)$ in interval $(\alpha^*, 1)$. Therefore conditions (5.19)-(5.21) are noncont radictory only in the case $m = 2$, $j = 1$ and $\theta_1^* = -1$, $\theta_2^* = \theta_m^* = 1$. In this case $l_1 = i$, $l_2 = n - i$, and the extremal polynomial is of the form

$$p_i(t) = (t+1)^i (t-1)^{n-i}. \tag{5.22}$$

Thus if $(t_1^*, \ldots, t_n^*)$ - is an extremal point in the problem (5.10), then $t_1^* = \cdots = t_i^* = -1$, $t_{i+1}^* = \ldots = t_n^* = 1$, and the extremal polynomial in (5.10) is of the form (5.22). The direct calculation shows us that

$$\frac{1}{B'_{n,i}} = \frac{n^2}{2\sqrt{(n-1i(n-i)}} \left(1 - \sqrt{\frac{n-i}{i(n-i)}}\right)^{i-1} \cdot$$

$$\left(1 + \sqrt{\frac{i}{(n-i)(n-1)}}\right)^{n-i-1}. \tag{5.23}$$

If now $p$ is a polynomial of the form (5.7) such that

$$\|p\|_{C[-1,1]} = |p(\xi)|, \quad \xi \in (t_i, t_{i+1}), \tag{5.24}$$

where $2 \le i \le n-1$, then its zeros satisfy the restrictions of the problem (5.10) and consequently

$$\frac{\|p'\|_C}{\|p\|_C} \ge \frac{|p'(\eta)|}{|p(\xi)|} \ge \frac{1}{B_{n,i}}, \quad \|\cdot\|_C = \|\cdot\|_{C[-1,1]}. \tag{5.25}$$

If $1 \le i \le n-2$ in (5.24) then by application of (5.25) to polynomial $p_1(t) = p(-t)$ we have

$$\frac{\|p'\|_C}{\|p\|_C} = \frac{\|p_1'\|_C}{\|p_1\|_C} \ge \frac{1}{B_{n,n-i}}.$$

Thus if $2 \le i \le n-2$ in (5.24) then

$$\frac{\|p'\|_C}{\|p\|_C} \ge \max\left(\frac{1}{B_{n,i}}, \frac{1}{B_{n,n-i}}\right). \tag{5.26}$$

Suppose that $i = 1$ or $i = n-1$ in (5.24). In order to obtain the estimation of the type (5.26) in this case we shall define on the set of points $(t_1, \dots, t_n)$ such that $t_1 < t_2 \le \dots \le t_n$ the function

$$G(t_1, t_2, \dots, t_n) = \frac{|p(\xi)|}{|p'(-1)|},$$

where $\xi$ is a unique point of the local extremum of the polynomial $p(t)$ of the form (5.7), belonging to interval $(t_1, t_2)$.

Consider the extremal problem

$$G(t_1, t_2, \dots, t_n) \longrightarrow \sup, \quad 1 \le t_1 < t_2 \le \dots \le t_n \le 1; \tag{5.27}$$

the value of the lower upper bound in (5.27) we denote by $B_{n,1}$.

If we shall study this problem similary to problem (5.10) we make sure that the extremal point $(t_1^*, \dots, t_n^*)$, $t_1^* < t_2^* \le \dots \le t_n^*$, in (5.27) exists, and if $\theta_1^* < \theta_2^* < \dots < \theta_m^*$ are the different zeros of the polynomial $p_* = p(t_1^*, \dots, t_n^*; t)$, then they satisfy the next necessary conditions:

$$\gamma \ge \frac{-1-\theta_1^*}{\alpha^* - \theta_1^*}; \tag{5.28}$$

$$\gamma = \frac{-1-\theta_k^*}{\alpha^* - \theta_k^*}, \quad k = 2, 3, \dots, m-1; \tag{5.29}$$

$$\gamma \le \frac{-1-\theta m^*}{\alpha^* - \theta_m^*}, \tag{5.30}$$

where $\alpha^*$ is a unique point of local extremum of the polynomial $p_*(t)$, belonging $(\theta_1^*, \theta_2^*)$, and

$$\gamma = \frac{p_*(-1)}{(\alpha^* + 1)p'_*(-1)}.$$

But the conditions (5.28)-(5.30) are noncontradictory only in the case $m = 2$, $\theta_1^* = -1$ and $\theta_m^* = 1$. Consequently $t_2^* = t_3^* = \ldots = t_n^* = 1$, and the extremal polynomial in (5.27) is of the form

$$p_*(t) = (1+t)(1-t)^{n-1}$$

and

$$\frac{1}{B_{n,1}} = \frac{n}{2}\left(1 - \frac{1}{n}\right)^{-n+1}.$$

If $i = 1$ in (5.24) then we obtain at once that

$$\frac{||p'||_{C[-1,1]}}{||p||_{C[-1,1]}} \geq \frac{|p'(-1)|}{p(\xi)} \geq \frac{1}{B_{n,1}}. \tag{5.31}$$

If $i = n - 1$, then for polynomial $p_1(t) = p(-t)$ we shall have $i = 1$ and in view of (5.31) (here and below $||\cdot||_C = ||\cdot||_{C[-1,1]}$ )

$$\frac{||p'||_C}{||p||_C} = \frac{||p_1'||_C}{||p_1||_C} \geq \frac{1}{B_{n,1}}.$$

Thus for $i = 1$ and for $i = n - 1$

$$\frac{||p'||_C}{||p||_C} \geq \frac{1}{B_{n,1}}. \tag{5.32}$$

Finally if the polynomial $p$ is of the form (5.7) and $-1 \leq t_1 \leq \ldots \leq t_n < 1$, then

$$\frac{p'(1)}{p(1)} = \sum_{k=1}^{n} \frac{1}{1 - t_k} \geq \frac{n}{2},$$

and if $-1 < t_1 \leq \ldots \leq t_n \leq 1$, then

$$\left\|\frac{p'(-1)}{p(-1)}\right\|_C = \sum_{k=1}^{n} \frac{1}{1 + t_k} \geq \frac{n}{2}.$$

Hence if $||p||_C = |p(1)|$ or $||p||_C = |p(-1)|$, then

$$\frac{||p'||_C}{||p||_C} \geq \left|\frac{p'(\pm 1)}{p(\pm 1)}\right| \geq \frac{n}{2} \geq A_n. \tag{5.33}$$

Taking into account the inequalities (5.5),(5.26) and (5.31)-(5.33), we see that to complete the proof of the theorem it remains to make sure that

$$\frac{1}{B_{n,1}} \geq A_n, \tag{5.34}$$

and for all $i = 2, 3, \ldots, n-2$

$$\max\{\frac{1}{B_{n,i}}, \frac{1}{B_{n,n-i}}\} \geq A_n = \begin{cases} \frac{1}{B_{n,n/2}}, & \text{if } n = 4, 6, \ldots; \\ \frac{1}{B_{n,(n-1)/2}}, & \text{if } n = 5, 7, \ldots. \end{cases} \tag{5.35}$$

For even $n \geq 4$ the inequality (5.34) is of the form

$$\frac{n}{2}\left(1 - \frac{1}{n}\right)^{-n+1} \geq \frac{n}{\sqrt{n-1}}\left(1 - \frac{1}{n-1}\right)^{(n-2)/2},$$

or, that is equivalent, of the form

$$n^{n-1} \geq 2(n-1)^{(n-1)/2}(n-2)^{(n-2)/2}. \tag{5.36}$$

But

$$n^{n-1} \geq n^{(n-1)/2}(n-2+2)^{(n-2)/2} >$$

$$(n-1)^{(n-1)/.2}\left((n-2)^{(n-2)/2} + \frac{n-2}{2}(n-2)^{(n-2)/2-1} \cdot 2\right) =$$

$$2(n-1)^{(n-1)/2}(n-2)^{(n-2)/2},$$

and so (5.36) and consequently (5.34) for even $n \geq 4$ are true. The proof of (5.34) for odd $n$ is only slight more difficult.

To prove (5.35) we note that

$$\max\{\frac{1}{B_{n,i}}, \frac{1}{B_{n,n-i}}\} \geq \left(\frac{1}{B_{n,i}} \cdot \frac{1}{B_{n,n-i}}\right)^{1/2} =$$

$$\frac{n}{2}\left(\frac{n^n}{(n-1)^{n-1}} \cdot \frac{1}{i(n-i)}\left(1 - \frac{1}{i}\right)^{i-1}\left(1 - \frac{1}{n-i}\right)^{n-i-1}\right)^{1/2}. \tag{5.37}$$

Let

$$g(t) = \ln\left(\frac{1}{t(n-t)}\left(1 - \frac{1}{t}\right)^{t-1}\left(1 - \frac{1}{n-t}\right)^{n-t-1}\right),$$

where $t \in (2, n-2)$. Then

$$g'(t) = \ln\frac{(t-1)(n-t)}{t(n-t-1)} < 0.$$

if $t \in (2, n/2)$. Consequently the function $g(t)$, and hence the function

$$\frac{1}{t(n-t)}\left(1-\frac{1}{t}\right)^{t-1}\left(1-\frac{1}{n-t}\right)^{n-t-1}$$

decrease in the interval $(2, n/2)$. Then for odd $n$ we have

$$\left(\frac{1}{B_{n.i}}\cdot\frac{1}{B_{n,n-i}}\right)^{1/2} \geq \frac{1}{B_{n.(n-1)/2}} = A_n,$$

and for even $n$ we have

$$\left(\frac{1}{B_{n,i}}\cdot\frac{1}{B_{n,n-i}}\right)^{1/2} \geq \frac{1}{B_{n,n/2}} = A_n.$$

In view of (5.37) the inequality (5.35) is proved.

The inequality (5.2) is proceed.

It immediately follows from our proof that inequality (5.2) becomes an equality for corresponding polynomials.

## 5.2 Inequalities for trigonometric polynomials

We shall prove the exact Turan's type inequality for trigonometric polynomials having all real zeros and present some its consequences.

**Theorem 5.2.1.** *Let $n = 1, 2, \ldots$ and trigonometric polynomial $\tau$ of oder $n$ has $2n$ real zeros inside period (calculating zeros we take into account their multiplicities). Then*

$$\|\tau'\|_C \geq \sqrt{\frac{n}{2}}\left(1-\frac{1}{2n}\right)^{n-1/2}\|\tau\|_C. \tag{5.38}$$

*For polynomial of the form*

$$\tau(x) = a\left(\sin\frac{x-\gamma}{2}\right)^{2n}, \quad a, \gamma \in R,$$

*and only for these polynomials the inequality (5.38) becomes an equality.*

At once we note the next

**Corollary 5.2.2.** *Under the conditions of theorem 5.2.1 the inequality*

$$\|\tau'\|_C \geq \sqrt{\frac{n}{2e}}\,\|\tau\|_C$$

*holds true.*

*We can not take in this inequality instead of the constant* $(2e)^{-1/2}$ *the large one if we want to have inequality for all* $n$.

This statement follows from theorem 5.2.1 since

$$\sqrt{\frac{1}{2}}\left(1-\frac{1}{2n}\right)^{n-1/2} \downarrow \sqrt{\frac{1}{2e}}, \qquad n \to \infty.$$

**Proof of the theorem 5.2.1.** Every trigonometric polynomial $\tau$ of oder $n$, having only real zeros can be written in the form

$$\tau(x) = a\prod_{k=1}^{2n} \sin\frac{x-x_k}{2}, \tag{5.39}$$

where $a \in R$, $x_1 \leq x_2 \leq \ldots \leq x_{2n} < x_1 + 2\pi$. We will consider $\tau(x)$ as a function of $a, x, x_1, x_2, \ldots, x_{2n}$.

For polynomials of the form (5.39) consider the next extremal problem:

$$\tau(x) \longrightarrow \sup, \quad \|\tau'\|_C = 1. \tag{5.40}$$

If $A_n$ is a value of the supremum in (5.40) then for each polynomial $\tau$ of the form (5.39) takes place the inequality

$$\|\tau'\|_C \geq A_n^{-1}\|\tau\|_C \tag{5.41}$$

with the exact for every $n$ constant $A_n^{-1}$.

To solve the problem (5.40) we will use the solution of the next auxiliary problem:

$$\begin{gathered}\tau(x) \longrightarrow \sup, \quad \tau'(y) = 1, \quad \tau''(y) = 0, \\ x_{2n-1} - 2\pi < y < x, \quad x_{2n} - 2\pi < x < x_1.\end{gathered} \tag{5.42}$$

We show that in the problem (5.40) supremum (denote it by $Q_n$) is attained. First we show that $Q_n < \infty$.

Let $\tau$ is a polynomial satisfies the restrictions of the problem (5.42). Since $\tau$ has inside the period exactly $2n$ zeros (taking into account their multiplicities) there exists a unique point $x' \in (x_{2n} - 2\pi, x_1)$ such that $\tau'(x') = 0$. Consequently $x'$ is a unique point of maximum of the polynomial

$\tau$ belonging to this interval. Let $y'$ be a nearest from the left to $y$ zero of the $\tau'(x)$. In view of the facts given above we have $y' \le x_{2n} - 2\pi$. Since together with $\tau$ the polynomial $\tau'$ has $2n$ zeros on the period then $y$ is a unique point in interval $(x', y')$ such that $\tau''(x)$ vanishes at this point. Thus $y$ is a point of maximum for $\tau'$ on $(x', y')$ and $\tau'(x)$ is less or equal to 1 at every point of this interval. Then for all $x \in (x_{2n} - 2\pi, x']$ we have

$$\tau(x) = \int_{x_{2n}-2\pi}^{x} \tau'(t)dt \le x - (x_{2n} - 2\pi) \le 2\pi. \tag{5.43}$$

So for all $x \in (x_{2n} - 2\pi, x')$

$$\tau(x) \le 2\pi \le \infty.$$

Concequently, $Q_n \le 2\pi < \infty$.

Next we prove that there exist a polynomial of the form (5.39) and points $y, x$, which realize the supremum in (5.42). By the definition of the supremum there exists a sequence $\{\tau_l\}_{l=1}^{\infty}$ of the polynomials

$$\tau_l(x) = a_l \prod_{k=1}^{2n} \sin \frac{x - x_{k,l}}{2}$$

$x_{1,l} \le x_{2,l} \le \ldots \le x_{2n,l} < x_{1,l} + 2\pi$, satisfying the restrictions of the problem (5.42) and such that

$$\tau_l(x^l) \longrightarrow Q_n, \quad l \to \infty, \tag{5.44}$$

where $x_l$ is the point of maximum of polynomial $\tau_l$ on the interval $(x_{2n} - 2\pi, x_{1,l})$. Really it follows from the first inequality (2,6) that in the case $x_{2n} - 2\pi = x$ for all sufficiently large $l$ the inequalities

$$\tau_l(x^l) \le x_{1,l} - (x_{2n,l} - 2\pi) \le Q_n/2,$$

hold true contrary to (5.44).

Polynomials $\tau_l$ are bounded by $Q_n$ on the interval $[x_{2n} - 2\pi + \epsilon, x_1 - \epsilon]$ (here $\epsilon$ is so small positive number that this interval is nonempty) for all sufficiently large $l$. So taking if it is nessesary the subsequence we may assume that on this interval, and consequently everywhere, the sequence $\{\tau_l\}_{l=1}^{\infty}$ converges uniformly to some polynomial $\tau$ (with zeros $x_1, \ldots, x_{2n}$). This polynomial $\tau$ satisfies evidently the restrictions of the problem (5.42) and

$$\max \tau(x) = \lim_{l \to \infty} \tau_l(x^l) = Q_n, \quad x \in [x_{2n} - 2\pi, x_1].$$

The existence of the extremal polynomial for problem (5.42) is proved.

Assume that extremal polymomial in (5.42) is of the form

$$\tau(x) = a \prod_{k=1}^{r} \left( \sin \frac{x - x_k}{2} \right)^{m_k}, \tag{5.45}$$

where $a \in R$, $r \leq 2n$, $x_1 < x_2 < \ldots < x_r < x_1 + 2\pi$, $m_1, \ldots, m_r \in N$,

$$\sum_{k=1}^{r} m_k = 2n.$$

Then this polynomial is also extremal in the problem (5.42) not only among all polynomials of the form (5.39) but among the polynomials of the form (5.45). We shall investigate the problem (5.42) for polynomials of the form (5.45).

We shall assume that inequalities

$$x_1 < \ldots < x_r < x_1 + 2\pi, \; x_{k-1} - 2\pi < y < x, \; x_r - 2\pi < x < x_1$$

determine the domain in the space $R^{r+3}$ where the variables $a$, $x$, $y$, $x_1, \ldots, x_r$ of this problem lie. We shall consider this problem as smooth extremal problem with restrictions of equality type. To investigate this problem we shall use the method of Lagrange fuctors.

The Lagrange's function for our problem we may write in the form

$$L = \lambda_0 \tau(x) + \lambda(\tau'(y) - 1) + \mu \tau''(y)$$

According to the Lagrange's method there exist factors $\lambda_0$, $\lambda, \mu$ which are not vanished simultaneously and such that coefficient $a$ and zeros $x_1, \ldots, x_r$ of the extremal polynomial as well as corresponding points $x$ and $y$ will satisfy the next necessary conditions

$$\frac{\partial L}{\partial a} = 0, \quad \frac{\partial L}{\partial x} = 0, \quad \frac{\partial L}{\partial y} = 0, \quad \frac{\partial L}{\partial x_j} = 0, \; j = 1, \ldots, r, \tag{5.46}$$

$$\tau'(y) = 1, \quad \tau''(y) = 0. \tag{5.47}$$

The first of three conditions in (5.46) we may write in the form

$$a^{-1} \lambda_0 \tau(x) + \lambda a^{-1} \tau'(y) + \mu a^{-1} \tau''(y) = 0, \tag{5.48}$$

$$\lambda_0 \tau'(x) = 0, \tag{5.49}$$

$$\lambda \tau''(y) + \mu \tau'''(y) = 0. \tag{5.50}$$

Taking into account condition (5.47) we obtain from (5.48)

$$\lambda_0 \tau(x) = -\lambda. \tag{5.51}$$

It follows from (5.50) and (5.47) that

$$\mu\tau'''(y) = 0. \tag{5.52}$$

Since under the conditions of problem we have $\tau'''(y) = 0$ then it follows from (5.52) that

$$\mu = 0. \tag{5.53}$$

If $\lambda_0 = 0$ then in view of (5.51) $\lambda = 0$ and all three Lagrange factors are equal to zero, but it is impossible. Consequently, $\lambda_0 \neq 0$, and we may assume that $\lambda_0 = 1$.

There are two possible cases: $\tau'(y) \neq 0$ or $\tau'(y) = 0$. If $\tau'(y) \neq 0$, then for all $j = 1, \ldots, r$

$$\frac{\partial}{\partial x}\tau(x) = -\frac{m_j}{2}\cot\frac{x - x_j}{2}\tau(x)$$

and since for all $y \neq x_j$

$$\tau'(y) = \frac{1}{2}\tau(y)\sum_{k=1}^{r} m_k \cot\frac{y - x_k}{2}.$$

We have

$$\frac{\partial}{\partial x_j}\tau'(y) = -\frac{m_j}{2}\cot\frac{y - x_j}{2}\tau'(y) + \frac{m_j}{2}\cdot\frac{1}{2}\sin^{-2}\frac{y - x_j}{2}\tau'(y).$$

so in view of (5.53) the last $r$ conditions (5.46) we may write in the form

$$-\cot\frac{x - x_j}{2}\tau(x) - \lambda\tau'(y)\cot y - x_j 2 + \lambda\,\tau(y)\,\frac{1}{2}\,\sin^{-2}\frac{y - x_j}{2} = 0. \tag{5.54}$$

Taking into account (5.47) and (5.52) the left part of every equation (5.54) we transform to the form

$$-\cot\frac{x - x_j}{2}\tau(x) + \tau(x)\cot\frac{y - x_j}{2} - \tau(x)\tau(y)\frac{1}{2}\sin^{-2}\frac{y - x_j}{2} =$$

$$\frac{\tau(x)}{\sin\frac{x-x_j}{2}\sin^2\frac{y-x_j}{2}}\left(\cos\frac{x - x_j}{2}\sin^2\frac{y - x_j}{2} + \right.$$

$$\left.\cos\frac{y - x_j}{2}\sin\frac{y - x_j}{2}\sin\frac{x - x_j}{2} - \frac{1}{2}\tau(y)\sin\frac{x - x_j}{2}\right) =$$

$$\frac{\tau(x)}{\sin\frac{x-x_j}{2}}\sin^{-2}\frac{y - x_j}{2}\left(\sin\frac{y - x_j}{2}\sin\frac{y - x}{2} + \frac{1}{2}\tau(y)\sin\frac{x - x_j}{2}\right).$$

Since

$$\tau(x) = \sin^{-1}\frac{x-x_j}{2}\sin^{-2}\frac{x-x_j}{2} \neq 0$$

(it follows from (5.54)) that all $x_1, \ldots, x_r$ are the solution of the equation with unknown $z$

$$2\sin\frac{y-x}{2}\sin\frac{y-z}{2} + \tau(y)\sin\frac{x-z}{2} = 0. \tag{5.55}$$

But equation (5.55) has exactly one root in any interval $[c, c+2\pi]$. Thus $x_1 = \ldots = x_r$, i.e. $r = 1$ (in the opposite case we obtain the contrary). So in extremal case (if $\tau(y) \neq 0$) it is necessary that the equality

$$\tau(x) = a\left(\sin\frac{x-\gamma}{2}\right)^{2n} \tag{5.56}$$

being satisfied. Now it is easy to calculate that the supremum in (5.40) is equal to the value

$$Q_n = \sqrt{\frac{2}{n}}\left(1 - \frac{1}{2n}\right)^{-n+1/2}.$$

Assume now that for extremal polynomial $\tau(x)$ and point $y$ be $\tau(y) = 0$, i.e. $y = x_r - 2\pi$. As above in this case we get once again that the points $x_1, \ldots, x_r$ must satisfy the equation (5.55) which in view of (5.51) and (5.47) and the fact that $\tau(y) = 0$, we may present in the form

$$\cot\frac{x-x_j}{2} = \cot\frac{y-x_j}{2}.$$

But this equation has no roots because $y \neq x (mod\ 2\pi)$. Thus actually $\tau(y) \neq 0$, supremum in the problem (5.40) is equal to $Q_n$ and extremal polynomials are of the form

$$\tau_*(x) = a\left(\sin\frac{x-\gamma}{2}\right)^{2n}.$$

Turn now to the basic problem (5.40). Let $\tau$ be arbitrary polynomial of order $n$ having only real zeros and such that $||\tau'||_C = 1$. Let also $||\tau||_C = \tau(x')$. By $y$ we denote the nearest from the left to the point $x'$ zero of the $\tau''(x)$ (it is evidently that $y \neq x'$). Let $\tau'(y) = \alpha > 0$ $(\alpha < 1)$. Then polynomial $\alpha^{-1}\tau(x)$ after the simple numeration of its zeros will satisfy the restrictions on the problem (5.42). Thus using the facts proved above we get

$$||\tau||_C = \alpha||\alpha^{-1}\tau||_C \leq \alpha\sqrt{\frac{2}{n}}\left(1 - \frac{1}{2n}\right)^{-n+1/2} \leq$$

$$\sqrt{\frac{2}{n}}\left(1-\frac{1}{2n}\right)^{-n+1/2}.$$

So supremum in the problem (5.40) is equal to

$$A_n = \sqrt{\frac{2}{n}}\left(1-\frac{1}{2n}\right)^{-n+1/2}$$

and is attained for polynomials $\tau_*$ only.

Taking into account (5.41) we see that

$$\|\tau'\| \geq \sqrt{\frac{2}{n}}\left(1-\frac{1}{2n}\right)^{-n+1/2} \|\tau\|_C.$$

The theorem is proved.

## 5.3 Inequalities for the second derivatives

In this section we shall present the analogs of the inequalities (5.1) and (5.38) for second derivatives of algebraic and trigonometric polynomials.

**Theorem 5.3.1** *For any algebraic polynomial $p(x) = \Sigma_{k=0}^{n} a_k x^k$ having all $n$ zeros inside [-1,1] the inequality*

$$\|p''\|_C \geq \min\{n, (n-1)n/4\}\|p\|_C \tag{5.57}$$

*holds true. If $n = 2,3,4,5$ then $\min\{n, n(n-1)/4\} = (n-1)n/4$ and inequality (5.57) becomes an equality for the polynomials of the form $p(x) = a(x \pm 1)^n$ and only for these polynomials. If $n \geq 6$ then $\min\{n, n(n-1)/4\} = n$ and for even $n = 2m$ inequality (5.57) becomes an equality for polynomials of the form $p(x) = a(x-1)^m(x+1)^m$, $a \in R$ only.*

**Proof.** If all $n$ zeros of the polynomial $p(x)$ belong to [-1,1], then $p(x)$ can be written in the form

$$p(x) = a\prod_{k=1}^{n}(x - x_k),$$

where $a \in R$, $-1 \leq x_1 \leq \ldots \leq x_n \leq 1$. For $x \neq x_k$, $k = 1, \ldots, n$ let

$$\Sigma(x) = \sum_{k=1}^{n}(x - x_k)^{-1}.$$

If $x \neq x_k,\ k = 1, \ldots, n$ then as it is easilly seen

$$p'(x) = p(x)\Sigma(x) \tag{5.58}$$

and

$$p''(x) = p'(x)\Sigma(x) + p(x)\Sigma'(x) = p(x)(\Sigma^2(x) + \Sigma'(x)). \tag{5.59}$$

First we assume that

$$||p||_C = |p(x_0)|, \tag{5.60}$$

where $x_0 \in (-1, 1)$. Then $p'(x_0) = 0$ and since it is evidently $p(x_0) \neq 0$. Consequently in view of (5.58) we have $\Sigma(x_0) = 0$. Then it follows from (5.59) that

$$|p''(x_0)| = |p(x_0)||\Sigma'(x_0)|. \tag{5.61}$$

Prove that $|\Sigma'(x_0)| \geq n$. Taking into account that $\Sigma(x_0) = 0$ and $|x_k| \leq 1$ for $k = 1, \ldots, n$ and using the inequality of Couchy-Bunjakovski we obtain

$$n = \sum_{k=1}^{n} \frac{x_0 - x_k}{x_0 - x_k} = \sum_{k=1}^{n} \frac{-x_k}{x_0 - x_k} \leq \left(\sum_{k=1}^{n} \frac{1}{(x_0 - x_k)^2}\right)^{1/2} \left(\sum_{k=1}^{n} x_k^2\right)^{1/2} \leq$$

$$\sqrt{|\Sigma'(x_0)|}\ \sqrt{n}.$$

Thus

$$\sqrt{n} \leq \sqrt{|\Sigma'(x_0)|}$$

and required assertion is proved.

It follows from this fact in view of (5.60) and (5.61) that

$$||p''||_C \geq |p''(x_0)| \geq n|p(x_0)| = n||p||_C. \tag{5.62}$$

Assume now that maximum of the $|p(x)|$ is attained at one of the end points of the interval [-1,1] (assume for definitness that $||p||_C = |p(1)|$). Using the relation (5.59) we obtain

$$|p''(1)| = |p(1)|(\Sigma^2(1) + \Sigma'(1)) =$$

$$|p(1)|\left(\left(\sum_{k=1}^{n} \frac{1}{1 - x_k}\right)^2 - \sum_{k=1}^{n} \frac{1}{(1 - x_k)^2}\right) = 2|p(1)| \sum_{i<j} \frac{1}{(1 - x_i)(1 - x_j)}.$$

Since there are $n(n-1)/2$ addends in the last sum and for all $i, j$

$$\frac{1}{(1 - x_i)(1 - x_j)} \geq \frac{1}{4},$$

we obtain

$$\sum_{i<j} \frac{1}{(1-x_i)(1-x_j)} \geq \frac{n(n-1)}{8}. \tag{5.63}$$

Therefore

$$\|p''\|_C \geq |p''(1)| \geq 2|p(1)| \left| \sum_{i<j} \frac{1}{(1-x_i)(1-x_j)} \right| \geq$$

$$2|p(1)| \frac{n(n-1)}{8} = \frac{n(n-1)}{4} \|p\|_C. \tag{5.64}$$

Taking into account the inequalities (5.62) and (5.64) we proving the inequality (5.57).

Direct calculation shows us that inequality (5.57) becomes an equality for according polynomials. It follows from presented proof that equality in (5.57) holds true for these polynomials only.

The analogoue of theorem 5.3.1 for trigonometric polynomials will be presented in the next theorem.

**Theorem 5.3.2** *If trigonometric polynomial $\tau(x)$ of order $n$, $n = 1, \ldots$ has (with account of their multiplisities) $2n$ zeros inside the interval $[0, 2\pi)$ then*

$$\|\tau''\|_C \geq \frac{n}{2} \|\tau\|_C. \tag{5.65}$$

*The inequality (5.65) becomes an equality for polynomials of the form*

$$\tau(x) = a\left(\sin \frac{x-\gamma}{2}\right)^{2n}, \quad a, \gamma \in R. \tag{5.66}$$

**Proof.** Let

$$\tau(x) = a \prod_{k=1}^{2n} \sin \frac{x - x_k}{2}, \tag{5.67}$$

where $a \in R$, $x_1 \leq x_2 \leq \ldots \leq x_{2n} < x_1 + 2\pi$ and for $x \neq x_k$, $k = 1, \ldots, 2n$,

$$\Sigma(x) = \sum_{k=1}^{2n} \cot \frac{x - x_k}{2}.$$

It is clear that for $x \neq x_k$

$$\Sigma'(x) = -\sum_{k=1}^{2n} \frac{1}{2} \sin^{-2} \frac{x - x_k}{2} \leq -n. \tag{5.68}$$

Besides that for the polynomials of the form (5.70)

$$\tau'(x) = \frac{1}{2}\tau(x)\Sigma(x),$$
$$\tau''(x) = \frac{1}{2}\tau'(x)\Sigma(x) + \frac{1}{2}\tau(x)\Sigma'(x). \tag{5.69}$$

Let

$$||\tau||_C = |\tau(x_0)|. \tag{5.70}$$

Then $\tau'(x_0) = 0$ and since $\tau(x_0) \neq 0$, we have in view of the second of the relations (5.72) that

$$\tau''(x_0) = \frac{1}{2}\tau(x_0)\Sigma'(x_0).$$

¿From this and also from (5.70) and (5.68) we obtain

$$||\tau''||_C \geq |\tau''(x_0)| = \frac{1}{2}|\tau(x_0)||\Sigma'(x_0)| \geq \frac{n}{2}|\tau(x_0)| = \frac{n}{2}||\tau||_C,$$

and equality (5.65) is proved.

It is easy to verify that for polynomials of the form (5.66) inequality (5.66) becomes an equality. Since the equality $|\Sigma'(x)| = n$ is possible only for polynomials of the form (5.66) the theorem is completely proved.

## 5.4 Turan's inequalities in the space $L_2$

Here we present some analogs of the inequalities from paragraphs 5.1 and 5.2 for algebraic and trigonometric polynomials in the spaces of $L_2$ type. We start from the trigonometric polynomials.

**Theorem 5.4.1** *For every trigonometric polynomial $\tau \in T_{2n-1}$, having only real zeros the inequality*

$$||\tau'||_{L_2[0,2\pi]} \geq \frac{\sqrt{n}}{2}||\tau||_{L_2[0,2\pi]} \tag{5.71}$$

*holds true. The inequality (5.71) is best possible in the sense that*

$$\sup_n \sup_\tau \frac{\sqrt{n}}{2} \frac{||\tau||_{L_2[0,2\pi]}}{||\tau'||_{L_2[0,2\pi]}} \tag{5.72}$$

*($\sup_\tau$ in (5.72) is calculated over the set of all polynomials $\tau \in T_{2n-1}$, having only real zeros).*

**Proof.** Let $t_1 \leq t_2 \leq \ldots \leq t_{2n} < t_1 + 2\pi$ are zeros of the polynomial $\tau$. Then

$$\tau(t) = a \prod_{k=1}^{2n} \sin \frac{t - t_k}{2}, \ a \in R,$$

and if

$$\Sigma(t) = \frac{1}{2} \sum_{k=1}^{2n} \cot \frac{t - t_k}{2},$$

as it was noted in section 5.2

$$\tau'(t) = \tau(t)\Sigma(t) \tag{5.73}$$

and

$$\tau''(t) = \tau'(t)\Sigma(t) + \tau(t)\Sigma'(t). \tag{5.74}$$

Integrating by parts and taking into account the relations (5.73) and (5.74) we obtain

$$||\tau'||^2_{L_2[0,2\pi]} = \int_0^{2\pi} (\tau'(t))^2 dt = \int_0^{2\pi} \tau'(t)\tau'(t)dt =$$

$$-\int_0^{2\pi} \tau(t)\tau''(t)dt = -\int_0^{2\pi} \tau(t)(\tau'(t)\Sigma(t) + \tau(t)\Sigma'(t))dt =$$

$$-\int_0^{2\pi} (\tau(t))^2 \Sigma'(t)dt - \int_0^{2\pi} (\tau'(t))^2 dt.$$

Since

$$-\Sigma'(t) = \frac{1}{4} \sum_{k=1}^{2n} \sin^{-2} \frac{t - t_k}{2} \geq \frac{n}{2},$$

we get the relations

$$2||\tau'||^2_{L_2[0,2\pi]} = 2 \int_0^{2\pi} (\tau'(t))^2 dt = -\int_0^{2\pi} (\tau(t))^2 \Sigma'(t)dt \geq$$

$$\frac{n}{2} \int_0^{2\pi} (\tau(t))^2 dt = \frac{n}{2} ||\tau||^2_{L_2[0,2\pi]}.$$

The inequality (5.71) is proved.

If $\tau_n(t) = \sin^{2n}(t/2)$ then, as it is easy to calculate,

$$\int_0^{2\pi} (\tau_n'(t))^2 dt = \frac{n^2}{4n-1} \int_0^{2\pi} (\tau_n(t))^2 dt, \tag{5.75}$$

it is evidently that (5.72) follows from (5.75).

The theorem is proved.

Note that actually for polynomials $\tau \in T_{2n+1}$ having only real zeros the next inequality

$$||\tau'||_{L_2[0,2\pi]} \geq \sqrt{\frac{n^2}{4n-1}}\, ||\tau||_{L_2[0,2\pi]} \tag{5.76}$$

holds true and this inequality is best possible already for every fixed $n$ . But its proof is more difficult and we omit it.

The analogue of the theorem 5.1.1 in the space $L_2[-1,1]$ is the next theorem.

**Theorem 5.4.2.** *Let an algebraic polynomial $p \in P_n$, $n = 1,2,\ldots$, has all zeros inside $[-1,1]$. Then*

$$||p'(\cdot)(1-(\cdot)^2)||_{L_2[-1,1]} \geq \sqrt{\frac{n}{2}}\, ||p||_{L_2[-1,1]}, \tag{5.77}$$

*and consequently*

$$||p'||_{L_2[-1,1]} \geq \sqrt{\frac{n}{2}}\, ||p||_{L_2[-1,1]} \tag{5.78}$$

*and*

$$||p'(\cdot)(1-(\cdot)^2)||_{L_2[-1,1]} \geq \sqrt{\frac{n}{2}}\, ||p(\cdot)(1-(\cdot)^2)||_{L_2[-1.1]}. \tag{5.79}$$

*The inequality (5.77) becomes an equality for every polynomials of the form*

$$p(x) = a(1-x)^k(1+x)^l,\ k+l=n,\ k,l \in N \tag{5.80}$$

*The inequalities (5.78) and (5.79) are best possible in the sense that*

$$\sup_n \sup_p \sqrt{\frac{n}{2}}\, \frac{||p||_{L_2[-1,1]}}{||p'||_{L_2[-1,1]}} = 1 \tag{5.81}$$

*and*

$$\sup_n \sup_p \sqrt{\frac{n}{2}}\, \frac{||p(\cdot)(1-(\cdot)^2)||_{L_2[-1,1]}}{||p'(\cdot)(1-(\cdot)^2)||_{L_2[-1,1]}} = 1 \tag{5.82}$$

*($\sup_p$ in (5.81) and (5.82) we calculate over all polynomials $p \in P_n$, having all zeros inside [-1,1]).*

**Proof.** If zeros $t_1, \ldots, t_n$ of a polynomial $p \in P_n$ are real then in view of the lemma 4.2 the identity

$$\int_{-1}^{1} (1-t^2)(p'(t))^2 dt = \frac{n}{2} \int_{-1}^{1} (p(t))^2 dt +$$
$$\frac{1}{2} \int_{-1}^{1} (p(t))^2 \sum_{k=1}^{n} \frac{1-t_k^2}{(t-t_k)^2} dt \tag{5.83}$$

holds true.

Under the conditions of the theorem we have $|t_k| \leq 1$ for $k = 1, \ldots, n$

$$\sum_{k=1}^{n} \frac{t - t_k^2}{(t-t_k)^2} \geq 0.$$

almost everywhere on $[-1, 1]$. From this and from (4.83) we immediately obtain (5.77).

Inequalities (4.78) and (5.79) follow from (5.77) and from obvious inequalities

$$\int_{-1}^{1} (p'(t))^2 dt \geq \int_{-1}^{1} (1-t^2)(p'(t))^2 dt,$$
$$\int_{-1}^{1} (p(t))^2 dt \geq \int_{-1}^{1} (1-t^2)(p(t))^2 dt.$$

For polynomials of the form (5.80)

$$\sum_{k=1}^{n} \frac{1-t_k^2}{(1-t_k)^2} \equiv 0,$$

and so in view of (5.83) the inequality (5.77) becomes an equality.

If $p_{2n}(t) = (1-t^2)^n$ then,

$$n \int_{-1}^{1} (p_{2n}(t))^2 dt = \left(1 - \frac{1}{n} + \frac{3}{16n^2}\right) \int_{-1}^{1} (p_{2n}')^2 dt$$

and consequently (5.71) holds true. Besides that

$$n \int_{-1}^{1} (1-t^2)(p_{2n}(t))^2 dt = \left(1 - \frac{1}{4n+2}\right) \int_{-1}^{1} (1-t^2)(p_{2n}'(t))^2 dt,$$

This relation yields us to (5.82). Theorem is proved.

Present one more proposition of this type.

**Theorem 5.4.3.** *Let all zeros $x_1, x_2, \ldots, x_n$ of the polynomial $p \in P_n$ are nonnegative and*

$$\sum_{k=1}^{n} \frac{1}{x_k} \geq \frac{1}{2}. \tag{5.84}$$

*Then*

$$\int_0^\infty e^{-x}(p'(x))^2 dx \geq \frac{1}{4} \int_0^\infty e^{-x} p^2(x) dx, \tag{5.85}$$

*The constant 1/4 in the right part of (5.85) is best possible in the sense that*

$$\sup_n \sup_p \frac{\int_0^\infty e^{-x} p^2(x) dx}{4\int_0^\infty e^{-x}(p'(x))^2 dx} = 1 \tag{5.86}$$

*($\sup_p$ we calculate over all polynomials $p \in P_n$, having real zeros which satisfy the conditions (5.84)).*

**Proof.** By means of integrating by parts we obtain

$$\int_0^\infty e^{-x}(p'(x))^2 dx = \int_0^\infty e^{-x} p'(t) dp(x) =$$

$$e^{-x} p'(x) p(x)|_0^\infty + \int_0^\infty e^{-x} p(x)(p'(x) - p''(x)) dx. \tag{5.87}$$

Taking into account the relations

$$p'(t) = p(t)\Sigma(t)$$

and

$$p''(t) = p'(t)\Sigma(t) + p(t)\Sigma'(t),$$

where

$$\Sigma(t) = \sum_{k=1}^{n} \frac{1}{t - x_k},$$

we obtain

$$\int_0^\infty e^{-x}(p'(x))^2 dx = -p'(0)p(0) + \int_0^\infty e^{-x} p'(x) p(x) dx -$$

$$\int_0^\infty e^{-x} p(x) p'(x) \Sigma(x) dx - \int_0^\infty e^{-x} p^2(x) \Sigma'(x) dx =$$

$$-p^2(0)\Sigma(0) + \int_0^\infty e^{-x} p(x) p'(x) dx -$$

$$\int_0^\infty e^{-x}(p'(x))^2 dx + \int_0^\infty e^{-x} p^2(x) |\Sigma'(x)| dx.$$

Thus,

$$2\int_0^\infty e^{-x}(p'(x))^2dx =$$

$$p^2(0)\sum_{k=1}^n \frac{1}{x_k} + \int_0^\infty e^{-x}p'(x)p(x)dx + \int_0^\infty e^{-x}p^2(x)|\Sigma'(x)|dx.$$

Besides that

$$\int_0^\infty e^{-x}p'(x)p(x)dx = \int_0^\infty e^{-x}p(x)dp(x) =$$

$$-p^2(0) + \int_0^\infty e^{-x}p(x)(p(x) - p'(x))dx,$$

i.e.

$$\int_0^\infty e^{-x}p'(x)p(x)dx = -\frac{p^2(0)}{2} + \frac{1}{2}\int_0^\infty e^{-x}p^2(x)dx,$$

and, consequently,

$$2\int_0^\infty (p'(x))^2dx = p^2(0)\left(\sum_{k=1}^n \frac{1}{x_k} - \frac{1}{2}\right) +$$

$$\frac{1}{2}\int_0^\infty e^{-x}p^2(x)dx + \int_0^\infty e^{-x}p^2(x)|\Sigma'(x)|dx \geq$$

$$\frac{1}{2}\int_0^\infty e^{-x}p^2(x)dx,$$

This completed the proof of the inequality (5.85).

The relation (5.86) follows from the next equality

$$\int_0^\infty e^{-x}x^{2n}dx = 2n(2n-1)\int_0^\infty e^{-x}x^{2n-2}dx =$$

$$\frac{4n-1}{n}\int_0^\infty e^{-x}((x^n)')^2dx.$$

The theorem is proved.

## 5.5 Turan's inequalities for complex polynomials

We shall present estimations from below for derivatives of complex algebraic polynomials having all zeros inside the disk $|z| \leq 1$.

**Theortem 5.5.1** *If all zeros of algebraic polynomial $p(z)$ of degree $n$ belong to the disk $|z| \leq 1$, then*

$$\max_{|z|=1} |p'(z)| \geq \frac{n}{2} \max_{|z|=1} |p(z)|. \tag{5.88}$$

*The inequality (5.87) becomes an equality for polynomials of form $p(z) = (\alpha + \beta z^n)/2$, where $|\alpha| = |\beta| = 1$.*

**Proof.** If all zeros of polynomial $p(z)$ belong to the disk $|z| \leq 1$, then

$$p(z) = c \prod_{j=1}^{n} (z - w_j),$$

where $|w_j| \leq 1$, $j = 1, \ldots, n$, $n, c \in C$. For the points $e^{i\theta}$, $0 \leq \theta < 2\pi$ , which are different from zeros $p(z)$ we have ( in view of $|w_j| \geq 1$, $j = 1, \ldots, n$)

$$Re\left(\frac{e^{i\theta} p'(e^{i\theta})}{n\ p(e^{i\theta})}\right) = \frac{1}{n} \sum_{j=1}^{n} Re\left(\frac{e^{i\theta}}{e^{i\theta} - w_j}\right) \geq \frac{1}{n} \sum_{j=1}^{n} \left(\frac{1}{2}\right) = \frac{1}{2}.$$

¿From this it follows that

$$\left|\frac{e^{i\theta} p'(e^{i\theta})}{np(e^{i\theta})}\right| \geq \left|1 - \frac{e^{i\theta} p(e^{i\theta})}{np(e^{i\theta})}\right|.$$

This inequality for such $\theta$ that $e^{i\theta} \neq w_j$, $j = 1, \ldots, n$ is equivalent to the inequality

$$\left|p'(e^{i\theta})\right| \geq \left|np(e^{i\theta}) - e^{i\theta} p'(e^{i\theta})\right|. \tag{5.89}$$

If $e^{i\theta} = w_j$ for some $j$ then (5.88) is trivial. Consequently if $|z| = 1$, then

$$|p'(z)| \geq |n\, p(z) - z\, p'(z)| \tag{5.90}$$

and moreover

$$|p'(z)| \geq n|p(z)| - |p'(z)|.$$

It follows from the last inequality (for $|z| = 1$)

$$|p'(z)| \geq \frac{n}{2} |p(z)|.$$

and so

$$\max_{|z|=1} |p'(z)| \geq \frac{n}{2} \max_{|z|=1} |p(z)|.$$

Theorem is proved.

Certainly the inequality (5.87) is equivalent to the inequality

$$\max_{|z|=1} |p'(z)| \geq \frac{n}{2} \max_{|z|\leq 1} |p(z)|.$$

**Tneorem 5.5.2** *If all zeros of the polynomial $p(z)$ of degree $n$ belong to the disk $|z| \leq 1$, then for every $q > 0$ the inequality*

$$n\left(\int_0^{2\pi} \left|p(e^{i\theta})\right|^q d\theta\right)^{1/q} \leq (A_q)^{1/q} \max_{|z|=1} |p'(z)|, \tag{5.91}$$

*where*

$$A_q = 2^{q+1}\sqrt{\pi}\, \Gamma((q+1)/2)/\Gamma(q/2+1) \tag{5.92}$$

*holds true. The inequality (5.90) becomes an equality for polynomials of the form $p(z) = \alpha z^n + \beta$, where $|\alpha| = |\beta|$.*

**Remark.** Lemitting in (5.90) $q \to \infty$ we obtain the inequality (5.87).

**Proof the theorem 5.5.2** Let $p(z)$ be a polynomial of degree $n$, and

$$Q(z) = z^n \overline{p(1/\overline{z})}.$$

It is clear that if $|z| = 1$, then $|Q(z)| = |p(z)|$. Besides that

$$p(z) = z^n \overline{Q(q/\overline{z})},$$

and as a simple calculation shows

$$p'(z) = nz^{n-1}\overline{Q(1/\overline{z})} - z^{n-2}\overline{Q'(1/\overline{z})}.$$

¿From this it follows the equality

$$z^{n-1}\overline{p'(1/\overline{z})} = n\, Q(z) - z\, Q'(z). \tag{5.93}$$

Since

$$\left|z^{n-1}\overline{p'(1/\overline{z})}\right| = |p'(z)|$$

for $|z| = 1$, then it follows from (5.92) that the equality

$$|p'(z)| \leq |nQ(z) - zQ'(z)|, \ |z| = 1 \tag{5.94}$$

holds true.

It is evident also that

$$\begin{aligned} n|Q(z)| = |nQ(z) - zQ'(z) + zQ'(z)| = \\ |nQ(z) - zQ'(z)| \cdot |1 + \omega(z)|, \end{aligned} \tag{5.95}$$

where

$$\omega(z) = zQ'(z)/(nQ(z) - zQ'(z)).$$

If we compare (5.93) and (5.94), we see that for $|z| = 1$ the equality

$$n|p(z)| = |p'(z)| \cdot |1 + \omega(z)| \tag{5.96}$$

holds true.

We shall show that if $p(z)$ is a polynomial of degree $n$ having all zeros inside disk $|z| \leq 1$, then the function $\omega(z)$ is analytical in this disk.

Since all zeros of the $p(z)$ lie in the disk $|z| \leq 1$ then by the Gauss-Luke theorem all zeros of the $p'(z)$ also lie in this disk. From this it follows that

$$z^{n-1}\overline{p'(1/\overline{z})} = nQ(z) - zQ'(z) \neq 0,$$

if $|z| < 1$. Moreover, if $\alpha$ ($|\alpha| = 1$) is a zero of the multiplicity $k$ ($k < n$) of the polynomial $nQ(z) - zQ'(z)$, then in view of (5.93) $\alpha$ is a zero of the multiplicity $k$ for $p'(z)$.

Since the unit disk, which contains all zeroz of the $p(z)$ is strictly convex we taking into account the Gauss-Luke theorem conclude that $\alpha$ is a zero of multiplicity $k + 1$ for $p(z)$ and consequently for $Q(z)$. So $\alpha$ is a zero of multiplisity of $k$ for $Q'(z)$. By defenition of $\omega(z)$ we make shure that this function is analytical in the disk $|z| \leq 1$.

Next, since

$$Q(z) = z^n\overline{Q(1/\overline{z})},$$

then $Q(z)$ has no zeros inside the disk $|z| \leq 1$. Then in view of the inequality (5.102)

$$|Q'(z)| \leq |nQ(z) - zQ'(z)|,$$

if $|z| = 1$. Thus, if $|z| = 1$, then $|\omega(z)| \leq 1$ and consequently the function $1 + \omega(z)$ has no zeros inside the disk $|z| < 1$. From this it follows that the function $(1 + \omega(z))^p$ will be regular in the disk $|z| < 1$ for every $p > 0$.

Now by Poisson's formula we have for $r_0 \in (0, 1)$ and for every $z$ ($|z| < r_0$) the equality

$$(1 + \omega(z))^p = \frac{1}{2\pi}\int_0^{2\pi} (1 + r_0 e^{i\theta})^p \, Re\frac{r_0 e^{i\theta} + \omega(z)}{r_0 e^{i\theta} - \omega(z)} d\theta.$$

Consequently

$$|1 + \omega(z)|^p \leq \frac{1}{2\pi}\int_0^{2\pi} \left|1 + r_0 e^{i\theta}\right|^p \, Re\frac{r_0 e^{i\theta} + \omega(z)}{r_0 e^{i\theta} - \omega(z)} d\theta.$$

Integrating the last inequality along the circle $|z| = r$ ($0 < r < r_0$) and taking into account that by theorem of the mean value

$$\int_0^{2\pi} Re\frac{z' + e^{i\theta}}{z' - e^{i\theta}} d\theta = 2\pi Re\frac{z' + \omega(0)}{z' - \omega(0)} = 2\pi.$$

we obtain

$$\int_0 2\pi|1+\omega(re^{it})|^p\,dt \le \int_0^{2\pi} |1+r_0e^{it}|^p\,dt.$$

Dicrecting $r \to r_0$ we obtain that for each $r_0 \in (0,1)$

$$\int_0 2\pi|1+\omega(r_0e^{it})|^p\,dt \le \int_0^{2\pi} |1+r_0e^{it}|^p\,dt.$$

¿From this when $r_0 \to 1$, we have

$$\int_0^{2\pi} |1+\omega(e^{it})|^p\,dt \le \int_0^{2\pi} |1+e^{it}|^p\,dt. \tag{5.97}$$

Now, taking into account (5.95) and (5.96) we obtain

$$n\left(\int_0^{2\pi} |P(e^{i\theta})|^p\,d\theta\right)^{1/p} \le \max_{|z|=1}|p'(z)|\left(\int_0^{2\pi} |1+w(e^{it})|\,dt\right)^{1/p} \le$$

$$\max_{|z|=1}|p'(z)|\left(\int_0^{2\pi} |1+e^{it}|\,dt\right)^{1/p}, \tag{5.98}$$

The last inequality is equivalent to (5.90). Theorem is proved.

The next theorem is a simple consequence of the theorem 5.5.2.

**Theorem 5.5.3.** *If all zeros of the polynomial $p(z)$ of degree $n$ lie on the circle $|z| = 1$ then for every $q > 0$*

$$\left(\int_0^{2\pi} |P(e^{i\theta})|^q\,d\theta\right)^{1/q} \le \frac{1}{2}(A_q)^{1/q}\max_{|z|=1}|p(z)|, \tag{5.99}$$

*where the constant $A_q$ is defined by equality (5.91).*

*The inequality (5.97) becomes an equality for the polynomials of the form $p(z) = \alpha z^n + \beta$, where $|\alpha| = |\beta| = 1$.*

To see that inequality (5.97) holds true it is sufficiently to recall that if all zeros of $p(z)$ lie on the circle $|z| = 1$ and consequently outside of the disk $|z| < 1$ then by theorem 4.5.4

$$\max_{|z|=1}|p'(z)| \le \frac{n}{2}\max_{|z|=1}|p(z)|.$$

To complete the proof of the theorem 5.5.3 we may use the theorem 5.5.2 and last inequality.

## 5.6 Commentaries

Sec.5.1. The opportunity of turning for algebraic polynomials which have only real zeros that being on segment $[-1,1]$ at first (it was remarked early) noticed P. Turan [1], who proved the inequality 1.1, D. Ered [1] with the help of rather delicate reasonings proved the inequality of Turan's type (5.2) with the exact constant for every $n$.

The ordered inequalities of Turan's type for trigonometric polynomials in uniform metric and also for trigonometric and algebraic polynomials in the spaces $L_p$, $0 \leq p \leq \infty$ were proved by Chzou Songping [1-3].

Sec.5.2. The inequality of Turan's type with exact constant for trigonometric polynomials having only real zeros (theorem 5.2.1) is proved by V.F. Babenko and S.A. Pichugov [2]. The method used by them under the proving of theorem allowed us to prove the row of other inequalities of such type (see the works of V.F. Babenko and S.A. Pichugov [3], V.F. Babenko and O.I. Severina [1], I.J. Tyrygin [2-4]).

Sec.5.3. The exact inequalities of type (5.1) for second derivatives of trigonometric polynomials having only real zeros and for algebraic polynomials which zeros are on the segment $[-1,1]$ (theorems 5.3.1 and 5.3.2) are proved by V.F. Babenko and S.A. Pichugov [3]. It is interest to note that the proofs of these theorems are more simple than the proofs of inequalities with the first derivatives.

Sec.5.4. Turan's type inequalities in the space $L_2$ on the segment $[-1,1]$ and on semiaxis exact in the class of all algebraic polynomials which zeros are on the segment $[-1,1]$ (in particular the theorems 5.4.2 and 5.4.3), are proved by A.K. Varma [1]. Theorem 5.4.1 is proved by V.F. Babenko.

Sec.5.5. Theorem 5.5.1 was proved by P. Turan [1]. Its generalization (theorem 5.5.2) was obtained by A.A. Malik [3]. Theorem 5.5.3 was established in the works of E.B. Saff and T. Sheil - Small [1].

## 5.7 Supplementary results

**1.** Let polynomial $p \in P_n$ has $n$ zeros on interval [-1,1]. Then

$$\|p'\|_{L_p[-1,1]} \geq c\sqrt{n}\,\|p\|_{L_p[-1,1]}, \quad 1 \leq p \leq \infty,$$

where $c$ is the absolute positive constant.

P.Turan ($p = \infty$) [2], Varma A.K., (p=2) [1], Chzou Songping [1].

If $0 < q < 1$ then also

$$\int_{-1}^{1} |p'(t)|^q dt \geq c\sqrt{n^q} \int_{-1}^{1} |p(t)|^q dt.$$

Chzou Songping [3].

**2.** If trigonometrical polynomial $\tau \in T_{2n+1}$ has with the calculation of its multiplicities $2n$ zeros on period and $h(\tau)$ is the minimal distance between its different zeros, then if

$$0 < 2h < \min\{h(\tau'), 2arcos(1/2)^{1/(2n)}\}$$

the following inequality

$$||\Delta_h^2 \tau||_C \geq 2(1 - \cos^{2n} h)||\tau||_C$$

is valid. This inequality turns into equality for polynomials

$$\tau(t) = a\left(\sin\frac{t-\gamma}{2}\right)^{2n}, a, \gamma \in R.$$

Babenko V.F., Pichugov S.A. [3].

**3.** For any trigonometical polynomial $\tau \in T_{2n+1}$ having $2n$ zeros with the calculation of its multiplicities on period $[0, 2\pi)$, the next exact inequality

$$\frac{||\tau'||_p}{||\tau||_C} \geq n\left(2B\left(\frac{(2n-1)p+1}{2}, \frac{p+1}{2}\right)\right)^{1/p}, \quad 1 \leq p < \infty,$$

holds true and for any $N$-function $\Phi$

$$\int_0^{2\pi} \Phi\left(\frac{|\tau(t)|}{||\tau||_C}\right) dt \geq \int_0^{2\pi} \Phi\left(\sin^{2n}\frac{1}{2}\right) dt.$$

Tyrygin I.J. [2]

**4.** Let polynomial $\tau \in T_{2n+1}$ has $2n$ zeros on period $[0, 2\pi)$ with the calculation of its multiplicities, $t_i$, $i = 1, \ldots, m$ are the different zeros of $\tau(t)$ on period and for every $i$ the point $y_i$ is the unique point of local extremum for polynomial $\tau$ in the interval $(t_i, t_{i+1})$.Then for $1 \leq p\infty$

$$\left(\sum_{i=1}^{m} |\tau(y_i)|^p\right)^{1/p} \geq ||\tau||_p \left(B\left(\frac{1}{2}, np + \frac{1}{2}\right)\right)^{1/p}$$

and

$$||\tau'||_p \geq n\left(2B\left(\frac{(2n-1)p+1}{2}, \frac{p+1}{2}\right)\right)^{1/p}\left(\sum_{i=1}^{m} |\tau(y_i)|^p\right)^{1/p}.$$

Tyrygin I.J. [3,4].

**5.** If polynomial $\tau \in T_{2n+1}$ has only real zeros, then for all $1 \le p \le q \le \infty$

$$\frac{\|\tau'\|_p}{\|\tau\|_q} \ge$$

$$n\,2^{\frac{1}{p}-\frac{1}{q}} \left(B\left(\frac{(2n-1)p+1}{2}, \frac{p+1}{2}\right)\right)^{1/p} \left(B\left(\frac{1}{2}, nq+\frac{1}{2}\right)\right)^{1/q}.$$

Inequality becomes an equality for polynomials $\tau(t)$ in the form

$$\tau(t) = c\left(\sin\frac{t-\gamma}{2}\right)^{2n}, \quad c, \gamma \in R, \ c \ne 0.$$

Tyrygin I.J. [3,4].

**6.** If all zeros of polynomial $p \in P_n$ are on the segment $[-1,1]$ then

$$\|p'\|_{L_2[-1,1]} \ge \frac{n}{4}|p(1)|$$

and there exists a polynomial $p_0 \in P_n$ for which $p_0(1) = 1$ and

$$\|p_0'\|_{L_2[-1,1]} = \frac{n}{4} + \frac{1}{8} + \frac{1}{8(2n-1)}.$$

Sabadosh I., Varma A.K. [1].

**7.** If all zeros of polynomial $p(z)$ of order $n$ are in disk $|z| \le k$, $k \le 1$, then

$$\max_{|z|=1} |p'(z)| \ge \frac{n}{k+1} \max_{|z|=1} |p(z)|.$$

The equality takes place for $p(z) = a(z+k)^n$.
Malik M.A. [1].

**8.** If all zeros $z_j$, $j = 1, \ldots, n$ of polynomial $p(z)$ of order $n$ are in the disk $|z| \le k$, $k \ge 1$, then

$$\max_{|z|=1} |p'(z)| \ge \frac{2}{k^n+1} \sum_{j=1}^{n} \frac{k}{k+|z_j|} \max_{|z|=1} |p(z)|.$$

The equality takes place for $p(z) = z^n + k^n$.
Aziz A. [2].

**9.** If all zeros of polynomial of order $n$

$$p(z) = \prod_{j=1}^{n} (z - x_j)$$

are in the semiplane $Re z \geq 1$, then next exact inequality

$$\max_{|z|=1} |p'(z)| \geq n \prod_{j=1}^{n} (1 + Re\, z_j)^{1/n} (\max_{|z|=1} |p(z)|)^{1-2/n}$$

is valid.

Chziro A., Rahman K.I.,Schmeisser D. [1].

**10.** If polynomial $p(z)$ of order $n$ is such that $p(z) = z^n p(1/z)$, then

$$\max_{|z|=1} |p'(z)| \geq \frac{n}{2} \max_{|z|=1} |p(z)|.$$

The inequality turns into equality for polynomial $p(z) = z^n + 1$.

Aziz A. [2].

# Chapter 6

# Inequalities for splines

The extremal properties of splines are investigated in this chapter. Though splines are glued from pieces of algedraic polynomials , they have a finite smoothness and their intrinsic nature principally differs from that of polynomials which are analytic functions.

There will be considered only splines of order $n$ with the minimal defect that are discontinuous only in their last $m$-th derivative being a piecewise constant. Such splines possess the most interesting properties of extremal and approximative character.

Collection of splines having the common fixed system of knots , i.e. of points an which last derivative may discontinue, is linear manifold which dimension is defined by the number of points of partition and boundary conditions. It turns out that subspaces of splines of minimal defect are the best approximation methods for many important classes of functions, moreover, unlike polynomials in series of cases the operator of spline-interpolation is optimal.

By the general purposefulness this chapter is closed to the chapter 3 and 4 in which were considered inner extremal properties of algebraic and trigonometric polynomials that are expressed through strict inequalities for derivatives and finite differences. Here subspaces of splines of the fixed partition are investigated in the view of that. At that time consideration is restricted by periodical splines which properties are studied much more profound and results have complete form.

## 6.1 Periodic splines of minimal defect

Let we have the partition of period $[0, 2\pi]$

$$\Delta_N = \{t_1 < t_2 < \ldots < t_N < 2\pi \leq t_1 + 2\pi = t_{N+1}\}.$$

The function $s(t)$ is called periodic spline-function of order $m = 1, 2, \ldots$ with minimal defect (or simply periodic spline) by partition $\delta_N$, if $s \in C^{m-1}$ and on each from intervals $(t_\nu, t_{\nu+1}), \nu = 1, 2, \ldots, N$, $s(t)$ coincides with the algebraic polynomial of order at most $m$. For fixed partition $\Delta_N$ a set of such splines is linear $n$-dimensional manifold. Denote it by $S_m(\Delta_N)$.

If $S_0(\Delta_N)$ is a set of all constant on each from intervals $(t_\nu, t_{\nu+1}), \nu = 1, 2, \ldots, N$ ,$2\pi$-periodic functions, then $S_m(\Delta_N)$ is a set of all $m$-th periodic integrals of the fuctions $s \in S_0(\Delta_N)$ which mean value is equal to zero on period.

Let for $m = 1, 2, \ldots$

$$B_m(t) = \frac{1}{\pi}\sum_{k=1}^{\infty} k^{-m}\cos(kt - m\pi/2)$$

is Bernully's functions. The function $B_m(t)$ we may define also as $(m-1)$-th $2\pi$-periodic integral with zero mean value on period from $2\pi$-periodic function $B_1(t)$ that equals to $(\pi - t)/(2\pi)$ for $t \in (0, 2\pi)$.

It is well known (see, for example,[11,p.36]) that for any function $f \in L_1^m$ the next integral representation

$$f(x) = \frac{1}{2\pi}\int_0^{2\pi} f(t)dt + \int_0^{2\pi} f^{(m)}(t)B_m(x-t)dt$$

takes place.

**Theorem 6.1.1.** *Any spline $s(t) \in S_m(\Delta_N)$, $m \in N$ may be represented by unique manner in the form*

$$s(t) = a + \sum_{k=1}^{N} a_k B_{m+1}(t - t_k), \tag{6.1}$$

*where $a$, $a_k \in R$,*

$$a_1 + a_2 + \ldots + a_N = 0. \tag{6.2}$$

*In this case*

$$a_k = s^{(m)}(t_k + 0) - s^{(m)}(t_k - 0), \quad k = 1, 2, \ldots, N. \tag{6.3}$$

**Proof.** Let $s \in S_m(\Delta_N)$,

$$s_h(t) = \frac{1}{2h}\int_{t-h}^{t+h} s(u)du, \quad 0 < h < \pi,$$

is Steklov's function with step $2h$ from the spline $s(t)$ and $\delta_h(t)$ is $2\pi$-periodic continuation to characteristic function of segment $[-h, h]$. As the function $s^{(m)}(t)$ is constant on intervals $(t_k, t_{k+1})$, $k = 1, 2, \ldots, N$, then for $h < \pi/n$ the function $s_h^{(m)}(t)$ is linear on intervals $(t_k - h, t_k + h)$ and constant on intervals $(t_{k-1} + h, t_k - h)$. Moreover for $t \in (t_k - h, t_k + h)$ $s_h^{(m+1)}(t) = (2h)^{-1}a_k$, where $a_k$ are defined by equalities (6.3). Therefore for $0 < h < \pi/n$

$$s_h^{(m+1)}(t) = \frac{1}{2h}\sum_{k=1}^{N} a_k \delta_h(t - t_k), \tag{6.4}$$

where $a_k$ are defined by equatlities (6.3)

$$0 = \int_0^{2\pi} s_h^{(m+1)}(t)dt = \sum_{k=1}^{N} a_k. \tag{6.5}$$

Moreover, taking into account that $B_1(t)$ is $2\pi$-periodic function which equals to $(\pi - t)/2\pi$ for $t \in (0, 2\pi)$ we obtain

$$\delta_h(t) = \pi^{-1}h + B_1(t + h) - B_1(t - h).$$

¿From here and from (6.2) and (6.5) we find

$$s_h^{(m+1)}(t) = \frac{1}{2h}\sum_{k=1}^{N} a_k[B_1(t - t_k + h) - B_1(t - t_k - h)]$$

and so as $B_{m+2}^{(m+1)}(t) = B_1(t)$ we obtain

$$s_h(t) = a + \frac{1}{2h}\sum_{k=1}^{N} a_k[B_{m+2}(t - t_k + h) - B_{m+2}(t - t_k - h)] =$$

$$a + \sum_{k=1}^{N} a_k B_{m+1,h}(t - t_k).$$

Therefore

$$s(t) = \lim_{h \to 0} s_h(t) = a + \sum_{k=1}^{N} a_k B_{m+1}(t - t_k).$$

The uniqueness of representation (6.1) follows from linear independence of functions 1, $B_k(t - t_1), B_k(t - t_2) \ldots, B_k(t - t_N)$.

**Proposition 6.1.2.** *Let $f \in L_1$ and $f_m(t)$, $m = 1, 2, \dots$ is m-th periodic integral from the function*

$$f(t) - \frac{1}{2\pi}\int_0^{2\pi} f(u)du$$

*while*

$$\int_0^{2\pi} f_m(t)dt = 0.$$

*The function $f(t)$ is orthogonal to subspace of splines $S_m(\Delta_N)$ if and only if*

$$f_{m+1}(t_k) = f_{m+1}(t_1), \quad k = 1, 2, \dots, N. \tag{6.6}$$

**Proof.** Since the set $S_m(\Delta_N)$ includes the constants then the condition $f \perp S_m(\Delta_N)$ is equivalent to condition

$$\int_0^{2\pi} f_{m+1}^{(m+1)}(t)s(t)dt = 0, \quad s \in S_m(\Delta_N).$$

With the help of integration by parts we verify that this condition is equivalent to condition

$$\int_0^{2\pi} f'_{m+1}(t)s(t)dt = 0, \quad s \in S_0(\Delta_N). \tag{6.7}$$

But any spline $s \in S_m(\Delta_N)$ is represented by unique manner in the form

$$s(t) = \sum_{k=1}^{N} c_k \psi_k(t),$$

where

$$\psi_k(t) = \begin{cases} 1 \ \ for \ \ t \in (t_k, t_{k+1}), \\ 0 \ \ for \ \ t \in (0, 2\pi) \setminus (t_k, t_{k+1}). \end{cases}$$

Therefore the condition (6.7) is equivalent to conditions

$$f_{m+1}(t_{k+1}) - f_{m+1}(t_k) = \int_0^{2\pi} f'_{m+1}(t)\psi_k(t)dt = 0,$$

that is equivalent to (6.6).

The spline of odd order

$$s(t) = s_{2\nu-1}(f, \Delta_N, t) \in S_{2\mu-1}(\Delta_N)$$

will be called interpolational one to the function $f(t)$ if

$$s_{2\nu-1}(f, \Delta_N, t_k) = f(t_k), \quad k = 1, 2, \ldots, N,$$

where $t_k$ are knots of spline, i.e. $t_k$ are the points of partition $\Delta_N$.

**Theorem 6.1.3.** *For any function $f \in C$ the spline $s_{2\nu-1}(f, \Delta_N, t)$ exists and is unique.*

**The proof** of existence and uniqueness of interpolational spline is usually conducted according to the next scheme. At first we establish that the spline which satisfies to zero interpolational conditions is identically equal to zero. From here owing to linear independence of basic functions (see (6.1)) it follows that uniform system of linear, with respect to parameters $a, a_1, \ldots, a_N$, equations has only zero solution. But then the corresponding system with arbitrary right part has the unique solution. In considered case if there exists the spline $s_* \in S_{2\nu-1}(\Delta_N)$ such that $s_*(t_k) = 0$, $k = 1, 2, \ldots, N$ and $s^*(t) \not\equiv 0$ then in view of proposition 6.1.2 $s_*^{(\nu)} \perp S_\nu(\Delta_N)$ and in particular $s_*^{(\nu)} \perp s_*^{(\nu)}$ that is impossible.

The interpolational splines of odd order possess interest extremal property.

**Theorem 6.1.4.** *Let $\Delta_N$ is arbitrary partition of period by points $t_i$, $y = \{y_i\}_{i=1}^N$ are arbitrary numbers and $s_{2\nu-1}(y, t)$ is spline from $S_{2\nu-1}(\Delta_N)$ such that $s_{2\nu-1}(y, t_i) = y_i$, $i = 1, 2, \ldots, N$. Then among all functions $f \in L_2^\nu$ such that $f(t_i) = y_i$ $i = 1, 2, \ldots, n$ the least value of norm $||f^{(\nu)}||_2$ is attained by spline $s_{2nu-1}(y, t)$, i.e.*

$$\min\{||f^{(\nu)}||_2 : f \in L_2^{(\nu)}, \; f(t_i) = y_i, \, i = 1, 2, \ldots, N\} = ||s_{2\nu-1}^{(\nu)}(y)||_2.$$

**Proof.** For the spline $s(t) = s_{2nu-1}(y, t)$ due to proposition 6.1.2 we have

$$(f^{(\nu)} - s^{(\nu)}) \perp S_{\nu-1}(\Delta_N)$$

and in particular

$$(f^{(\nu)} - s^{(\nu)}) \perp s^{(\nu)}(t).$$

But then

$$\|f^{(\nu)} - s^{(\nu)}\|_2 = \int_0^{2\pi} (f^{(\nu)}(t) - s^{(\nu)}(t))^2 dt =$$

$$\int_0^{2\pi} \left(f^{(\nu)}(t)\right)^2 dt - \int_0^{2\pi} \left(s^{(\nu)}(t)\right)^2 dt - 2\int_0^{2\pi} (f^{(\nu)}(t) - s^{(\nu)}(t))\, s^{(\nu)}(t) dt =$$

$$\int_0^{2\pi} \left(f^{(\nu)}(t)\right)^2 dt - \int_0^{2\pi} \left(s^{(\nu)}(t)\right)^2 = \|f^{(\nu)}\|_2^2 - \|s^{(\nu)}\|_2^2$$

i.e.

$$\|f^{(\nu)}\|_2^2 = \|s^{(\nu)}\|_2^2 + \|f^{(\nu)} - s^{(\nu)}\|_2^2 \geq \|s^{(\nu)}\|_2^2.$$

The function $\gamma \in L_\infty^m$ is called perfect spline of order $m$ with knots at points

$$t_1 < t_2 < \ldots < t_{2k} < 2\pi \leq t_1 + 2\pi = t_{2k+1}, \tag{6.8}$$

if for $t \in (t_\nu, t_{\nu-1})$

$$\gamma^{(m)}(t) = \varepsilon(-1)^\nu, \quad |\varepsilon| = 1, \quad \nu = 1, 2, \ldots, 2k.$$

Denote by $\Gamma_{2n,m}$ a set of all perfect splines of order $m$ having at most $2n$ knots on period .

It is easy to se that the $(\alpha, \beta)$-perfect splines are the natural generalization of ordinary perfect splines.

A function $\gamma \in L_\infty^m$ will be called $(\alpha, \beta)$-perfect spline of order $m$ with the knots at points $t_\nu$ (see (6.8)) if for $t \in [t_{2\nu-1}, t_{2\nu})$, $\nu = 1, 2, \ldots, k$, $\gamma^{(m)} = \alpha > 0$ and for $t \in [t_{2\nu}, t_{2\nu+1})$, $\nu = 1, 2, \ldots, k$, $\gamma^{(m)} = -\beta$, $\beta > 0$.

Denote by $\Gamma_{2n,m}(\alpha, \beta)$ a set of all $(\alpha, \beta)$-perfect splines of order to $m$ having at most $2n$ knots on period,. It is clear that $\Gamma_{2n,m}(1, 1) = \Gamma_{2n,m}$.

**Proposition 6.1.5.** *Any spline $\gamma \in \Gamma_{2n,m}(\alpha, \beta)$ is represented by unique manner in the form*

$$\gamma(t) = a + (\alpha + \beta) \sum_{\nu=1}^{2k} (-1)^\nu B_{m+1}(t - t_\nu), \quad k \leq n, \tag{6.9}$$

*where $t_\nu$, $\nu = 1, 2, \ldots, 2k$ are the points of some partition (6.8). For these points next condition*

$$(\alpha + \beta) \sum_{\nu=1}^{2k} (-1)^\nu t_\nu + 2\pi\alpha = 0. \tag{6.10}$$

*is valid. In particular if* $\gamma \in \Gamma_{2n,m}$ *then*

$$\gamma(t) = a + 2\sum_{\nu=1}^{2k} (-1)^\nu B_{m+1}(t - t_\nu),$$

*and*

$$\sum_{\nu=1}^{2k} (-1)^\nu t_\nu = -\pi.$$

This statement immediately follows from the theorem 6.1.1 and evident equalities

$$0 = \int_0^{2\pi} \gamma^{(m)}(t)dt = \alpha \sum_{\nu=1}^{k} (t_{2\nu} - t_{2\nu-1}) - \beta \sum_{\nu=1}^{k} (t_{2\nu+1} - t_{2\nu}) =$$

$$(\alpha + \beta) \sum_{\nu=1}^{2k} (-1)^\nu t_\nu + 2\pi\alpha.$$

Note that the functions from the set $\Gamma_{2n,1}(\alpha, \beta)$ are uniformly continuous. From Arzell's theorem it follows that the set

$$\Gamma^0_{2n,1}(\alpha, \beta) = \{\gamma : \ \gamma \in \Gamma_{2n,1}(\alpha, \beta), \ \gamma \perp 1\}$$

is compact in the space $C$, moreover only polygonal function from the set $\Gamma^0_{2n,1}(\alpha, \beta)$ may be the limit for the sequence of polygonal functions from this set.

The sets $\Gamma_{2n,m}$ and $\Gamma_{2n,m}(\alpha, \beta)$ unlike $S_m(\Delta_N)$ aren't linear manifolds. The peculiarity of perfect splines doesn't allow to consider the problems of existence and uniqueness in view of theorem 6.1.3. However we may decide these problems for perfect splines having zeros at fixed points.

**Theorem 6.1.6.** *Let* $\alpha, \beta > 0$; $n, m \in N$ *and*

$$\Delta_{2n} = \{x_1 < x_2 < \ldots < x_{2n} < 2\pi \le x_1 + 2\pi = x_{2n+1}\}.$$

*Then there exist two and only two perfect* $(\alpha, \beta)$*-splines* $\gamma_1, \gamma_2 \in \Gamma_{2n,m}(\alpha, \beta)$ *such that*

$$\gamma_i(x_\nu) = 0, \ \ i = 1, 2, \ \ \nu = 1, 2, \ldots, 2n,$$

*and for* $t \in [x_1, x_1 + 2\pi)$

$$sgn\, \gamma_i(t) = (-1)^i sgn \prod_{\nu=1}^{2n+1} (t - x_\nu).$$

**Proof.** Let $f \in C_m$ is arbitrary function for which

$$sgn f^{(m)} = (-1)^{m+1} sgn \prod_{\nu=1}^{2n+1} (t - x_\nu)$$

and $s_0 \in S_{m-1}(\Delta_{2n})$ is the best $(\alpha, \beta)$-approximation spline (see section 1.1) for function $f$ in the space $L_1$. As for any spline $s \in S_{m-1}(\Delta_{2n})$ the difference $f^{(m-1)}(t) - s^{(m-1)}(t)$ on each from intervals $(x_\nu, x_{\nu+1})$ has at most one zero, then the difference $f(t) - s(t)$ has at most $2n$ zeros on period. According to the criterion of best $(\alpha, \beta)$-approximation element by subspace (theorem 1.1.9) the function

$$h(t) = \alpha\, sgn(f(t) - s_0(t))_+ - \beta\, sgn(f(t) - s_0(t))_-$$

is orthogonal to the subspace $S_{m-1}(\Delta_{2n})$, moreover

$$||f - s_0||_{1;\alpha,\beta} = \int_0^{2\pi} f(t)h(t)dt. \tag{6.11}$$

¿From proposition 6.1.2 it follows that if $h_m(t)$ is $m$-th periodic integral from function $h(t)$ such that $h_m(x_1) = 0$ then next condition $h \perp S_{m-1}(\Delta_{2n})$ is equivalent to conditions $h_m(x_\nu) = 0$, $\nu = 1, 2, \ldots, 2n$. It is clear that $h_m \in \Gamma_{2n,m}(\alpha, \beta)$ since $h(t)$ and hence $h_m(t)$ change sign on period at most $2n$ times. Thus

$$sgn\, h_m(t) = \varepsilon\, sgn\, f^{(m)}(t), \quad |\varepsilon| = 1.$$

Besides that, in view of (6.11)

$$(-1)^m \int_0^{2\pi} f^{(m)}(t)h_m(t)dt = \int_0^{2\pi} f(t)h_m^{(m)}(t)dt =$$

$$\int_0^{2\pi} f(t)h(t)dt = ||f - s_0||_{1;\alpha,\beta},$$

thus

$$sgn\, h_m(t) = (-1)^m sgn f^{(m)}(t) = -sgn \prod_0^{2\pi} (t - x_\nu),$$

and we may put that $\gamma_1(t) = h_m(t)$.

To prove the existence of $\gamma_2(t)$ it sufficies to change a sign of the function $f(t)$.

Now prove the uniqueness. If there exist two perfect splines $g_1, g_* \in \Gamma_{2n,m}(\alpha, \beta)$ sutisfying the conditions of theorem, then under the suitable choice of $\lambda > 0$, $\lambda \neq 0$ the function $\delta(t) = \gamma_1(t) - \lambda\gamma_*(t)$ will have at least $2n+1$ different zeros on period $[0, 2\pi)$. But then in view of Rolle's theorem the function $\delta^{(m-1)}$ will possess the same property too. But in other hand $\delta^{(m)}(t)$ has only finite number of zeros and changes sign at most $2n$ times on period, therefore $\delta^{(m-1)}(t)$ changes sign at most $2n$ times on period.

**Theorem 6.1.7.** *Let $n, m \in N$,*

$$\Delta_h = \{z_1 < x_2 < \ldots < x_k < 2\pi \leq x_1 + 2\pi = x_{k+1}\},$$

*$\rho_1, \rho_2, \ldots, \rho_k$ are integral positive numbers such that*

$$\sum_{\nu=1}^{k} \rho_k = 2n. \tag{6.12}$$

*Then there exist the perfect splines $\gamma_i \in \Gamma_{2n,m}(\alpha, \beta)$ sutisfying the conditions*

$$\gamma_i^{(j)}(x_\nu) = 0, \;\; j = 0, 1, \ldots, \rho_\nu - 1, \;\; \nu = 1, 2, \ldots, k, \;\; i = 1, 2,$$

*and such that for $t \in (x_1, x_{k+1})$ the equalities*

$$sgn\, \gamma_i(t) = (-1)^i sgn \prod_{\nu=1}^{k+1} (t - x_\mu)^{\rho_\nu}$$

*are valid.*

**Proof.** Let $\varepsilon > 0$ and $\varepsilon < \min_\nu (x_{\nu+1} - x_\nu)/2$. On each of intervals $(x_\nu - \varepsilon, x_\nu + \varepsilon,\ \nu = 1, 2, \ldots, k$ select any $\rho_\nu$ different points. In view of previous theorem there exists the perfect spline $\gamma_\varepsilon \in \Gamma_{2n,m}(\alpha, \beta)$ vanishing at all of those points and such that

$$(-1)^{\rho_1+1}\gamma_\varepsilon(t) > 0, \quad t \in (x_1 + \varepsilon, x_2 - \varepsilon).$$

As the set $\Gamma_{2n,m}(\alpha, \beta)$ is locally compact in $C$ then there exists the spline $\gamma_* \in \Gamma_{2n,m}(\alpha, \beta)$ and sequence $\varepsilon_n$, $\varepsilon_n \to 0$ $(n \to \infty)$ such that

$$||\gamma_{\varepsilon_n} - \gamma_*||_C \to 0, \quad n \to \infty.$$

It is clear that $\gamma_*^{(j)}(x_\nu) = 0,\ j = 1, 2, \ldots, \rho_\nu - 1,\ \nu = 1, 2, \ldots, k$ and

$$sgn\, \gamma_*(t) = -sgn \prod_{\nu=1}^{k+1} (t - x_\nu),$$

so we may put $\gamma_1(t) = \gamma_*(t)$. By analogy we may establish the existence of spline $\gamma_2$. As in previous theorem the uniqueness of splines $\gamma_1(t)$ and $\gamma_2(t)$ is established with the help of Rolle's theorem.

Note that if $\alpha = \beta$ then for splines $\gamma_1(t)$ and $\gamma_2(t)$ that appeared in theorems 6.1.6 and 6.1.7 the equality $\gamma_1(t) = -\gamma_2(t)$ takes place.

¿From the theorem 6.1.7 it follows that for any partition $\Delta_k$ and any collection of numbers $\rho = \{\rho_\nu\}_{\nu=1}^k$ for which the condition (6.12) holds true there exists unique perfect spline $\gamma_m(\Delta_k, \rho, t) \in \Gamma_{2n,m}$ having $\rho_\nu$-divisible zeros at points $x_\nu$, $\nu = 1, 2, \ldots, k$ and being positive on interval $(x_1, x_2)$.

Let

$$\{t_1 < t_2 < \ldots < t_{2n} \leq 2\pi < t_1 + 2\pi = t_{2n+1}\} := \Delta_{2n}$$

are knots of perfect spline, that is defined by unique manner.

Denote by $s(f, \Delta_k, \rho, t)$ the spline from $S_{m-1}(\Delta_{2n})$ such that

$$s^{(j)}(f, \Delta_k, \rho, t_\nu) = f^{(j)}(x_\nu), \ \ j = 0, 1, \ldots, \rho_{\nu-1}, \ \ \nu = 1, 2, \ldots, k.$$

**Theorem 6.1.8.** *Let $n, m \in N$. For any function $f \in C^{m-1}$ the interpolational spline $s(f, \Delta_k, \rho, t)$ in the set $S_{m-1}(\Delta_{2n})$ exists and is unique. Moreover for all $t$*

$$\sup\{|f(t) - s(f, \Delta_k, \rho, t)| : \ f \in W_\infty^m\} = |\gamma_m(\Delta_k, \rho, t)|. \tag{6.13}$$

*and consequently for each monotonous norm (i.e. such that from inequality $|f(t)| \leq |g(t)|$ $(t \in (0, 2\pi))$ it follows that inequality $\|f\| \leq \|g\|$ holds true) the inequality*

$$\sup\{\|f - s(f, \Delta_k, \rho)\| : \ f \in W_\infty^m\} = \|\gamma_n(\Delta_k, \rho)\|$$

*is valid. In particular for all $p \in [0, \infty]$*

$$\sup\{\|f - s(f, \Delta_k, \rho)\|_p : \ f \in W_\infty^m\} = \|\gamma_n(\Delta_k, \rho)\|_p.$$

**Proof.** To prove the existence and uniqueness it is sufficiently to show that if $s_0 \in S_{m-1}(\Delta_{2n})$ and

$$s_0^{(j)}(t_\nu) = f^{(j)}(x_\nu), \ \ j = 0, 1, \ldots, \rho_{\nu-1}, \ \ \nu = 1, 2, \ldots, k. \tag{6.14}$$

then $s_0(t) \equiv 0$. Suppose that it is not true, i.e. there exists the spline $s \in S_{m-1}(\Delta_{2n})$ sutisfying the conditions (6.14) and such that $s(t_*) \neq 0$, $t_* \neq x_\nu$. Then for

$$\lambda = \gamma(t_*)/s(t_*) \ \ (\gamma(t) = \gamma_m(\Delta_k, \rho, t))$$

the difference $\gamma(t) - \lambda s(t)$ has zeros with multiplicity at least $\rho_\nu$ at points $x_\nu$, $\nu = 1, 2, \ldots, k$ and zero at point $t_*$. Hence on period $[0, 2\pi)$ this difference has at least $2n + 1$ zeros with the calculation of their multiplicities. But then in view of Rolle's theorem function

$$\gamma^{(m-1)}(t) - \lambda s^{(m-1)}(t)$$

has on period at least $2n+1$ zeros with the calculation of their multiplicities.

On the other hand the function $\gamma^{(m-1)}$ changes by turns the strict monotony on intervals of partition $\Delta_{2n}$. Thus for any spline $s \in S_{m-1}(\Delta_{2n})$ the difference $\gamma^{(m-1)}(t) - \lambda s^{(m-1)}(t)$ has on $[0, 2\pi)$ at most $2n$ zeros (at most one zero on each from intervals of partition $\Delta_{2n}$) but that contradicts to previous supposition.

Now prove (6.13). If the correlation (6.13) is not true, then there exist function $f \in W_\infty^m$ and point $t^*$ such that

$$|f(t^*) - s(t^*)| > |\gamma(t^*)|,$$

where

$$s(t) = s(f, \Delta_k, \rho, t), \quad \gamma(t) = \gamma_m(\Delta_k, \rho, t).$$

But then for

$$\lambda = \frac{\gamma(t^*)}{f(t^*) - s(t^*)}, \quad |\lambda| < 1$$

the difference

$$\delta(t) = \gamma(t) - \lambda(f(t) - s(t)),$$

and hence the function $\delta^{(m-1)}$ must have on period at least $2n$ zeros with the calculation of their multiplicities.

Besides that on each from intervals $(t_\nu, t_{\nu+1})$, $\nu = 1, 2, \ldots, 2n$ of partition $\Delta_{2n}$

$$\gamma^{(m)}(t) - \lambda f^{(m)}(t) = (-1)^\nu \varepsilon - \lambda f^{(m)}(t), \quad |\varepsilon| = 1$$

Therefore

$$(-1)^\nu \varepsilon - |\lambda| < \gamma^{(m)}(t) - \lambda f^{(m)}(t) < (-1)^\nu \varepsilon + |\lambda|,$$

thus $\gamma^{(m-1)}(t) - \lambda f^{(m-1)}(t)$ changes by turns strict monotony on intervals of partition $\Delta_{2n}$ and therefore for any spline $s \in S_{m-1}(\Delta_{2n})$ the difference $\gamma(t) - \lambda(f(t) - s(t))$ has on period $[0, 2\pi)$ at most $2n$ zeros, but that contradicts to previous.

Denote by $S_{2n,m}$ the set of $S_m(\Delta_{2n})$ by equidistant partition

$$\Delta_{2n} = \overline{\Delta}_{2n} = \{\nu\pi/n\}_{\nu=1}^{2n}$$

and by $s_{2n,m}(f,t)$ the spline from $S_{2n,m}$ interpolating $f(t)$ at points $\nu\pi/n$, $\nu = 1, 2, \ldots, 2n$ for odd $m$ and at points $(2\nu - 1)\pi/(2n)$, $\nu = 1, 2, \ldots, 2n$ for even $m$.

The interpolation knots of the spline $s_{2n,m}(t)$ are zeros of perfect spline $\varphi_{n,m+1} \in \Gamma_{2n,m+1}$ and hence in view of theorem 6.1.8 interpolational spline $s_{2n,m}(f,t)$ exists and is unique. Moreover the next statement follows from this theorem.

**Corollary 6.1.9.** *At each point t the next equality*

$$\sup\{|f(t) - s_{2n,m-1}(f,t)| : f \in W_\infty^m\} = |\varphi_{n,m}(t)|$$

*is valid and in the space with monotonous norm the following equalities*

$$\sup\{\|f - s_{2n,m-1}(f)\| : f \in W_\infty^m\} = \|\varphi_{n,m}\|,$$

*in particular*

$$\sup\{\|f - s_{2n,m-1}(f)\|_p : f \in W_\infty^m\} = \|\varphi_{n,m}\|_p, \ p \in [1, \infty],$$

*hold true.*

Denote by $s_{2n,m,*}(f,t)$ the spline from $S_{2n,m}$ such that

$$s_{2n,m,*}^{(j)}\left(f, \frac{2\nu\pi}{n}\right) = f^{(j)}\left(\frac{2\nu\pi}{n}\right), \ j = 0, 1, \ \nu = 1, 2, \ldots, n,$$

if the number $m$ is even and

$$s_{2n,m,*}^{(j)}\left(f, \frac{(4\nu - 1)\pi}{2n}\right) = f^{(j)}\left(\frac{(4\nu - 1)\pi}{2n}\right), \ j = 0, 1, \ \nu = 1, 2, \ldots, n,$$

if $m$ is odd.

The interpolation knots of spline $s_{2n,m,*}(f,t)$ are twice multiplle zeros of perfect spline $\varphi_{n,m}(t) + \|\varphi_{n,m}\|_\infty$ or $\varphi_{n,m}(t) - \|\varphi_{n,m}\|_\infty$. According to theorem 6.1.8 the spline $s_{2n,m,*}(f,t)$ exists and is unique and for it the next statement takes place.

**Corollary 6.1.10.** *Let $n \in N$, $m = 2, 3, \ldots$. For any space $X$ with monotonous norm $\|\cdot\|_X = \|\cdot\|$ the next equality*

$$\sup\{\|f - s_{2n,m-1,*}(f)\| : f \in W_\infty^m\} = \|\varphi_{n,m} + \|\varphi_{n,m}\|_\infty\|$$

*holds true.*

## 6.2 Estimates for derivatives in the space $C$

Present now a few inequalities for splines by equidistant partition after the model of inequalities from section 3.1 for trigonometric polynomials.

Begin from the direct analog of Bernstein's inequality.

**Theorem 6.2.1.** *Let $n, m \in N$. For any spline $s \in S_{2n,m}$ the exact inequalities*

$$\frac{||s^{(k)}||_\infty}{||\varphi_{n,m-k}||_\infty} \leq \frac{||s||_\infty}{||\varphi_{n,m}||_\infty}, \quad k = 1, 2, \ldots, m \tag{6.15}$$

*are valid.*

The next statement, that will be repeatedly used furthermore, are required to us.

Define for $s \in S_{2n,m}$ and $\varepsilon > 0$

$$\varphi_\varepsilon(t) = \left(\varepsilon + \frac{||s||_\infty}{||\varphi_{n,m}||_\infty}\right) \varphi_{n,m}(t). \tag{6.16}$$

**Lemma 6.2.2.** *For any $\tau$ the difference*

$$\delta_\varepsilon(t) = \varphi_\varepsilon(t) - s(t + \tau)$$

*has on period of function $\varphi_\varepsilon(t)$ two changes of sign.*

Really, the $m$-th derivative of $\delta_\tau^{(m)}(t)$ is the difference of two piece-constant functions with the knots $\nu\pi/n$, $\nu = 1, 2, \ldots, 2n$ and $\nu\pi/n - \tau$, $\nu = 1, 2, \ldots, 2n$ correspondingly and consequently changes sign on period $[0, 2\pi)$ at most $2n$ times. Hence in view of Rolle's theorem this fact takes place and for $\delta(t)$.

On the other hand for all $t$

$$\min_t \varphi_\varepsilon(t) < s(t) < \max_t \varphi_\varepsilon(t).$$

¿From here and from properties of function $\varphi_{n,m}(t)$ it follows that on each segment of monotony for function $\varphi_{n,m}(t)$ (and hence for function $\varphi_\varepsilon(t)$)) the difference $\delta_\tau(t)$ changes sign at least one time, that together with previous fact proves lemma.

**Proof of theorem 6.2.1.** In view of lemma 6.2.2 the functions $s(t)$ and $\varphi_\varepsilon(t)$ satisfy the conditions of lemma 1.4.1 and proposition 1.4.2. For any $\xi$ there exists $\eta$ such that $s(\xi) = \varphi_\varepsilon(\eta)$. In view of proposition 1.4.2

$$|s'(\xi)| < |\varphi'_\varepsilon(\eta)| \leq \|\varphi'_\varepsilon\|_\infty,$$

consequently taking into account the arbitrariness of $\xi$

$$\|s'\|_\infty \leq \|\varphi_\varepsilon\|_\infty = \left(\varepsilon + \frac{\|s\|_\infty}{\|\varphi_{n,m}\|_\infty}\right)\|\varphi'_{n,m}\|_\infty.$$

Taking into account now the arbitrariness of $\varepsilon > 0$ we obtain

$$\frac{\|s'\|_\infty}{\|\varphi_{n,m-1}\|_\infty} \leq \frac{\|s\|_\infty}{\|\varphi_{n,m}\|_\infty}, \quad k = 1, 2, \ldots, m,$$

i.e. inequality (6.15) is proved for $k = 1$. The general case is easy infered with the help of induction.

It remains to note that inequality becomes an equality for splines $s(t) = a\varphi_{n,m}(t)$.

By analogy from proposition 1.4.5 we may obtain

**Proposition 6.2.3.** *Let $n, m \in N$ and $\alpha \in R$. For any spline $s \in S_{2n,m}$ the exact inequality*

$$\frac{\|s^{(k)} + \alpha s^{(k-1)}\|_\infty}{\|\varphi_{n,m-k} + \alpha\varphi_{n,m-k+1}\|_\infty} \leq \frac{\|s\|_\infty}{\|\varphi_{n,m}\|_\infty}, \quad k = 1, 2, \ldots, m \tag{6.17}$$

*is valid.*

Note also that in view of theorem 6.2.1 the next estimate

$$\frac{\|s^{(k)}\|_\infty}{\|\varphi_{n,m-k}\|_\infty} \leq \frac{\min\{\|s - \lambda\|_\infty : \lambda \in R\}}{\|\varphi_{n,m}\|_\infty}, \quad k = 1, 2, \ldots, m$$

holds true.

**Corollary 6.2.4.** *For any constant sign spline $s \in S_{2n,m}$ the exact inequality*

$$\frac{\|s^{(k)}\|_\infty}{\|\varphi_{n,m-k}\|_\infty} \leq \frac{\|s\|_\infty}{2\|\varphi_{n,m}\|_\infty}, \quad k = 1, 2, \ldots, m \tag{6.18}$$

*holds true.*

The next statememt intensifies the theorem 6.2.1.

**Theorem 6.2.5.** *Let $n, m \in N$ and $k = 1, 2, \dots, m$. For any spline $s \in S_{2n,m}$ the exact inequalities*

$$\frac{||s^{(k)}||_\infty}{||\varphi_{n,m-k}||_\infty} \leq \frac{||\Delta_h^k(s)||_\infty}{||\Delta_h^k \varphi_{n,m}||_\infty} \leq \frac{||s||_\infty}{||\varphi_{n,m}||_\infty} \tag{6.19}$$

*holds true for $h \in (0, \pi/(2kn))$ and $h \in (0, \pi/(4kn))$.*

**Proof.** Let $\delta_{1,h}(t)$ is even $2\pi$-periodic function which equals to 1 for $|t| < h$ and 0 for $t \in [h, \pi]$, and $\delta_{k,h}(t)$ is a convolution of $k$ functions $\delta_{1,h}(t)$, i.e.

$$\delta_{k,h} = \delta_{k-1,h} * \delta_{1,h}.$$

It is clear that for $f \in L_1^1$

$$\Delta_h^1(f, t) = \int_0^{2\pi} \delta_{1,h}(t-u) f'(u) du = (\delta_{1,h} * f')(t).$$

Therefore if $f \in L_1^k$ then

$$\Delta_h^k(f, t) = (\delta_{k,h} * f^{(k)})(t) = \int_0^{2\pi} \delta_{k,h}(t-u) f^{(k)}(u) du. \tag{6.20}$$

The functions $s^{(k)}(t)$ and

$$\psi_\varepsilon(t) = \left(\varepsilon + \frac{||s^{(k)}||_\infty}{||\varphi_{n,m-k}||_\infty}\right) \varphi_{n,m-k}(t), \;\; \varepsilon \geq 0,$$

in view of lemma 6.2.2 satisfy the conditions of lemma 1.4.1. Let $||s^{(k)}||_\infty = s^{(k)}(t_*)$ and for clearness $s^{(k)}(t_*) > 0$ and $\varphi_{n,m-k}(t') = ||\varphi_{n,m-k}||_\infty$. From lemma 1.4.1 it follows that for $t \in (-\pi/n, \pi/n)$

$$\psi_0(t' - t) \leq s^{(k)}(t_* - t). \tag{6.21}$$

Moreover, owing to (6.21)

$$||\Delta_h^k(s)||_\infty \geq \int_{-\pi}^{\pi} \delta_{k,h}(t) s^{(k)}(t_* - t) dt = \int_{-\pi/n}^{\pi/n} \delta_{k,h}(t) s^{(k)}(t_* - t) dt. \tag{6.22}$$

For $h \in (o, \pi/(2kn))$ the bearer of narrowing for function $\delta_{k,h}(t)$ on period $[-\pi, \pi)$ is contained in interval $(-\pi/n, \pi/n)$ and $\delta_{k,h}(t) \geq 0$.Taking into account this fact and the correlations (6.22) and (6.23) we obtain

$$||\Delta_h^k(s)||_\infty \geq \int_{-\pi/n}^{\pi/n} \delta_{k,h}(t) \psi_0(t' - t) dt =$$

$$\frac{\|s^{(k)}\|_\infty}{\|\varphi_{n,m-k}\|_\infty}\int_{-\pi/n}^{\pi/n}\delta_{k,h}(t)\varphi_{n,m-k}(t'-t)dt =$$

$$\frac{\|s^{(k)}\|_\infty}{\|\varphi_{n,m-k}\|_\infty}|\Delta_h^k(\varphi_{n,m}(t'))| = \frac{\|s^{(k)}\|_\infty}{\|\varphi_{n,m-k}\|_\infty}\|\Delta_h^k(\varphi_{n,m})\|_\infty.$$

The first inequality from (6.19) is proved.

Due to theorem 6.2.1 for $\nu = k-1, k$

$$\|s^{(\nu)}\|_\infty \le \|\varphi^{(\nu)}\|_\infty, \tag{6.23}$$

where the function $\varphi(t) = \varphi_\varepsilon(t)$ is defined by (6.16). ¿From the (6.24) and lemma 6.2.2 it follows that the functions $s^{(k-1)}$ and $\varphi^{(k-1)}$ satisfy the conditions of comparison theorem for rearrangements 1.4.6 and moreover to the conditions of lemma 1.4.1 and corollaries from it.

Denote by $r(f;\alpha;\beta,t)$ the rearrangement of narrfwing on segment $[\alpha,\beta]$ of function $f(t) \ge 0$ and we shall consider that $r(f;\alpha;\beta,t) = 0$ for $t > \beta-\alpha$.

Let $(\alpha_i,\beta_i)$ are composing intervals of the set

$$E = \{t:\ t\in(t_0,t_0+2\pi),\ s^{(k)} \ne 0\},$$

where $t_0$ is zero of $\varphi(t)$ and $(\alpha,\beta)$ is one from the constant sign segments of function $\varphi^{(k)}(t)$. Prove now that for all $t \ge 0$

$$\int_0^t r(|s^{(k)}|;\alpha_i;\beta_i,u)du \le \int_0^t r(|\varphi^{(k)}|;\alpha;\beta,u)du. \tag{6.24}$$

In view of inequality (6.23) for $\nu = k$

$$r(|s^{(k)}|;\alpha_i;\beta_i,0) = \max_{\alpha_i\le t\le\beta_i}|s^{(k)}(t)| \le \|s^{(k)}\|_\infty <$$

$$\|\varphi^{(k)}\|_\infty = r(|\varphi^{(k)}|;\alpha;\beta,0), \tag{6.25}$$

and owing to (6.23) for $\nu = k-1$

$$\int_0^{\beta_i-\alpha_i} r(|s^{(k)}|;\alpha_i;\beta_i,u)du = |\int_{\alpha_i}^{\beta_i} s^{(k)}(t)dt = \left|s^{(k-1)}(\beta_i) - s^{(k-1)}(\alpha_i)\right| \le$$

$$2\|s^{(k-1)}\|_\infty < 2\|\varphi^{(k-1)}\|_\infty = \left|\int_\alpha^\beta \varphi^{(k)}(t)dt\right| =$$

$$\int_0^{\pi/n} r(|\varphi^{(k)}|;\alpha;\beta,t)dt. \tag{6.26}$$

Prove now that if

$$r(|s^{(k)}|;\alpha_i;\beta_i,\tau)=r(|\varphi^{(k)}|;\alpha;\beta,\tau)=a, \tag{6.27}$$

then

$$|r'(|s^{(k)}|;\alpha_i;\beta_i,\tau)|\le|r'(|\varphi^{(k)}|;\alpha;\beta,\tau)|. \tag{6.28}$$

Let for clearness $s^{(k)}(t)>0$ for $\alpha_i<t<\beta_i$, points $\alpha_i'$ and $\beta_i'$, $\alpha_i<\alpha_i',\beta_i'<\beta_i$ are such that $\beta_i'-\alpha_i'=\tau$ and $s^{(k)}(\alpha_i')=s^{(k)}(\beta_i')=a$, and points $\alpha'$ and $\beta'$, $\alpha<\alpha'<\beta'<\beta$ are such that $\beta'-\alpha'=\tau$ and

$$|\varphi^{(k)}(\alpha')|=|\varphi^{(k)}(\beta')|=a.$$

Due to proposition 1.4.2

$$|s^{(k+1)}(\alpha_i')|<\left|\varphi^{(k+1)}(\alpha')\right|$$

and

$$\left|s^{(k+1)}(\beta_i')\right|<\left|\varphi^{(k+1)}(\beta')\right|.$$

Besides that in view of proposition 1.3.2

$$\left|r'(|s^{(k)}|;\alpha_i;\beta_i,\tau)\right|=\left(\frac{1}{s^{(k+1)}(\alpha_i')}-\frac{1}{s^{(k+1)}(\beta_i')}\right)^{-1}.$$

Therefore

$$\left|r'(|s^{(k)}|;\alpha_i;\beta_i,\tau)\right|\le\left|\frac{1}{\varphi^{(k+1)}(\alpha')}-\frac{1}{\varphi^{(k+1)}(\beta')}\right|^{-1}=$$

$$\left|r'(|\varphi^{(k+1)}|;\alpha;\beta,\tau)\right|$$

¿From the correlations (6.25), (6.27) and (6.28) it follows that the difference

$$r(|\varphi^{(k)}|;\alpha;\beta,t)-r(|s^{(k)}|;\alpha_i;\beta_i,t)$$

changes sign on $(0,\infty)$ (from plus to minus) at most one time. From here and from (6.26) it follows that inequality (6.24) is true.

Note also that from proposition 1.4.2 and lemma 1.4.1 it follows that if $[\alpha_i,\beta_i]\subset[\alpha,\beta]$ then for all $t\in(\alpha_i,\beta_i)$

$$|s^{(k)}(t)|\le|\varphi^{(k)}(t)|. \tag{6.29}$$

Let now

$$||\Delta_h^k(s)||_\infty=|\Delta_h^k(s,t_0)|=\left|\int_{t_0-\pi}^{t_0+\pi}\delta_h^k(t-t_0)s^{(k)}(t)dt\right| \tag{6.30}$$

and

$$\|\Delta_h^k(\varphi)\|_\infty = \left| \int_{t_1-\pi}^{t_1+\pi} \delta_h^k(t-t_1)\varphi^{(k)}(t)dt \right|, \quad \alpha \le t_1 \le \beta. \tag{6.31}$$

Without the loss of community we may consider that $t_1 = t_0$. Two cases are possible:

1) exactly one from intervals $I_i = (\alpha_i, \beta_i)$ intersects with interval $(\alpha, \beta)$;

2) there are at least two such intervals $I_i, I_{i+1}, \dots, I_{i+l}$.

In the first case with the help of proposition 1.3.3 and owing to (6.30) we obtain

$$\|\Delta_h^k(s)\|_\infty \le \int_0^{2\pi} r(|s^{(k)}|; I_i, t) r(\delta_h^k, t)dt.$$

¿From here, from (6.24) and from proposition 1.3.14 we find

$$\|\Delta_h^k(s)\|_\infty \le \int_0^{2\pi} r(|\varphi^{(k)}|; I_i, t) r(\delta_h^k, t)dt.$$

and in view of (6.31) and properties of the function $\varphi(t)$ the right part of last inequality has the form

$$\left| \int_{t_1-\pi}^{t_1+\pi} \delta_h^k(t-t_1)\varphi^{(k)}(t)dt \right| = |\Delta_h^k(\varphi), t_1)| = \|\Delta_h^k(\varphi)\|_\infty.$$

Thus in first case

$$\|\Delta_h^k(s)\|_\infty \le \|\Delta_h^k(\varphi)\|_\infty = \left( \varepsilon + \frac{\|s\|_\infty}{\|\varphi_{n,m}\|_\infty} \right) \left\|\Delta_h^k(\varphi_{n,m})\right\|_\infty.$$

In second case

$$\|\Delta_h^k(s)\|_\infty = \left| \int_{t_0-\pi}^{t_0+\pi} \delta_h^k(t-t_0) s^{(k)}(t)dt \right| =$$

$$\sum_{\nu=i}^{i+l} \int_{\alpha_\nu}^{\beta_\nu} \delta_h^k(t-t_0)|s^{(k)}(t)|dt. \tag{6.32}$$

If $(\alpha_\nu, \beta_\nu) \subset (\alpha, \beta)$ then due to (6.19)

$$\int_{\alpha_\nu}^{\beta_\nu} \delta_h^k(t-t_0)|s^{(k)}(t)|dt \le \int_{\alpha_\nu}^{\beta_\nu} \delta_h^k(t-t_0)|\varphi^{(k)}(t)|dt. \tag{6.33}$$

If $\alpha \in (\alpha_i, \beta_i)$ then in view of lemma 1.4.1

$$|s^{(k)}(\beta_i - t)| < |\varphi^{(k)}(t)|, \ \ t \in (\alpha, \beta),$$

and hence

$$r(|s^{(k)}|;\alpha;\beta_i,t) \le r(|\varphi^{(k)}|;\alpha;\beta_i,t), \;\; t\in(\alpha,\beta_i). \tag{6.34}$$

By analogy we can verify that if $\beta \in (\alpha_{i+l},\beta_{i+l})$ then

$$r(|s^{(k)}|;\alpha_{i+l};\beta,t) \le r(|\varphi^{(k)}|;\alpha_{i+l};\beta,t), \;\; t\in(\alpha_{i+l},\beta). \tag{6.35}$$

Now we use the proposition 1.4.2. For $h \in (0,\pi/(2kn))$ the bearer of function $\delta_h^k(t-t_0)$ is contained at the interval $[\alpha,\beta]$ and therefore in view of (6.34) and (6.35)

$$\int_{\alpha_i}^{\beta_i} \delta_h^k(t-t_0)|s^{(k)}(t)|dt = \int_{\alpha}^{\beta_i} \delta_h^k(t-t_0)|s^{(k)}(t)|dt \le$$

$$\int_0^{\beta_i-\alpha} r(\delta_h^k(\cdot-t_0);\alpha;\beta_i,t) r(|s^{(k)}|;\alpha;\beta_i,t)dt \le$$

$$\int_0^{\beta_i-\alpha} r(\delta_h^k(\cdot-t_0);\alpha;\beta_i,t) r(|\varphi^{(k)}|;\alpha;\beta_i,t)dt, \tag{6.36}$$

and

$$\int_{\alpha_{i+l}}^{\beta_{i+l}} \delta_h^k(t-t_0)|s^{(k)}(t)|dt = \int_{\alpha_{i+l}}^{\beta} \delta_h^k(t-t_0)|s^{(k)}(t)|dt \le$$

$$\int_0^{\beta-\alpha_{i+l}} r(\delta_h^k(\cdot-t_0);\alpha_{i+l};\beta,t) r(|s^{(k)}|;\alpha_{i+l};\beta,t)dt \le$$

$$\int_0^{\beta-\alpha_{i+l}} r(\delta_h^k(\cdot-t_0);\alpha_{i+l};\beta,t) r(|\varphi^{(k)}|;\alpha_{i+l};\beta,t)dt \tag{6.37}$$

and due to (6.33) for $\nu$ so that $(\alpha_i,\beta_i)\subset(\alpha,\beta)$

$$\int_{\alpha_\nu}^{\beta_\nu} \delta_h^k(t-t_0)|s^{(k)}(t)|dt \le$$

$$\int_0^{\beta_\nu-\alpha_\nu} r(\delta_h^k(\cdot-t_0);\alpha_\nu;\beta_\nu,t) r(|\varphi^{(k)}|;\alpha_\nu;\beta_\nu,t)dt. \tag{6.38}$$

Moreover, if $a=t_0<t_1<\ldots<t_n=b$ then for any summable on $[a,b]$ functions $f(t)$ and $g(t)$

$$\sum_{\nu=1}^{n}\int_0^{t_\nu-t_{\nu-1}} r(|f|;t_{\nu-1};t_\nu,t) r(|g|;t_{\nu-1};t_\nu,t)dt \le$$

$$\int_0^{b-a} r(|f|;a;b,t)r(|g|;a;b,t)dt.$$

¿From here and from correlations (6.32) and (6.36)-(6.38) it follows that

$$\left\|\Delta_h^k(s)\right\|_\infty \leq \int_0^{\beta-\alpha} r(\delta_h^k(\cdot - t_0);\alpha;\beta,t)r(|\varphi^{(k)}|;\alpha;\beta,t)dt =$$

$$\int_{t_0-\pi}^{t_0+\pi} \delta_h^k(t-t_0)\varphi^{(k)}(t)dt = \left\|\Delta_h^k(\varphi)\right\|_\infty =$$

$$\left(\varepsilon + \frac{||s||_\infty}{||\varphi_{n,m}||_\infty}\right)\left\|\Delta_h^k(\varphi_{n,m})\right\|_\infty.$$

Thus for each $\varepsilon > 0$ and $h \in (0, \pi/(4kn))$

$$||\Delta_h^k(s)||_\infty \leq$$

$$\left(\varepsilon + \frac{||s||_\infty}{||\varphi_{n,m}||_\infty}\right)\left\|\Delta_h^k(\varphi_{n,m})\right\|_\infty.$$

that due to arbitrariness of $\varepsilon > 0$ gives us second inequality (6.19).

Inequalities become in equalities for $s(t) = a\varphi_{n,m}(t)$.

**Corollary 6.2.6.** *Let $n, m \in N$, $k = 1, 2, \ldots, m-1$ and $h_n = \pi/(2n)$. For each spline $s \in S_{2n,m}$ the next exact inequalities*

$$\frac{||s^{(k)}||_\infty}{||\varphi_{n,m-k}||_\infty} \leq \frac{||\Delta_{h_n}^k(s)||_\infty}{2^k||\varphi_{n,m}||_\infty} \leq \frac{||s||_\infty}{||\varphi_{n,m}||_\infty}. \tag{6.39}$$

*are valid.*

Really, in view of theorem 6.2.5

$$\frac{||s^{(k)}||_\infty}{||\varphi_{n,m-k}||_\infty} \leq \frac{||\Delta_{h_n}^1(s^{(k-1)})||_\infty}{2||\varphi_{n,m-k+1}||_\infty}.$$

But

$$\Delta_{h_n}^1(s^{(k-1)}) \in S_{n,m-k+1}$$

and hence

$$\frac{||\Delta_{h_n}^1(s^{(k-1)})||_\infty}{2||\varphi_{n,1}||_\infty} \leq \frac{||\Delta_{h_n}^2(s^{(k-2)})||_\infty}{2^2||\varphi_{n,2}||_\infty} \leq \cdots \frac{||\Delta_{h_n}^k(s)||_\infty}{2^k||\varphi_{n,m}||_\infty}.$$

The next follows from inequality

$$||\Delta_{h_n}^k(s)||_\infty \leq 2^k||s||_\infty. \tag{6.40}$$

Now we adduce the making more precise of inequality (6.1) that is analogous to the theorem 3.1.3.

Consider the mapping $F : R \to R_2$ which is defined for $m > 1$ by means of formula

$$F(t) = (\varphi_{1,m}(t), \varphi_{1,m-1}(t)).$$

The image $F(R)$ of the set $R$ by such mapping is the curve on the plane $R_2$ restricting the convex central-symmetric and symmetric as to coordinate axises body $A$. The function of Minkowski

$$\psi_A(u_1, u_2) = \inf\{t; \, t > 0, \, t^{-1}(u_1, u_2) \in A\}$$

is the monotonous norm in $R_2$. The monotony of the norm we understand so, that from the inequalities $|u_1'| \leq |u_1''|$ and $|u_2'| \leq |u_2''|$ it follows that $\psi_A(u_1', u_2') \leq \psi_A(u_1'', u_2'')$. It is clear that

$$\psi_A(\varphi_{1,m}(t), \varphi_{1,m-1}(t)) \equiv 1. \tag{6.41}$$

**Theorem 6.2.7.** *For any spline $s \in S_{2n,m}$, $n = 1, 2, \ldots$, $m = 2, 3, \ldots$ for all $t$ the exact inequality*

$$\psi_A(n|s(t)|, |s'(t)|) \leq n\frac{||s||_\infty}{||\varphi_{1,m}||_\infty} \tag{6.42}$$

*takes place.*

Note that in terms of Minkowski's functions we may formulate theorem 3.1.3 concerning trigonometric polynomials . As mapping $F$ in this case we will consider the $F(t) = (\cos t, \sin t)$, the set $A$ will be the unit disk , and corresponding monotonous norm will be Euclid's norm in $R_2$.

**The proof of the theorem 6.2.7.** Let again

$$\varphi(t) = \left(\varepsilon + \frac{||s||_\infty}{||\varphi_{n.m}||_\infty}\right)\varphi_{n,m}(t), \;\; \varepsilon > 0.$$

Fix $t$ and choose the point $y$ from conditions $s(t) = \varphi_\varepsilon(y)$ and $s'(t)\varphi_\varepsilon'(y) \geq 0$. In view of lemma 6.2.2 the functions $s(t)$ and $\varphi_\varepsilon(t)$ satisfy the conditions of proposition 1.4.2 and due to this proposition

$$|s'(t)| \leq |\varphi_\varepsilon'(y)|.$$

But then from the monotony of the norm $\psi_A(u_1, u_2)$ it follows that

$$\psi_A(n|s(t)|, |s'(t)|) \le \psi_A(n|\varphi_\varepsilon(y)|, |\varphi'_\varepsilon(y)|).$$

At the limit for $\varepsilon \to 0$ we arrive to inequality

$$\psi_A(n|s(t)|, |s'(t)|) \le \psi_A(n|\varphi_0(y)|, |\varphi'_0(y)|) =$$

$$\frac{||s||_\infty}{||\varphi_{n,m}||_\infty}\psi_A(n|\varphi_{n,m}(y)|, |\varphi'_{n,m}(y)|) =$$

$$\frac{||s||_\infty}{||n^{m-1}\varphi_{n,m}||_\infty}\psi_A(n|\varphi_{1,m}(y)|, |\varphi'_{1,m}(y)|) = \frac{||s||_\infty}{||n^{m-1}\varphi_{n.m}||_\infty} = n\frac{||s||_\infty}{||\varphi_{1,m}||_\infty}.$$

For each $t$ inequality (6.42) becomes an equality for $s(y) = a\varphi_{n,m}$.

The theorem 6.2.7 we may formulate in such equivalent form.

**Theorem 6.2.7'.** *For any spline $s \in S_{2n,m}$, $n = 1, 2, \ldots$, $m = 2, 3, \ldots$ for any numbers $\alpha$ and $\beta$ the exact inequality*

$$||\alpha s + \beta s'||_\infty \le \frac{||s||_\infty}{||\varphi_{n,m}||_\infty} \left\|\alpha\varphi_{n,m} + \beta\varphi'_{n,m}\right\|_\infty . \tag{6.43}$$

*takes place.*

If now $p \in [1, \infty)$ then passing in both parts (6.43) to upper bounds by $\alpha$ and $\beta$ such that

$$|\alpha|^{p'} + |\beta|^{p'} \le 1, \;\; 1/p + 1/p' = 1,$$

we obtain

$$\left|||\alpha s|^p + |\beta s'|^p\right||_\infty^{1/p} \le \frac{||s||_\infty}{||\varphi_{n,m}||_\infty} \left\||\alpha\varphi_{n,m}|^p + |\beta\varphi'_{n,m}|^p\right\|_\infty^{1/p} .$$

Note that for $\alpha = 1$, $\beta = 0$ the inequality becomes an equality (6.15) for $k = 1$.

Present still one making more precise of the theorem 6.2.1.

**Theorem 6.2.8.** *Let $n, m \in N$,*

$$s_i = s\left(\frac{i\pi}{n}\right), \;\; s_{i-1/2} = s\left(\frac{(2i-1)\pi}{2n}\right), \;\; i = 0, \pm 1, \ldots,$$

*and*

$$\Delta a_i = a_i - a_{i-1}, \;\; \Delta^k a_i = \Delta\Delta^{k-1} a_i$$

*For any spline $s \in S_{2n,m}$ for odd $m$*

$$\frac{\|s^{(k)}\|_\infty}{\|\varphi_{n,m-k}\|_\infty} \le \frac{\max_i |\Delta^k(s_i)|}{2^k\|\varphi_{n,m}\|_\infty} \le \frac{|\max_i s_i|}{\|\varphi_{n,m}\|_\infty}, \quad k = m, m-1 \tag{6.44}$$

*and for even $m$*

$$\frac{\|s^{(k)}\|_\infty}{\|\varphi_{n,m-k}\|_\infty} \le \frac{\max_i |\Delta^k(s_{i-1/2})}{2^k\|\varphi_{n,m}\|_\infty} \le \frac{|\max_i s_{i-1/2}|}{\|\varphi_{n,m}\|_\infty}, \quad k = m, m-1. \tag{6.45}$$

**Proof.** The proof of this proposition we demonstrate for cases when $m = 2$ and $m = 3$.

Let as before $\delta_{k,h}$ is the convolution of $k$ functions $\delta_{1,h}(t)$ where $\delta_{1,h}(t)$ is the even $2\pi$-periodic function which equals to one for $|t| \le h$ and equals to zero for $h < |t| \le \pi$.Thus

$$\delta_{k,h} = \int_{t-h}^{t+h} \delta_{k-1,h}(u)du.$$

For $h = \pi/(2n)$ the function

$$B_{n,m}(t) = c_m \delta_{m,h}(t),$$

where

$$c_m = \left( \sum_{|k| \le mh/2} \delta_{m,h}(kh) \right)^{-1}$$

is called B-spline of order $m$. It is easy to show that for odd $m$ any spline $s \in S_{2n,m}$ is represented by unique manner in the form

$$s(t) = \sum_{k=1}^{2n} a_k B_{n,m}\left(t - \frac{k\pi}{n}\right),$$

and for even $m$ in the form

$$s(t) = \sum_{k=1}^{2n} a_k B_{n,m}\left(t - \frac{(2k-1)\pi}{2n}\right).$$

Let at first $m = 3$. Taking into account that

$$B_{n,3}(0) = \frac{2}{3}, \quad B_{n,3}(\pm h) = \frac{1}{6},$$

it is easy to see that

$$s(ih) = \frac{1}{6}\left(c_{i-1} + 4c_i + c_{i+1}\right) = c_i + \frac{1}{6}\Delta^2 c_{i+1}.$$

¿From here it follows that

$$s(t) = \sum_{i=1}^{2n}\left(\sum_{k=0}^{\infty}\left(-\frac{1}{6}\right)^k \Delta^{2k} s_{k+i}\right) B_{n,3}\left(t - \frac{i\pi}{n}\right). \tag{6.46}$$

With the help of direct verification it is easy to see

$$s''\left(\frac{i\pi}{n}\right) = \frac{n^2}{\pi^2}\Delta^2 c_{i+1}. \tag{6.47}$$

Therefore

$$\|s''\|_\infty = \max_i \left|s''\left(\frac{i\pi}{n}\right)\right| = \frac{n^2}{\pi^2}\left|\sum_{k=0}^{\infty}\left(-\frac{1}{6}\right)^k \Delta^{2k+2} s_{k+i+1}\right| \le$$

$$\frac{n^2}{\pi^2}\max_i \sum_{k=0}^{\infty}\left(\frac{1}{6}\right)^k \left|\Delta^{2k+2} s_{k+i+1}\right| \le$$

$$\frac{n^2}{\pi^2}\sum_{k=0}^{\infty}\left(\frac{1}{6}\right)^k \max_i \left|\Delta^{2k+2} s_{i+k+1}\right| = \frac{n^2}{\pi^2}\sum_{k=0}^{\infty}\left(\frac{1}{6}\right)^k \max_i \left|\Delta^{2k+2} s_i\right|.$$

Besides that

$$\max_i \left|\Delta^{2k+2} s_i\right| \le 4^k \max_i |\Delta^2 s_i|. \tag{6.48}$$

Thus

$$\|s''\|_\infty \le \frac{n^2}{\pi^2}\max_i |\Delta^2 s_i| \sum_{k=0}^{\infty}\left(\frac{2}{3}\right)^k = \frac{3n^2}{\pi^2}\max_i |\Delta^2 s_i|,$$

that is equivalent to first equality from (6.44) for $m = 3$ and $k = 2$.

¿From (6.46) it follows that for $t \in (i\pi/n, (i+1)\pi/n)$

$$s'''(t) = \frac{n^3}{\pi^3}\Delta^3 c_{i+2}.$$

Using this fact by analogy with previous we obtain the inequality (6.44) for $m = k = 3$.

By the same way we may obtain the inequalities (6.44) for $m = 2$ using the equality

$$s(t) = \sum_{i=1}^{2n}\left(\sum_{k=0}^{\infty}\left(-\frac{1}{8}\right)^k \Delta^{2k} s_{k+i-1/2}\right) B_{n,2}\left(t - \frac{(2i-1)\pi}{2n}\right). \tag{6.49}$$

With the help of most difficult reasonings the justice of theorem 6.2.8 for all $m$ was established in the work of Y.N.Subbotin [2].

The representations (6.46) and (6.47) are useful and in other problems.

For example from (6.46) it immediately follows that for any function $f \in C$

$$s_{2n,3}(f,t) == \sum_{i=1}^{2n} \left( \sum_{k=0}^{\infty} \left(-\frac{1}{8}\right)^k \Delta^{2k} f_{k+i} \right) B_{n,2}\left(t - \frac{i\pi}{n}\right).$$

and from (6.48) it follows that for any function $f \in C$

$$s_{2n,2}(f,t) = \sum_{i=1}^{2n} \left( \sum_{k=0}^{\infty} \left(-\frac{1}{8}\right)^k \Delta^{2k} f_{k+i-1/2} \right) B_{n,2}\left(t - \frac{(2i-1)\pi}{2n}\right),$$

where

$$s_i = s\left(\frac{i\pi}{n}\right), \quad s_{i-1/2} = s\left(\frac{(2i-1)\pi}{2n}\right), \quad i = 0, \pm 1, \ldots.$$

Besides that from (6.46) and (6.47) it follows that for $s \in S_{2n,3}$

$$\|s\|_\infty \le \max_i |s_i| + \frac{1}{2}\left|\Delta^2 s_{i+1}\right| \le 3 \max_i |s_i|$$

and for $s \in S_{2n,2}$

$$\|s\|_\infty \le \max_i |s_{i-1/2}| + \frac{1}{4}\left|\Delta^2 s_{i-1/2}\right| \le 2 \max_i |s_{i-1/2}|.$$

Let

$$f_h(t) = \frac{1}{2h}\delta_{1,h} * f(t) = \frac{1}{2h}\int_{t-h}^{t+h} f(t-u)du$$

is the Steklov's function,

$$g_0(t) = \delta_{1,h}(t), \ g_1(t) = \frac{1}{2}\left(1 - \frac{t}{h}\right), \quad for \ \ t \in (0,h)$$

and for $t \in (-h, 0)$ $g_1(t) = g_(t)$ and

$$g_m(t) = \int_{\gamma_m} g_{m-1}(u)du,$$

where $\gamma_m$ is chuse such that

$$\int_{-h}^{h} g_m(t)dt = 0.$$

Put moreover for $\nu = 0, 1, 2, \ldots$

$$G_{2\nu}(t) = g_{2\nu}(t) - g_{2\nu}(h), \quad G_{2\nu+1}(t) = g_{2\nu+1}(t).$$

Note that

$$G_n(\pm h) = 0, \quad m = 1, 2, \ldots, \quad G'_m(t) = g_{m-1}(t), \; m = 2, 3, \ldots.$$

Consider the operator

$$A_{h,m}(f,t) = f(t) + 2h \sum_{\nu=0}^{[(m-1)/2]} g_{2\nu}(h) f_h^{(2\nu)}(t). \tag{6.50}$$

**Lemma 6.2.9.** *For any function $f \in L_1^m$ the next equality*

$$A_{h,m}(f,t) = \int_{-h}^{h} G_m(u) f^{(m)}(t-u)du \tag{6.51}$$

*holds true.*

**Proof.** Let $m = 1$. Then from (6.50) we get that

$$A_{h,1}(f,t) = f(t) - f_h(t). \tag{6.52}$$

Using the integrating by parts we have

$$\int_0^h G_1(u) f'(t-u)du = \frac{1}{2} f(t) - \frac{1}{2h} \int_0^h f(t-u)du$$

and

$$\int_{-h}^0 G_1(u) f'(t-u)du = \frac{1}{2} f(t) - \frac{1}{2h} \int_{-h}^0 f(t-u)du.$$

¿From this and from (6.52) we obtain (6.51) for $m = 1$.

Now let the statement of lemma is true for $m = 1, 2, \ldots, k-1$. Consider two cases:

1)k=2j. In view of (6.50)

$$A_{h,2j}(f,t) = A_{h,2j-1}(f,t)$$

and hence, taking into account the supposition of induction, we obtain

$$A_{h,2j}(f,t) = \int_{-h}^{h} G_{2j-1}(u) f^{(2j-1)}(t-u)du =$$

$$\int_{-h}^{h} f^{(2j-1)}(t-u)dG_{2j}(u) = \int_{-h}^{h} f^{(2j)}(t-u)G_{2j}(u)du,$$

as so $G_{2j}(\pm h) = 0$.

2)k=2j+1. With the help of (6.50) and taking into account the supposition of induction, we obtain

$$A_{h,2j+1}(f,t) = A_{h,2j} + 2hg_{2j}(h)f_h^{(2j)} =$$

$$\int_{-h}^{h} f^{(2j)}(t-u)G_{2j}(u)du + g_{2j}(h)\int_{-h}^{h} f^{(2j)}(t-u)du =$$

$$\int_{-h}^{h} g_{2j}(u)f^{(2j)}(t-u)du = \int_{-h}^{h} G_{2j}(u)f^{(2j)}(t-u)du.$$

In the lemma 6.2.9 $h = \pi/(2n)$ and $f(t) = \varphi_{n,m}(t)$ if $m$ is odd and $f(t) = \varphi_{n,m}(t+h)$ if $m$ is even. So we obtain the next proposition.

**Lemma 6.2.10.** *For $n, m \in N$ and $h = \pi/(2n)$ the equalities*

$$\int_{-h}^{h} |G_m(t)|dt + 2\sum_{\nu=1}^{[(m-1)/2]} |g_{2\nu}(h)|||\varphi_{n,m-2\nu+1}||_\infty = ||\varphi_{n,m}||_\infty$$

*are valid.*

**Theorem 6.2.11.** *Let $n, m \in N$, $k = 1, 2, \ldots, m$ and $h_n = \pi/(2n)$. For any spline $s \in S_{2n,m}$ the exact inequality*

$$\frac{||s^{(k)}||_\infty}{||\varphi_{n,m-k}||_\infty} \leq q(1-\eta_{m,k})\frac{||\Delta(h_n{}^{k+2}(s)||_\infty}{2^{k+2}||\varphi_{n,m}||_\infty} + \eta_{m,k}\frac{||\Delta(h_n{}^{k}(s)||_\infty}{2^{k}||\varphi_{n,m}||_\infty} \quad (6.53)$$

*is valid. Here*

$$\eta_{m,k} = \left(\frac{2}{\pi}\right)^k \frac{||\varphi_{1,m}||_\infty}{||\varphi_{1,m-k}||_\infty}.$$

*Note that*

$$\eta_{1,1} = 1,\ \eta_{2,1} = \frac{1}{2},\ \eta_{3,1} = \frac{2}{3},\ \eta_{4,1} = \frac{16}{25}, \ldots,$$

$$\eta_{2,2} = \frac{1}{2},\ \eta_{3,2} = \frac{5}{12}\ldots,$$

$$\eta_{i,1} < \eta_{i+1,1},\ i = 1, 2, \ldots,\ \lim_{m\to\infty} \eta_{m,k} = \left(\frac{2}{\pi}\right)^k.$$

**Proof.** In view of lemma 6.1.10 for any spline $s \in S_{2n,m}$

$$\|s'\|_\infty = \left\| A_{h,m} - 2h \sum_{\nu=0}^{[(m-2)/2]} g_{2\nu}(h) s_h^{(2\nu+1)} \right\|_\infty \leq$$

$$\|A_{h,m-1}\|_\infty + 2h \sum_{\nu=0}^{[(m-2)/2]} |g_{2\nu+1}(h)| \, \|s_h^{(2\nu)}\|_\infty \leq$$

$$\|s^{(m)}\|_\infty \int_{-h}^{h} |G_{m-1}(t)| dt + 2h \sum_{\nu=1}^{[(m-1)/2]} |g_{2\nu}(h)| \, \|s_h^{(2\nu+1)}\|_\infty \leq$$

$$\|s^{(m)}\|_\infty \int_{-h}^{h} |G_{m-1}(t)| dt + \sum_{\nu=1}^{[(m-1)/2]} |g_{2\nu}(h)| \, \|\Delta s_h^{(2\nu)}\|_\infty.$$

Puting $h = h_n$ and applying the theorem 6.2.6 we have

$$\|s'\|_\infty \leq \frac{n}{\pi} \frac{\|\Delta s\|_\infty}{+} \int_{-h_n}^{h_n} |G_{m-1}(t)| dt \frac{\|\Delta^{m-1} s\|_\infty}{2^{m-1} \|\varphi_{1,m-1}} +$$

$$\sum_{\nu=1}^{[(m-2)/2]} |g_{2\nu}(h)| \frac{\|\Delta^{2\nu+1} s\|_\infty \, \|\varphi_{m-\nu}\|_\infty}{2^{2\nu} \|\varphi_{m-1}\|_\infty}.$$

¿From here and from evidental inequality

$$\|\Delta^k s\|_\infty \leq 2^{k-2} \|\Delta^2 s\|_\infty$$

we obtain

$$\|s'\|_\infty \leq \frac{n}{\pi} \|\Delta s\|_\infty +$$

$$\int_{-h_n}^{h_n} |G_{m-1}(t)| dt \frac{\|\Delta^2 s\|_\infty}{4 \|\varphi_{1,m-1}} + \sum_{\nu=1}^{[(m-2)/2]} |g_{2\nu}(h)| \frac{\|\Delta^2 s\|_\infty \, \|\varphi_{m-\nu}\|_\infty}{4 \|\varphi_{m-1}\|_\infty}$$

Applying now the statement of lemma 6.2.10, we get

$$\|s'\|_\infty \leq \frac{n}{\pi} \|\Delta s\|_\infty + \left( \|\varphi_{n,m-1}\|_\infty - \frac{2n}{\pi} \|\varphi_{n,m}\|_\infty \right) \|\Delta^2 s\|_\infty,$$

that is equivalent to the statement of theorem for $k = 1$. For $k \geq 2$ the statement of theoren we can easy prove with the help of induction.

## 6.3 Inequalities in the spaces $L_1$ and $L_2$

At first we present the analog of inequality (6.15) in the space $L_1$.

**Theorem 6.3.1.** *Let $n, m \in N$ and $k = 1, 2, \ldots, m$. For any spline $s(t) \in S_{2n,m}$ the exact inequalities*

$$\frac{\bigvee_0^{2\pi}(s^{(m)})}{\bigvee_0^{2\pi}(\varphi_{n,0})} \leq \frac{||s||_1}{||\varphi_{n,m}||_1}; \tag{6.54}$$

*and*

$$\frac{||s^{(k)}||_1}{||\varphi_{n,m-k}||_1} \leq \frac{||s||_1}{||\varphi_{n,m}||_1} \tag{6.55}$$

*hold true.*

**Proof.** If for $t \in (\nu\pi/n, (\nu+1)\pi/n)$ $s^{(m)}(t) = a_\nu$, $\nu = 1, 2, \ldots, 2n$, then (puting $h_n = \pi/(2n)$)

$$\bigvee_0^{2\pi}(s^{(m)}) = \sum_{\nu=1}^{2n} |a_\nu - a_{\nu-1}| = \frac{n}{\pi}\int_0^{2\pi} |\Delta_{h_n}^1(s^{(m)}, t)|dt =$$

$$\frac{\bigvee_0^{2\pi}(\varphi_{n,0})\left\|\Delta_{h_n}^1(s^{(m)})\right\|_1}{2||\varphi_{n,0}||_1},$$

i.e.

$$\frac{\bigvee_0^{2\pi}(s^{(m)})}{\bigvee_0^{2\pi}(\varphi_{n,0})} = \frac{\left\|\Delta_{h_n}^1(s^{(m)})\right\|_1}{2||\varphi_{n,0}||_1}, \tag{6.56}$$

and consequently

$$\frac{\bigvee_0^{2\pi}(s^{(m)})}{\bigvee_0^{2\pi}(\varphi_{n,0})} \leq \frac{||s^{(m)}||_1}{||\varphi_{n,0}||_1}. \tag{6.57}$$

Besides due to Stain's inequality (theorem 1.7.4)

$$\left(\frac{||s^{(k)}||_1}{||\varphi_{1,m-k+1}||_\infty \bigvee_0^{2\pi}(s^{(m)})}\right)^{1/(m-k+1)} \leq$$

$$\left(\frac{||s||_1}{||\varphi_{1,m+1}||_\infty \bigvee_0^{2\pi}(s^{(m)})}\right)^{1/(m+1)},$$

or that is the same

$$\left(\frac{||s^{(k)}||_1 \bigvee_0^{2\pi}(\varphi_{n,0})}{\bigvee_0^{2\pi}(s^{(m)})||\varphi_{n,m-k}||_1}\right)^{1/(m-k+1)} \leq \left(\frac{||s||_1 \bigvee_0^{2\pi}(\varphi_{n,0})}{\bigvee_0^{2\pi}(s^{(m)})||\varphi_{n,m}||_1}\right)^{1/(m+1)}. \tag{6.58}$$

¿From here (for $k = m$) and from (6.58) the next inequality immediately follows

$$\bigvee_0^{2\pi}(\varphi_{n,0})||s||_1 \geq \bigvee_0^{2\pi}(s^{(m)})||\varphi_{n,m}||_1,$$

which is equivalent (6.55). Applying now (6.55) in (6.59) we obtain

$$\frac{\|s^{(k)}\|_1}{\|\varphi_{n,m-k}\|_1} \le \left(\frac{\|s\|_1}{\|\varphi_{n,m}\|_1}\right)^{(m-k+1)/(m+1)} \left(\frac{\bigvee_0^{2\pi}(s^{(m)})}{\bigvee_0^{2\pi}\varphi_{n,0}}\right)^{k/(m+1)} \le$$

$$\left(\frac{\|s\|_1}{\|\varphi_{n,m}\|_1}\right)^{(m-k+1)/(m+1)} \left(\frac{\|s\|_1}{\|\varphi_{n,m}\|_1}\right)^{k/(m+1)} = \frac{\|s\|_1}{\|\varphi_{n,m}\|_1}.$$

Using now the method of finite differences, we can intensify theorem 6.3.1 and in particular we can obtain the analog of theorem 6.2.6.

**Theorem 6.3.2.** *Let $n, m \in N$, $k = 1, 2, \ldots, m$ and $h = \pi/(2n)$. For any spline $s \in S_{2n,m}$ the exact inequalities*

$$\frac{\bigvee_0^{2\pi}(s^{(m)})}{\bigvee_0^{2\pi}(\varphi_{n,0})} \le \frac{\|\Delta_{h_n}^{m+1}(s)\|_1}{2^{m+1}\|\varphi_{n,m}\|_1} \tag{6.59}$$

*and*

$$\frac{\|s^{(k)}\|_1}{\|\varphi_{n,m-k}\|_1} \le \frac{\|\Delta_{h_n}^{k} s\|_1}{2^k\|\varphi_{n,m}\|_1} \tag{6.60}$$

*hold true.*

**Proof.** Taking into account the definition of the operator $A_{h,m}(f,t)$ (see the previous section) we have

$$\|s'\|_1 \le \|A_{h,m}(f')\|_1 + 2h \sum_{\nu=0}^{[(m-1)/2]} |g_{2\nu}(h)| \|s_h^{(2\nu+1)}\|_1. \tag{6.61}$$

But owing to lemma 6.2.9 we establish that

$$A_{h,m}(s', t) = \int_{-h}^{h} G_m(u) ds^{(m)}(t-u)$$

and consequently

$$\|A_{h,m}\|_1 \le \bigvee_0^{2\pi} s^{(m)} \int_{-h}^{h} |G_m(u)| d(u). \tag{6.62}$$

Besides that

$$2h\|s_h^{2\nu+1}\|_1 = \left\|\Delta_h^1(s^{(2\nu)})\right\|_1. \tag{6.63}$$

Comparing the correlations (6.61)-(6.63) and (6.56) we obtain

$$||s'||_1 \leq 2n \frac{||\Delta_h^1(s^{(2\nu)})||_1}{||\varphi_{n,0}||_1} \int_{-h}^{h} |G_m(u)|d(u) +$$

$$\sum_{\nu=0}^{[(m-1)/2]} |g_{2\nu}(h)|||\Delta_h^1(s^{(2\nu)})||_1. \tag{6.64}$$

As $\Delta_{h_n}^1(s^{(k)}) \in S_{2n,m-k}$ so

$$\left\|\Delta_{h_n}^1(s^{(k)})\right\|_1 \leq ||\varphi_{n,m-k}||_1 \frac{\Delta_{h_n}^1(s)}{||\varphi_{n,m}||_1} = 4n||\varphi_{n,m-k+1}||_\infty \frac{\Delta_{h_n}^1(s)}{||\varphi_{n,m}||_1}$$

and

$$||s'||_1 \leq$$

$$2n \frac{\Delta_{h_n}^1(s)}{||\varphi_{n,m}||_1} \left( \int_{-h_n}^{h_n} |G_m(t)|dt + 2 \sum_{\nu=0}^{[(m-1)/2]} |g_{2\nu}(h_n)|||\varphi_{n,m-2\nu+1}||_\infty \right).$$

Applying now lemma 6.3.1 we find that

$$||s'||_1 \leq 2n \frac{\Delta_{h_n}^1(s)}{||\varphi_{n,m}||_1} ||\varphi_{n,m}||_\infty$$

or

$$\frac{||s'||_1}{||\varphi_{n,m-1}||_1} \leq \frac{||\Delta_{h_n}^k s||_1}{2||\varphi_{n,m}||_1}.$$

Thus the inequality (6.59) for $k = 1$ is valid. Suppose that this inequality is valid for $k = l$. Then

$$\frac{||s^{(l+1)}||_1}{||\varphi_{n,m-l-1}||_1} \leq \frac{||\Delta_{h_n}^k(s)||_1}{2^l||\varphi_{n,m}||_1}$$

and

$$\frac{||\Delta_{h_n}^1(s')|_1}{||\varphi_{n,m-l-1}||_1} \leq \frac{||\Delta_{h_n}^1(\Delta_{h_n}^l(s))||_1}{2||\varphi_{n,m}||_1} = \frac{||\Delta_{h_n}^{l+1}(s)||_1}{2||\varphi_{n,m}||_1},$$

i.e.

$$\frac{||s^{(l+1)})||_1}{||\varphi_{n,m-l-1}||_1} \leq \frac{||\Delta_{h_n}^{l+1}(s)||_1}{2^{l+1}||\varphi_{n,m}||_1}.$$

So the inequality (6.59) is true for $k = 1, 2 \ldots, m$ and, in particular,

$$\frac{||\Delta_{h_n}^1(s^{(m)})||_1}{||\varphi_{n,0}||_1} \leq \frac{||\Delta_{h_n}^{m+1}(s^{(m)})||_1}{||2^m \varphi_{n,m}||_1}.$$

¿From here and from the inequality (6.56) we immediately obtain the inequality (6.60).

The theorem 6.3.2 also allows some intensification, namely the next theorem is true.

**Theorem 6.3.3.** *Let* $n = 1, 2, \dots$, $m = 2, 3, \dots$ *and* $k = 1, 2, \dots, m$. *Then for* $h_n = \pi/(2n)$ *and*

$$\theta_{m,k} = \left(\frac{2}{\pi}\right)^k \frac{\|\varphi_{n,m+1}\|_\infty}{\|\varphi_{n,m+1-k}\|_\infty},$$

*the next exact inequalities*

$$\frac{\|s^{(k)}\|_1}{\|\varphi_{n,m-k}\|_1} \le \frac{1}{2^k\|\varphi_{n,m}\|_1}\left(\theta_{m,k}\|\Delta^k_{h_n}(s)\|_1 + \frac{1}{4}(1-\theta_{m,k})\|\Delta^{k+2}_{h_n}(s)\|_1\right) \tag{6.65}$$

*and*

$$\frac{\bigvee_0^{2\pi}(s^{(m)})}{\bigvee_0^{2\pi}(\varphi_{n,0})} \le$$

$$\frac{1}{2^{m+1}\|\varphi_{n,m}\|_1}\left(\theta_{m,m+1}\|\Delta^{m+1}_{h_n}(s)\|_1 + \frac{1}{4}(1-\theta_{m,m+1})\|\Delta^{m+3}_{h_n}(s)\|_1\right) \tag{6.66}$$

*are valid.*

**Proof.** From the inequalities (6.64) and (6.59) taking into account that $g_0(h) = 1/(2h)$ we obtain

$$\|s'\|_1 \le 4n\frac{\Delta^{m+1}_{h_n}(s)}{2^{m+1}\|\varphi_{n,m}\|_1}\int_{-h_n}^{h_n}|G_m(t)|dt+$$

$$+2\sum_{\nu=1}^{[(m-1)/2]}|g_{2\nu}(h_n)|\frac{\Delta^{2\nu+1}_{h_n}(s)}{2^{2\nu}\|\varphi_{n,m}\|_1} + \frac{1}{h_n}\|\Delta^1_{h_n}(s)\|_1 \le$$

$$4n\frac{\Delta^3_{h_n}(s)}{8\|\varphi_{n,m}\|_1}\left(\int_{-h_n}^{h_n}|G_m(t)|dt + 2\sum_{\nu=1}^{[(m-1)/2]}|g_{2\nu}(h_n)|\|\varphi_{n,m-2\nu+1}\|_\infty\right)+$$

$$\frac{n}{\pi}\|\Delta^1_{h_n}(s)\|_1.$$

¿From here and from the lemma 6.2.10 we infer that

$$\|s'\|_1 \le 4n\frac{\Delta^3_{h_n}(s)}{8\|\varphi_{n,m}\|_1}\left(\|\varphi_{n,m}\|_\infty - \frac{2n}{\pi}\|\varphi_{n,m+1}\|_\infty\right) + \frac{n}{\pi}\|\Delta^1_{h_n}(s)\|_1 =$$

$$\frac{\Delta^3_{h_n}(s)}{8||\varphi_{n,m}||_1}\left(||\varphi_{n,m-1}||_1 - \frac{2n}{\pi}||\varphi_{n,m}||_1\right) + \frac{n}{\pi}||\Delta^1_{h_n}(s)||_1,$$

or that is the same

$$\frac{||s'||_1}{||\varphi_{n,m-1}||_1}\frac{1}{2||\varphi_{n,m}||_1}\left(\theta_{m,1}||\Delta^1_{h_n}(s)||_1 + \frac{1}{4}(1-\theta_{m,1})||\Delta^3_{h_n}(s)||_1\right),$$

i.e. the inequality (6.65) for $k = 1$ is valid. With the help of induction we verify that it is true for all $k = 1, 2, \ldots, m$. In particular from (6.65) it follows that when $k = m$ then

$$\frac{||\Delta^1_{h_n}(s^{(m)})||_1}{||\varphi_{n,0}||_1} \leq$$

$$\frac{1}{2^m||\varphi_{n,m}||_1}\left(\theta_{m,m}||\Delta^m_{h_n}(s)||_1 + \frac{1}{4}(1-\theta_{m,m})||\Delta^{m+2}_{h_n}(s)||_1\right)$$

and to complete the proof it remains to use the inequality (6.56).

The analogs of theorem 6.2.5 and corollary 6.2.6 take place and in the metric $L_2$.

**Theorem 6.3.4.** *For all $n, m \in N$, $k = 1, 2, \ldots, m$ and any spline $s \in S_{2n,m}$ the next exact inequalities*

$$\frac{||s^{(k)}||_2}{||\varphi_{n,m-k}||_2)} \leq \frac{||\Delta^k_h(s)||_2}{||\Delta^k_h(\varphi_{n,m})||_2)} \leq \frac{||s||_2}{||\varphi_{n,m}||_2)} \tag{6.67}$$

*are valid for $h \in (0, \pi/(4kn))$ and $h \in (0, \pi/(8kn))$ correspondingly.*

*Besides if $h_n = \pi/(2n)$ then*

$$\frac{||s^{(k)}||_2}{||\varphi_{n,m-k}||_2} \leq \frac{||\Delta^k_{h_n}(s)||_2}{2^k||\varphi_{n,m}||_2} \leq \frac{||s||_2}{||\varphi_{n,m}||_2}. \tag{6.68}$$

**Proof.** From the theorem 6.1.1 it follows that any spline $s \in S_{2n,m}$ may be represented by unique manner in the form

$$s(t) = c + \sum_{\nu=1}^{2n} c_\nu B_{m+1}\left(\frac{\nu\pi}{n} - t\right), \tag{6.69}$$

where

$$\sum_{\nu=1}^{2n} c_\nu = 0$$

and $B_m(t)$ is Bernully's function.

Denote by $s_{m+1}(t)$ the $(m+1)$-th periodic integral of function

$$s_0(t) = s(t) - \frac{1}{2\pi}\int_0^{2\pi} s(t)dt = \sum_{\nu=1}^{2n} c_\nu B_{m+1}\left(\frac{\nu\pi}{n} - t\right),$$

i.e.

$$s_{m+1} = B_{m+1} * s_0$$

and put

$$s_*(t) = \sum_{\nu=1}^{2n} c_\nu s_{m+1}\left(\frac{\nu\pi}{n} + t\right).$$

It is clear that $s_* \in S_{2n,2m+1}$. In view of (6.69)

$$\int_0^{2\pi} s^{(k)}(u-t)s^{(k)}(u)du =$$

$$\int_0^{2\pi} s^{(k)}(u)\sum_{\nu=1}^{2n} c_\nu B_{m+1}\left(\frac{\nu\pi}{n} + t - u\right)du = s_*^{(2k)}(t). \tag{6.70}$$

By analogy we can establish the next equalities

$$\int_0^{2\pi} \Delta_h^k(s, u-t)\Delta_h^k(s, u)du == \Delta_h^{2k}(s_*, t) \tag{6.71}$$

and

$$\int_0^{2\pi} s_0(u-t)s_0(u)du = s_*(t). \tag{6.72}$$

Note now that for any function $f \in L_2$ the relation

$$||f||_2^2 = \max_t \int_0^{2\pi} f(u-t)f(u)du \tag{6.73}$$

is valid.

Really due to Holder's inequality for any $t$

$$\int_0^{2\pi} f(u-t)f(u)du \le ||f||_2||f(\cdot - t)||_2 = ||f||_2^2$$

and moreover

$$\max_t \int_0^{2\pi} f(u-t)f(u)du \ge \int_0^{2\pi} f(u)f(u)du = ||f||_2^2.$$

¿From (6.70) and (6.73) it follows that

$$||s^{(k)}||_2^2 = |s_*^{(2k)}(0)| = \left\| s_*^{(2k)} \right\|_\infty . \tag{6.74}$$

The first inequality (6.19) gives us when $h \in (0, \pi/(4kn))$

$$\frac{||s_*^{(2k)}||_\infty}{||\varphi_{n,2m-2k+1}||_\infty} \leq \frac{||\Delta_h^{2k}(s_*)||_\infty}{||\Delta_h^{2k}(\varphi_{n,2m+1})||_\infty}, \tag{6.75}$$

and owing to (6.55) and (6.73)

$$||\Delta_h^{2k}(s_*)||_\infty = \max_t |\Delta_h^{2k}(S_*, t)| =$$

$$\max_t \left| \int_0^{2\pi} \Delta_h^k(s, u-t)\Delta_h^k(s, u)du \right| = ||\Delta_h^k(s)||_2^2. \tag{6.76}$$

Comparing the correlations (6.74)-(7.76) and taking into account that for $m = 1, 2, \ldots$

$$\left\| \Delta_h^{2k}(\varphi_{n,2m+1}) \right\|_\infty = \frac{1}{4n} \left\| \Delta_h^k(\varphi_{n,m}) \right\|_2^2 \tag{6.77}$$

and

$$||\varphi_{n,2m+1}||_\infty = \frac{1}{4n} \left\| \Delta_h^k \varphi_{n,m} \right\|_2^2 \tag{6.78}$$

we obtain the first inequality in (7.64).

Due to second inequality in (6.19) for $h \in (0, \pi/(8kn))$ we establish that

$$\frac{||\Delta_h^{2k}(s_*)||_\infty}{||\Delta_h^{2k}(\varphi_{n,2m+1})||_\infty} \leq \frac{||s_*||_\infty}{||\varphi_{n,2m+1}||_\infty}. \tag{6.79}$$

Comparing now the correlations (6.72), (6.73), (6.76) and (3.79) and taking into account (6.60) and (6.78) we obtain

$$\frac{||\Delta_h^k(s)||_2}{||\Delta_h^k(\varphi_{n,m})||_2} \leq \frac{||s||_2}{||\varphi_{n,m}||_2}.$$

Inequalities (6.68) immediately follow from the corollary 6.2.6.

The sign of equaliy in (6.67) and (6.68) takes place for $s(t) = a\varphi_{n,m}(t)$.

## 6.4 Inequalities of different metrics

At present section the extremal properties of splines are expressed by inequalities connecting their characteristics in metrics of different functional spaces.

**Proposition 6.4.1** *Let $n, m \in N$ and $k = 1, 2, \ldots, m$. For any $s \in S_{2n,m}$ the following exact inequalities hold true:*

$$\frac{\|s^{(k)}\|_1}{\|\varphi_{n,m-k}\|_1} \le \frac{\|\Delta^{k-1}_{\pi/(2n)}(s)\|_\infty}{2^{k-1}\|\varphi_{n,m}\|_\infty} \le \frac{\|s\|_\infty}{\|\varphi_{n,m}\|_\infty}; \tag{6.80}$$

$$\frac{\|s^{(k)}\|_1}{\|\varphi_{n,m-k}\|_1} \le \frac{\|\Delta^{k-1}_h(s)\|_\infty}{\|\Delta^{k-1}_h(\varphi_{n,m})\|_\infty}, \quad h \in \left(0, \frac{\pi}{2kn}\right); \tag{6.81}$$

*and for $1 \le k + l \le m + 1$*

$$\frac{\|\Delta^{k-1}_h(s^{(l)})\|_1}{\|\Delta^{k-1}_h(\varphi_{n,m-l})\|_1} \le \frac{\|s\|_\infty}{\|\varphi_{n,m}\|_\infty}, \quad h \in \left(0, \frac{\pi}{4kn}\right). \tag{6.82}$$

**Proof.** Any spline with $2n$ knots on the period $[0, 2\pi)$ changes it sign at most $2n$ times. Hence if $h_n = \pi/(2n)$ then

$$\|\Delta^{k-1}_{h_n}(s^{(k)})\|_1 = \bigvee_0^{2\pi} \Delta^{k-1}_{h_n}(s^{(k-1)}) \le 4n \left\|\Delta^{k-1}_{h_n}(s^{(k-1)})\right\|_\infty$$

and

$$\|s^{(k)}\|_1 \le 4n\|s^{(k-1)}\|_\infty. \tag{6.83}$$

¿From here and from the corollary 6.2.6 we find that

$$\frac{\|s^{(k)}\|_1}{\|\varphi_{n,m-k}\|_1} = \frac{\|s^{(k)}\|_1}{4n\|\varphi_{n,m-k+1}\|_\infty} \le \frac{\|s^{(k-1)}\|_\infty}{\|\varphi_{n,m-k+1}\|_\infty} \le$$

$$\frac{\|\Delta^{k-1}_{h_n}((s)\|_\infty}{\|2^{k-1}\varphi_{n,m}\|_\infty} \le \frac{\|s\|_\infty}{\|\varphi_{n,m}\|_\infty}.$$

By the same way, using first inequality in (6.19), we can obtain the inequality (6.81).

Moreover, from the representation (6.1) and from the equality (6.20) it follows that

$$\Delta^{k-1}_h(s^{(m-k+1,t)}) = \sum_{\nu=1}^{2n} a_\nu \delta_{k,h}\left(t - \frac{\nu\pi}{n}\right), \tag{6.84}$$

and hence for $h \in (0, \pi/(2kn))$ the bearers of the functions $\delta_{k,h}(t - \nu\pi/n)$ don't intersect and therefore the function (6.84) has at most $2n$ changes of sign on period. In view of Rolle's theorem for $l \leq m-k+1$ the function $\Delta_h^{k-1}(s^{(l)},t)$ has at most $2n$ changes of sign too. Consequently

$$\left\|\Delta_h^{k-1}(s^{(l)})\right\|_1 \leq 4n \left\|\Delta_h^{k-1}(s^{(l-1)})\right\|_\infty,$$

and to prove (6.82) it remains to use the second inequality in (6.19). To substitute in (6.80)-(6.82) metric $L_1$ by metric $L_p$ it's necessary to attract the device of rearrangements.

**Theorem 6.4.2.** *Let $n, m \in N$. For any spline $s \in S_{2n,m}$ the following exact inequalities hold true:*

$$|s^{(k)} - \lambda| \prec \frac{||\Delta_{\pi/(2n)}^{k-1}(s)||_\infty}{2^{k-1}||\varphi_{n,m}||_\infty} |\varphi_{n,m-k} - \lambda|; \tag{6.85}$$

$$|s^{(k)} - \lambda| \prec \frac{||\Delta_h^{k-1}(s)||_\infty}{||\Delta_h^{k-1}\varphi_{n,m}||_\infty} |\varphi_{n,m-k} - \lambda|, \quad 0 < h \leq \frac{\pi}{2(k-1)n}, \tag{6.86}$$

*and for $l \leq k+j \leq m$*

$$|\Delta_h^{k-1} s^{(j)} - \lambda| \prec \frac{||s||_\infty}{||\varphi_{n,m}||_\infty} |\Delta_h^{k-1}(\varphi_{n,m-j}) - \lambda|, \quad 0 < h \leq \frac{\pi}{4kn}. \tag{6.87}$$

**Proof.** Put

$$f(t) = \frac{s^{(k-1)}(t)}{||s^{(m)}||_\infty}.$$

Then owing to theorem 1.4.7 for any $\mu \in R$

$$|f' - \mu| \prec \left| \left( \frac{||f^{(k-1)}||_\infty}{||\varphi_{n,m-k}||_\infty} \right)^{(m-1)/m} \varphi_{n,m-k} - \mu \right|,$$

or that is the same

$$\left| \frac{s^{(k)}}{||s^{(m)}||_\infty} - \mu \right| \prec \left| \left( \frac{||s||_\infty}{||s^{(m)}||_\infty ||\varphi_{n,m-k+1}||_\infty} \right)^{\frac{m-k}{m-k+1}} \varphi_{n,m-k} - \mu \right|.$$

Puting now

$$\mu = \lambda ||s^{(m)}||_\infty,$$

we find

$$|s^{(k)} - \lambda| \prec \left| \left( \frac{||s^{(k-1)}||_\infty}{||||\varphi_{n,m-k+1}||_\infty} \right)^{\frac{m-k}{m-k+1}} \varphi_{n,m-k} ||s^{(m)}||_\infty^{\frac{1}{m-k+1}} - \lambda \right|.$$

¿From here and from the theorem 6.2.1 we obtain

$$|s^{(k)} - \lambda| \prec \left| \frac{||s^{(-1)}||_\infty}{||\varphi_{n,m-k+1}||_\infty} ||\varphi_{n,m-k} - \lambda \right|,$$

and to complete the proof of inequalities (6.85) and (6.86) we must apply theorem 6.2.5 and corollary 6.2.6.

Further for $0 < h \leq \pi/(2kn)$ and for any $a, b$ and $\tau$ the function

$$a\delta_{k,h}(t) + a\delta_{k,h}(t - \tau)$$

either does not change sign on period (if $a, b \geq 0$) or changes sign equally two times. From here and from equality (6.84) it follows that if

$$\xi(t) = \Delta_h^k(\varphi_{n,m}, t) - \Delta_h^k(s, t - \tau),$$

then $\xi^{(m-k+1)}(t)$ changes sign on period at most $2n$ times and the same fact takes place also for the function

$$\xi^{(j)}(t) = \Delta_h^k(\varphi_{n,m-j}, t) - \Delta_h^k(s^{(j)}, t - \tau).$$

Besides that in view of theorem 6.2.5

$$\frac{||\Delta_h^k(||s^{(j-1)})||_\infty}{||\Delta_h^k(\varphi_{n,m-j+1})||_\infty} \leq \frac{||s||_\infty}{||\varphi_{n,m}||_\infty},$$

so the functions $g(t) = \Delta_h^k(s^{(j)}, t)$ and

$$\varphi_\varepsilon(t) = \left( \frac{||s||_\infty}{||\varphi_{n,m}||_\infty} + \varepsilon \right) \Delta_h^k(\varphi_{n,m-j+1}, t), \quad \varepsilon > 0$$

satisfy the conditions of the theorem 1.4.5. Due to this theorem for any $\lambda \in R$

$$|g' - \lambda| \prec |\varphi_\varepsilon' - \lambda|$$

and owing to the arbitrariness of $\varepsilon > 0$

$$|g' - \lambda| \prec |\varepsilon_0' - \lambda|,$$

that is equivalent to (6.87).

Using the inequalities (6.85)-(6.87) and theorem 1.3.11 we establish the next proposition.

**Proposition 6.4.3.** *Let $n, m \in N$, $k = 1, 2, \ldots, m$ and $p \in [1, \infty]$. For any $s \in S_{2n,m}$ the following exact inequalities hold true:*

$$\frac{||s^{(k)}||_p}{||\varphi_{n,m-k}||_p} \leq \frac{||\Delta^{k-1}_{\pi/(2n)}(s)||_\infty}{2^{k-1}||\varphi_{n,m}||_\infty} \leq \frac{||s||_\infty}{||\varphi_{n,m}||_\infty}; \tag{6.88}$$

$$\frac{||s^{(k)}||_p}{||\varphi_{n,m-k}||_p} \leq \frac{\Delta^{k-1}_h(s)||_\infty}{||\Delta^{k-1}_h(\varphi_{n,m}||_\infty)}, \quad h \in \left(0, \frac{\pi}{2kn}\right); \tag{6.89}$$

*and for $1 \leq k + l \leq m$*

$$\frac{||\Delta^{k-1}_h(s^{(l)})||_p}{||\Delta^{k-1}_h(\varphi_{n,m-l})||_p} \leq \frac{||s||_\infty}{||\varphi_{n,m}||_\infty} \quad h \in \left(0, \frac{\pi}{4kn}\right). \tag{6.90}$$

Comparing the inequalities (6.85), (6.86) and the theorem 1.6.3 we establish the next exact estimates for $L_1$-norms of derivatives of the function $\tilde{s}(t)$ which are conjugate with the functions $s \in S_{2n,m}$.

**Proposition 6.4.4.** *Let $n, m \in N$, $m = 2, 3, \ldots$ and $k = 1, 2, \ldots, m$. For any $s \in S_{2n,m}$ the following exact inequalities hold true:*

$$\frac{||\tilde{s}^{(k)}||_1}{||\tilde{\varphi}_{n,m-k}||_1} \leq \frac{||s||_\infty}{||\varphi_{n,m}||_\infty}$$

*and for $0 < h < \pi/(2kn)$*

$$\frac{||\tilde{s}^{(k)}||_1}{||\tilde{\varphi}_{n,m-k}||_1} \leq \frac{||\Delta^{k-1}_h(s)||_\infty}{||\Delta^{k-1}_h(\varphi_{n,m})||_\infty},$$

*Inequalities become the equalities for*

$$s(t) = a\varphi_{n,m}(t).$$

Note that for derivatives of conjugate trigonometric polynomials in the chapter 3 we proved more general inequalities.

The next statement immediately follows from the inequality (6.85) and from the theorem 6.4.4.

**Theorem 6.4.5.** *Let $n, m \in N$, $m = 2, 3, \dots$ and $k = 1, 2, \dots, m$. For any $s \in S_{2n,m}$ the following exact inequalities hold true:*

$$\frac{\|\tilde{s}^{(k)}\|_1}{\|\tilde{\varphi}_{n,m-k}\|_1} \le \frac{\|\Delta^{k-1}_{\pi/(2n)}(\tilde{s})\|_\infty}{2^{k-1}\|\varphi_{n,m}\|_\infty} \le \frac{\|s\|_\infty}{\|\varphi_{n,m}\|_\infty}$$

*and for $1 \le k + l \le m$*

$$\frac{\|\Delta^{k-1}_h(\tilde{s}^{(l)})\|_1}{\|\Delta^{k-1}_h(\tilde{\varphi}_{n,m-l})\|_1} \le \frac{\|\Delta^{k-1}_{\pi/(2n)}(s)\|_\infty}{2^{k-1}\|\varphi_{n,m}\|_\infty} \le \frac{\|s\|_\infty}{\|\varphi_{n,m}\|_\infty}\ h \in \left(0, \frac{\pi}{4kn}\right).$$

**Theorem 6.4.6.** *Let $n, m \in N$ and $k = 1, 2, \dots, m$. For any spline $s \in S_{2n,m}$ and any $N$-function $\Phi$ the next exact inequality*

$$\int_0^{2\pi} \Phi(|\Delta^{k-1}_{\pi/(2n)}(s, t)|)dt \ge \int_0^{2\pi} \Phi\left(\frac{2^{k-1}\|s^{(k)}\|_1}{\|\varphi_{n,m-k}\|_1}|\varphi_{n,m}(t)|\right)dt,$$

*is valid and consequently for all $p \in [1, \infty]$*

$$\frac{|s^{(k)}\|_1}{\|\varphi_{n,m-k}\|_1} \le \frac{\|\Delta^{k-1}_{\pi/(2n)}(s)\|_p}{2^{k-1}\|\varphi_{n,m}\|_p} \le \frac{\|s\|_p}{\|\varphi_{n,m}\|_p}.$$

**Proof.** Let $s \in S_{2n,m}$, $s_h$ is the Steklov's function with the step $h$ of function $s$ and

$$f_h(t) = s_h(t)/\|s_h^{(m+1)}\|_1.$$

It is clear that $s_h \in W_1^{m+1}$ and $f'(t)$ changes sign on period at most $2n$ times. But then due to corollary 1.5.7 for any number $a > 0$ and any $N$-function $\Phi$ the inequality

$$\int_0^{2\pi} \Phi(a|f_h(t)|)dt \ge n(A_h)^{1/m} \int_0^{2\pi} \Phi(aA_h|\varphi_{n,m}(t)|/(4n))dt,$$

holds true. Here

$$A_h = 4\|f'\|_1/\|\varphi_{n,m-1}\|_1.$$

Puting $a = \vee_0^{2\pi}(s^{(m)})$ and passing to limit by $h \to 0$ we obtain

$$\int_0^{2\pi} \Phi(|s(t)|)dt \ge$$

$$n\left(\frac{4\|s'\|_1}{\|\varphi_{n,m-1}\|_1 \bigvee_0^{2\pi}(s^{(m)})}\right)^{1/m}\int_0^{2\pi}\Phi\left(\frac{\|s'\|_1}{n\|\varphi_{n,m-1}\|_1}|\varphi_{n,m}(t)|\right)dt =$$

$$\left(\frac{4\|s'\|_1 \bigvee_0^{2\pi}(\varphi_{n,0})}{\|\varphi_{n,m-1}\|_1 \bigvee_0^{2\pi}(s^{(m)})}\right)^{1/m}\int_0^{2\pi}\Phi\left(\frac{\|s'\|_1}{n\|\varphi_{n,m-1}\|_1}|\varphi_{n,m}(t)|\right)dt.$$

¿From here and from the theorem 6.3.1 we infer that for any spline $s \in S_{2n,m}$

$$\int_0^{2\pi}\Phi(|s(t)|)dt \geq \int_0^{2\pi}\Phi\left(\frac{\|s'\|_1}{n\|\varphi_{n,m-1}\|_1}|\varphi_{n,m-1}|\right)dt$$

and so $\Delta_{h_n}^{k-1} \in S_{2n,m}$, $h_n = \pi/(2n)$ then

$$\int_0^{2\pi}\Phi(|\Delta_{h_n}^{k-1}(s,t)|)dt \geq \int_0^{2\pi}\Phi\left(\frac{\|\Delta_{h_n}^{k-1}(s')\|_1}{\|\varphi_{n,m-1}\|_1}|\varphi_{n,m}(t)|\right)dt.$$

Moreover owing to the theorem 6.3.2

$$\frac{\|\Delta_{h_n}^{k-1}(s',t)\|_1}{\|\varphi_{n,m-1}\|_1} \geq 2^{k-1}\frac{\|s^{(k)}\|_1}{\|\varphi_{n,m}\|_1}$$

and this fact with previous one proves the theorem.

In conclusion of paragraph we shall give exact estimates of spline values and their derivatives through $L_2$-norms of splines.

**Theorem 6.4.7.** *Let $n, m \in N$, $k = 0, 1, \ldots, n$, $x \in R$. Then for arbitrary spline $s \in S_{2n,m}$ the following inequality takes place (when $k = 0$ we think, that $s^{(0)} = s(t) - \frac{1}{2\pi}\int_0^{2\pi} s(u)du$ )*

$$|s^{(k)}(t)| \leq \|s_{2m+1}^{(m+1)}(B_{m+1-k}(x-\cdot))\|_2\|s\|_2,$$

*where $s_{2m+1}(f)$ is the spline from $S_{2n,2m+1}$ interpolating the function $f$ at points $k\pi/n$, $k \in Z$ (the nodes of spline ). This inequality becomes an equality for*

$$s(x) = s_{2m+1}^{(m+1)}(B_{m+1-k}(x-\cdot),t) \in S_{2n,m}.$$

*Besides that for arbitrary $s \in S_{2n,m}$ the following exact inequality takes place*

$$|s(x)| \leq \|s_{2m+1}^{(m+1)}(B_{m+1}(x-\cdot)) + 1/(2\pi)\|_2\|s\|_2.$$

**Proof.** We have

$$s^{(k)}(x) = \int_0^{2\pi} B_1(x-t)s^{(k+1)}(t)dt =$$

$$\int_0^{2\pi} \left(B_1(x-t) - s_{2m+1}^{(m-k)}(B_{m+1-k}(x-\cdot),t)\right) s^{(k+1)}(t)dt.$$

Since the fuunction

$$B_1(x-t) - s_{2m+1}^{(m-k)}(B_{m+1-k}(x-\cdot),t) =$$

$$(B_{m+1-k}(x-t) - s_{2m+1}(B_{m+1-k}(x-\cdot),t))^{(m-k)}$$

is orthogonal to splines from $S_{2n,m-k-1}$ and $s^{(k+1)} \in S_{2n,m-k-1}$. That is why

$$s^{(k)}(x) = \int_0^{2\pi} s_{2m+1}^{(m-k)}(B_{m+1-k}(x-\cdot),t)s^{(k+1)}(t)dt$$

and after $(k+1)$-multiply integration by parts

$$|s^{(k)}(x)| = \left|\int_0^{2\pi} s_{2m+1}^{(m+1)}(B_{m+1-k}(x-\cdot),t)s(t)dt\right|.$$

¿From here, using inequality of Couchy-Bunyakowski, we obtain

$$|s^{(k)}(x)| \le \left\|s_{2m+1}^{(m+1)}(B_{m+1-k}(x-\cdot))\right\|_2 \|s\|_2.$$

In other hand, since $s = s_{2m+1}^{(m+1)}(B_{m+1-k}(x-\cdot)) \in S_{2n,m}$ and

$$|s^{(k)}(x)| = |s_{2m+1}^{(m+1)}(B_{m+1-k}(x-\cdot),x)| =$$

$$\left|\int_0^{2\pi} s_{2m+1}^{(m+2+k)}(B_{m+1-k}(x-\cdot),t)s_{2m+1}^{(m-k)}(B_{m+1-k}(x-\cdot),t)dt\right| =$$

$$\left|\int_0^{2\pi} s_{2m+1}^{(m+1)}(B_{m+1-k}(x-\cdot),t)s_{2m+1}^{(m+1)}(B_{m+1-k}(x-\cdot),t)dt\right| =$$

$$\left\|s_{2m+1}^{(m+1)}(B_{m+1-k}(x-\cdot))\right\|_2 \|s\|_2.$$

For arbitrary spline $s \in S_{2n,m}$ the following equality takes place

$$s(x) = \int_0^{2\pi} \left(s_{2m+1}^{(m+1)}(B_{m+1}(x-\cdot),t) + \frac{1}{2\pi}\right) s(t)dt$$

¿From this equality taking into account inequality of Couchy-Bunyakowski we obtain the second inequality of theorem.

The second inequality becomes an equality for the spline

$$s(t) = s_{2m+1}^{(m+1)}(B_{m+1}(x-\cdot),t) + \frac{1}{2\pi}.$$

## 6.5 Inequalities for generalized splines

Let $m = 1, 2, \ldots$

$$\Im_m(x) = \sum_{k=0}^{m} a_k x^{m-k} \tag{6.91}$$

is arbitrary polynomial of order $m$ having real coefficients and

$$\Im_m(D) = \sum_{k=0}^{m} a_k D^{m-k}, \quad D = \frac{d}{dx} \tag{6.92}$$

is the corresponding differential operator.

The function $s \in C^{m-1}$ is called periodic (corresponding to the operator $\Im_m(D)$) with $2n$ equidistant knots on period if $\Im_m(D)(s) \in S_{2n,0}$. The set of all such splines denote by $S_{2n,\Im_m}$.

It is clear that if $\Im_m(x) = x^m$ then $S_{2n,\Im_m} = S_{2n,m}$.

By $\varphi_{n,\Im_m}(t)$ denote the $2\pi/n$-periodic function having zero mean value on period such that

$$\Im_m(D)\varphi_{n,\Im_m}(t) = \varphi_{n,0}(t).$$

For $\Im_m(x) = x^m$ we have $\varphi_{n,\Im_m}(t) = \varphi_{n,m}$.

The next statement is a generalization of theorem 6.2.1.

**Theorem 6.5.1.** *Let $m, n \in N$, the polynomial $\Im_m$ of the form (6.91) has only real zeros and*

$$\Im_m(x) = \Im_k(x)\Im_{m-k}(x), \quad k = 1, 2, \ldots, m-1.$$

*Then for any $s \in S_{2n,\Im_m}$ the following next inequality*

$$\frac{||\Im_k(D)s||_\infty}{||\varphi_{n,\Im_{m-k}}||_\infty} \le \frac{||s||_\infty}{||\varphi_{n,\Im_m}||_\infty} \tag{6.93}$$

*holds true. In particular*

$$||\Im_m(D)s||_\infty \le \frac{||s||_\infty}{||\varphi_{n,\Im_m}||_\infty}. \tag{6.94}$$

*The inequalities (6.93) and (6.94) aren't improved on the set $S_{2n,\Im_m}$.*

**Proof.** If the function $f(t)$ is differentiable at the point $t$ then

$$f'(t) + af(t) = e^{-at}(f(t)e^(at))'. \tag{6.95}$$

¿From here and from Rolle's theorem it follows that for any piece-wise, differential function $f$

$$\nu(f) \leq \nu(f' + af) \tag{6.96}$$

(remind that $\nu(f)$ is the number of sign changes for $2\pi$-periodic function $f$ on period).

Let $\Im_m$ is arbitrary polynomial of the form (6.91) having real zeros $\beta_j$, $j = 1, 2, \ldots$. Then

$$\Im_m(x) = a_0 \prod_{j=1}^{m} (x - \beta_j)$$

and consequently the operator $\Im_m(D)$ is the composition of $m$ operators of first order $D - \beta_j$, $j = 1, 2, \ldots$. Therefore from (6.96) it follows that for any function $f \in C^{m-1}$ such that $f^m(t)$ is the piece-continuous function the inequality

$$\nu(\Im_m(D)f) \geq \nu(f) \tag{6.97}$$

is valid.

As for any numbers $\lambda$ and $\tau$ and any $s \in S_{2n,0}$

$$\nu\left(\varphi_{2n,0}(\cdot) - \lambda s(\cdot - \tau)\right) \leq 2n \tag{6.98}$$

so for $s \in S_{2n,\Im_m}$

$$\nu\left(\varphi_{n,\Im_m}(\cdot) - \lambda s(\cdot - \tau)\right) \leq 2n. \tag{6.99}$$

The function $\varphi_{n,\Im_m}$ has period $2\pi/n$ and if

$$\varphi_{n,\Im_m}(t_0) = \min_t \varphi_{n,\Im_m}(t),$$

then on the segment $[t_0, t_0 + 2\pi]$ this function has exact two intervals of monotony. Really if it is not true, then there exists the number $c$ such that

$$\nu(\varphi_{n,\Im_m} + c) \geq 4n.$$

But then due to (6.97)

$$\nu\left(\Im_m(D)(\varphi_{n,\Im_m} + c)\right) \geq 4n,$$

that is impossible because

$$\Im_m(D)(\varphi_{n,\Im_m} + c) = \varphi_{n,0} + c_1,$$

where $c_1$ is some constant.

¿From above it follows that for any $\varepsilon > 0$ the functions $s(t)$ and

$$\varphi_\varepsilon(t) = K_\varepsilon\, \varphi_{n,\Im_m}(t),$$

where

$$K_\varepsilon = \varepsilon + \frac{||s||_\infty}{||\varphi_{n,\Im_m}||_\infty},$$

satisfy the conditions of the lemma 1.4.1 and hence to the conditions of the proposition 1.4.5. Therefore for any $\beta \in R$

$$K_\varepsilon \min_t (D-\beta)\varphi_{n,\Im_m}(t) < (D-\beta)s(t) < K_\varepsilon \max_t (D-\beta)\varphi_{n,\Im_m}(t).$$

Thus if

$$\Im_{m-1}(x) = a_0 \sum_{j=1}^{m} (x-\beta_j),$$

then function

$$f = (D-\beta)s \in S_{2n,\Im_{m-1}}$$

and

$$K_\varepsilon\, (D - beta_1)\varphi_{n,\Im_m}(t) = K_\varepsilon\, \varphi_{n,\Im_{m-1}}(t)$$

satisfy the conditions of proposition 1.4.5. With the help of induction we obtain that if

$$\Im_k(x) = \prod_{j=1}^{k} (x-\beta_j),$$

then

$$K_\varepsilon \min_t \Im_k(D)\varphi_{n,\Im_m}(t) < \Im_k(D)s(t) < K_\varepsilon \max_t \Im_k(D)\varphi_{n,\Im_m}(t).$$

Due to arbitrariness of $\varepsilon$ the last inequality is equivalent to inequality (6.93).

**Theorem 6.5.2.** *Let $n, m \in N$ and a polynomial $\Im(x)$ of the form (6.91) has only real zeros and may be represented in the form*

$$\Im_m(x) = \Im_k(x)\Im_{m-k}(x),$$

*where $\Im_k(x) = x\Im_{k-1}(x)$, $\Im_{k-1} \in P_{k-1}$, $\Im_{m-k} \in P_{m-k}$. Then for any $\Im$-spline $s \in S_{2n,\Im_m}$ and any $\lambda \in R$ the exact inequalities*

$$|\Im_k(D)s - \lambda| \prec \left| \frac{||s||_\infty}{||\varphi_{n,\Im_m}||_\infty} \varphi_{n,\Im_{m-k}} - \lambda \right| \tag{6.100}$$

*are valid. Consequently for any $p \in [1,\infty]$*

$$||\Im_k(D)s - \lambda||_p \le \frac{||s||_\infty}{||\varphi_{n,\Im_m}||_\infty} ||\varphi_{n,\Im_{m-k}} - \lambda||_p. \tag{6.101}$$

**Proof.** Let $f(t) = \Im_{k-1}(D)s(t)$ and

$$g_\varepsilon(t) = \left(\frac{||s||_\infty}{||\varphi_{n,\Im_m}||_\infty} + \varepsilon\right)\Im_{k-1}(D)\varphi_{n,\Im_m}(t).$$

Under the proving of the theorem 6.5.1 we established that for any $\tau$ function $g_\varepsilon^{(\nu)}(t) - f^{(\nu)}(t-\tau)$, $\nu = 1, 2$ has at most $2n$ changes of sign on period. Besides that due to theorem 6.5.1 $|f(t)| \le ||g_\varepsilon||_\infty$ and $|f'(t)| \le ||g'_\varepsilon||_\infty$, i.e the functions $f(t)$ and $f'(t)$ satisfy the conditions of the theorem 1.4.6 and thanks to this one we can write

$$|f' - \lambda| \prec |g'_\varepsilon - \lambda| \;\; \forall \varepsilon > 0.$$

Therefore

$$|f' - \lambda| \le |g'_0 - \lambda|,$$

that is equivalent to inequality (6.100).

The inequlity (6.101) immediately follows from (6.100) and from the theorem 1.3.11.

Note that theorem 6.5.1 we can generalize on the case when polynomial $\Im_m$ has both real and complex zeros.

**Theorem 6.5.3.** *Let $n, m \in N$ and a polynomial $\Im(x)$ of the form (6.91) has only real zeros and is represented in the form*

$$\Im_m(x) = \Im_k(x)\Im_{m-k}(x),$$

*where $\Im_k \in P_k$, $\Im_{m-k} \in P_{m-k}$. Then for*

$$n > \max_k |Im\, x_k|$$

*for any $\Im$-spline $s \in S_{2n,\Im_m}$ the exact inequality*

$$\frac{||\Im_k(D)s||_\infty}{||\varphi_{n,\Im_{m-k}}||_\infty} \le \frac{||s||_\infty}{||\varphi_{n,\Im_m}||_\infty} \tag{6.102}$$

*holds true. In particular*

$$||\Im_m(D)s||_\infty \le \frac{||s||_\infty}{||\varphi_{n,\Im_m}||_\infty}. \tag{6.103}$$

At last consider the analogous problems for functions wich are convolution of $CVD$-kernel with the splines of zero order. With the kernel $K(t)$ we will connect the number $\theta = \theta(K)$ which equals to 1 if $K \perp 1$ and equals to 0 in opposite case.

**Theorem 6.5.4.** *Let $n \in N$ and a kernel $K$ is CVD-kernel. For any $s \in S_{2n,0}$, $s \perp \theta(K)$ the next exact inequality*

$$||s||_\infty \leq \frac{||K * s||_\infty}{||K * \varphi_{n,0}||_\infty}$$

*is valid.*

**Proof.** Let $g(t) = s(t)/||s||_\infty$. Show that

$$||K * g||_\infty \geq ||K * \varphi_{n,0}||_\infty. \tag{6.104}$$

Suppose that it is not true. Then taking into account the equality

$$K * \varphi_{n,0}\left(t + \frac{k\pi}{n}\right) = (-1)^k K * \varphi_{n,0}(t)$$

we shall have

$$\nu(K * \varphi_{n,0} - \varepsilon K * g) = \nu(K * (\varphi_{n,0} - \varepsilon g)) \geq 2n, \quad |\varepsilon| = 1.$$

Then for small $h > 0$ and $|\varepsilon| = 1$

$$\nu((K * (\varphi_{n,0} - \varepsilon g))_h) = \nu(K * (\varphi_{n,0} - \varepsilon g)_h) \geq 2n,$$

where $f_h(t)$ is the Steklov's functiom of $f(t)$. As $K \in CVD$ from (6.105) we obtain

$$\nu((\varphi_{n,0} - \varepsilon g)_h) \geq 2n. \tag{6.105}$$

While $||g||_\infty = ||\varphi_{n,0}||_\infty = 1$ then under the suitable choice of $\varepsilon = \pm 1$ the difference $\varphi_{n,0}(t) - \varepsilon g(t)$ identically equals to zero at least on one interval $k\pi/n, (k+1)\pi/n$ and hence

$$\nu(\varphi_{n,0} - \varepsilon g) \leq 2n - 2.$$

But then for small $h > 0$

$$\nu((\varphi_{n,0} - \varepsilon g)_h) \leq 2n - 2,$$

that contradicts to (6.106). Thus we proved the inequality (6.104) and the theorem.

As we will show in the next chapter the important applications of Bernstein's type inequalities for splines and $\Im$-splines connect with receiving the estimates from below for the diameter of periodical classes of functions. The substitution in the definition of these splines the set $S_{2n,0}$ by the most general set essentially extend the sphere of its application.

Let on the segment $[0, \pi/n]$ the positive measurable and limited function $\psi(t)$ is given. Continue this function with the help of equality $\psi(t+\pi/n) = -\psi(t)$ on segment $[\pi/n, 2\pi/n)$ and then with the period $2\pi/n$ on all real axis. Denote continued function by $\psi_{n,0}(t)$. For $j = 1, 2, \dots, 2n$ define the $2\pi$-periodic function $\psi_j(t)$ on the segment $[0, 2\pi$ by equalities

$$\psi_i(t) = \begin{cases} |\psi_{n,0}(t)|, & (j-1)\pi/n \le t < j\pi/n, \\ 0, & t \in [0, 2\pi) \setminus [(j-1)\pi/n, j\pi/n). \end{cases}$$

The set of all functions in the form

$$g(t) = \sum_{j=1}^{2n} c_j \psi_j(t), \tag{6.106}$$

where $c_j$ are arbitrary numbers, makes in $L_\infty$ linear manifold with dimension $2n$. Denote it by $S_{2n,0}(\psi)$. For $c_j = (-1)^{j+1}$ , $j = 1, 2, \dots, 2n$ the function $g(t)$ coincides with the function $\psi_{n,0}$. If $\psi(t) \equiv 1$ for $t \in [0, \pi/n)$ them $S_{2n,0(\psi)} = S_{2n,0}$.

**Theorem 6.5.5.** *Let $n \in N$ and $K \in CVD$. For any function $g \in S_{2n,0}(\psi)$, $g \perp \theta(K)$ the exact inequality*

$$\frac{\|g\|_\infty}{\|\psi_{n,0}\|_\infty} \le \frac{\|K * g\|_\infty}{\|K * \psi_{n,0}\|_\infty} \tag{6.107}$$

*is valid.*

**Proof.** It is sufficiently to prove that if

$$\|K * g\|_\infty \le \|K * \psi_{n,0}\|_\infty, \tag{6.108}$$

then

$$\|g\|_\infty \le \|\psi_{n,0}\|_\infty. \tag{6.109}$$

That is equivalent to the fact that for any coefficient $c_j$ in the representation (6.107) for the function $g(t)$ the inequality $|c_j| \le 1$ is valid.

Suppose the opposite fact i.e. that the inequality (6.109) holds true for function $g(t)$ but among its coefficients there exists the coefficient $c_k$ for wich $|c_k| > 1$. Then puting $\lambda = \pm |c_k|^{-1}$ we see that

$$-\|K * \psi_{n,0}\|_\infty = \min_t K * \psi_{n,0}(t) \le \lambda K * g(t) \le$$

$$\max_t K * \psi_{n,0}(t) = \|K * \psi_{n,0}\|_\infty$$

and consequently

$$\nu(\lambda K * g - K * \psi_{n,0}) \geq 2n. \tag{6.110}$$

At the same time

$$\lambda K * g - K * \psi_{n,0} = K * (\lambda g - \psi_{n,0}).$$

Under the suitable choice for sign of the number $\lambda$ the function $\lambda g(t) - \psi_{n,0}(t)$ is equal to zero for $t \in [(k-1)\pi/n, k\pi/n)$ and on each from the rest of intervals $[(i-1)\pi/n, i\pi/n)$ it has constant sign. Then for such $\lambda$

$$\nu(\lambda g - \psi_{n,0}) \leq 2n - 2. \tag{6.111}$$

Passing to Steklov's functions $(\lambda K * g - K * \psi_{n,0})_h$ and $(\lambda g - \psi_{n,0})_h$ by the similar way as it was done in the proof of theorem 6.5.4 with the help of (6.111) and (6.112) we obtain the contradiction with the fact that $K \in CVD$.

Let $\omega(t)$ is convex from below modulus of continuity. For $t \in [0, \pi/n]$ we put

$$\psi^\omega(t) = \frac{1}{2}\min\{\omega(2t), \omega\left(\frac{2\pi}{n} - 2t\right)\}.$$

The corresponding function $\psi^\omega_{n,0}$ usually denote by $f_{n,0}(\omega, t)$. Let $f_{n,m}(\omega, t)$ is $m$-th periodic integral of the function $f_{n,0}(\omega, t)$ having zero mean value on period. Denote by $S_{2n,m}(\omega)$ the set of all functions $f \in C^m$ such that $f^{(m)} \in S_{2n,0}(\psi^\omega)$.

¿From the theorem 6.5.5 the justice of the next statement immediately follows.

**Corollary 6.5.6.** *Let $n, m \in N$. For any function $f \in S_{2n,m}(\omega)$ the exact inequality*

$$\|f^{(m)}\|_\infty \leq 2\frac{\|f\|_\infty}{\|f_{n,m}(\omega)\|_\infty}\omega\left(\frac{\pi}{n}\right)$$

*is valid.*

## 6.6 Commentaries

Sec. 6.1. General facts concerning with the foundations of spline-functions theory are stated here. In detail one can familiarize oneself with these problems in a few well-known monographs of N.P.Korneichuk [22], D.Alberg, E.Nilson and D.Wolsh [1], S.B.Stechkin and Y.N.Subbotin [1] and etc.

Sec. 6.2. By another way the theorem 6.2.1 was proved by Tihomirov [2]. Theorem 6.2.5 was proved by V.F. Babenko [15]. He also proved the theorem 6.2.7 which gave the analog of theorem 3.1.2 and made more precise the theorem 6.2.1. Theorem 6.2.8 was proved by Y.N. Subbotin [2].

Sec. 6.3. With the using of Stein's method the inequalities 6.3.1 was obtained by Y.N. Subbotin [4]. Theorems 6.3.2 and 6.3.5 had been proving by V.F.Babenko and A.A.Ligun during the writing of this book. Theorem 6.3.6 was proved by V.F. Babenko and S.A.Pichugov.

Sec. 6.4. Theorem 6.4.2 was proved by A.A. Ligun [4] and V.F. Babenko [15] with the help comparison theorem for rearrangements.

Theorem 6.4.4 was proved by V.F. Babenko [13].

Theorem 6.4.6 was proved by A.A. Ligun [10].

Sec.6.5. The some inequalities of Bernstein's type for generalized splines were proved by another way by V.T. Schevaldin [1], S.I. Novikov [1], Nguen Tkhy Tkhiey Hoa [2]. Theorems 6.5.2 and 6.5.4 were obtained by V.F. Babenko [15].

Theorem 6.5.6 in the case when $K = B_m$ for the classes of differentiable functions was proved by V.I. Ruban [1].

## 6.7 Supplementary results

**1.** Let $f \in C^k$, $k = 1, 2, \ldots$ is fixed and $\{s_{2m+1}(f,t)\}$, $m = k, k+1, \ldots$ is sequence of $2\pi$-periodic splines of order $2m+1$ with the defect $k$ by equidistant fixed partition

$$\Delta_N = \{0 = t_0 < t_1 \ldots < t_N = 2\pi\},$$

which defined by the unique manner with the help of conditions

$$s_{2m+1}^{(j)}(f, t_i) = f^{(j)}(t_i), \;\; j = 0, 1, \ldots, k, \;\; i = 1, 2, \ldots, N.$$

Then uniformly by $t$

$$\lim_{m \to \infty} s_{2m+1}^{(\nu)}(f, t) = T_N^{(\nu)}(f, t),$$

where $T_N(f,t)$ is trigonometric polynomial

$$T_N(f,t) = \sum_{\mu=0}^{[N/2]} (a_\mu \cos \mu t + b_\mu \sin \mu t),$$

interpolating the function $f(t)$ at the same nodes $t_i$ and with the same multiplicity

$$T_N^{(j)}(f, t_i) = f^{(j)}(t_i), \;\; j = 0, 1, \ldots, k, \;\; i = 0, 1, \ldots, N;$$

the uniqueness of polynomial is provide for even $N$ by condition of minimization for the sum $a_{N/2}^2 + b_{N/2}^2$.

Kvade V., Collatz L. [1], Shoenberg I. [1].

**2.** Let $T_N$ is the subspace of trigonometric polynomials in the form

$$\sum_{\mu=0}^{[(N-1)/2]} (a_\mu \cos \mu t + b_\mu \sin \mu t) + \frac{1+(-1)^\mu}{2} a_{N/2} \cos \frac{Nt}{2}.$$

Then

$$\lim_{m\to\infty} E(f, S_{2n,m})_p = E(f, T_n)_p.$$

Velikin V.L. [2]

In this article the statements characterizing the limited tie of trigonometric and spline-approximation in the case of arbitrary linear approximation methods and also in the case of the best one-sided approximations were proved.

**3.** Let $n, m \in N$ the polynomial $L_m$ of the form (5.89) has only real zeros and

$$L_m(x) = L_k(x) L_{m-k}(x), \quad k = 1, 2, \ldots, m.$$

If

$$n > 2 \max_{1\le k \le y_m} Im\, y_k,$$

then for any $L$-spline $s \in S_{2n,L_m}$ the exact inequality

$$\frac{||L_k(D)s||_p}{||\varphi_{n,L_{m-k}}||_p} \le \frac{||s||_p}{||\varphi_{n,L_m}||_p}, \quad p = 1, 2, \infty.$$

is valid. In particular

$$\frac{||L_r(D)s||_p}{(2\pi)^{1/p}} \le frac ||s||_p ||\varphi_{n,L_m}||_p.$$

It, in addition $L_k(x)$ is such that $L_k(0) = 0$ than for any $s \in S_{2n,L_m}$ and any $\lambda \in R$

$$|L_k(D)s - \lambda| \prec \left| \frac{||s||_\infty}{||\varphi_{n,L_m}||_\infty} \varphi_{n,L_{m-k}} - \lambda \right|$$

and consequently, for any $p \in [1, \infty]$

$$\frac{||L_k(D)s||_p}{||\varphi_{n,L_{m-k}}||_p} \le \frac{||s||_\infty}{||\varphi_{n,L_{m-k}}||_\infty}.$$

If polynomial $L_m(x)$ has only real zeros, then for any $s \in S_{2n,L_m}$ and any $q \in [1,\infty]$

$$\frac{V_0^{2\pi}(L_k(D)s)}{V_0^{2\pi}(\varphi_{n,L_{m-k}})} \leq \frac{||s||_q}{||\varphi_{n,L_{m-k}}||_q},$$

and consequently, if $L_k(0) = 0$, then

$$\frac{||L_k(D)s||_1}{||\varphi_{n,L_{m-k}}||_1} \leq \frac{||s||_q}{||\varphi_{n,L_{m-k}}||_q},$$

Babenko V.F. and Ligun A.A.[1]

**4.** For all $n, m \in N$, $l \geq m-1$ and $p \in [1,\infty]$

$$E(W_p^m, S_{2n,l})_1 = ||\varphi_{n,m}||_{p'}, \;\; 1/p + 1/p' = 1.$$

Ligun A.A. [3].
More general: for all $\alpha, \beta > 0$

$$E(W_p^m, S_{2n,m})_{1;\alpha,\beta} = \inf_{\lambda} ||\varphi_{n,m}(\cdot;\alpha,\beta) - \lambda||_{p'}, \;\; 1/p + 1/p' = 1.$$

This result was obtained for $\max\{\alpha,\beta\} = \infty$ by Doronin V.G. and Ligun A.A. [2] and for $\max\{\alpha,\beta\} \leq \infty$ by Babenko V.F. [12].

**5.** For all $n, m \in N$ and $p \in [1,\infty]$

$$\sup_{f \in W_\infty^m} ||f' - s'_{2n,m-1}(f)||_p = ||\varphi_{n,m-1}||_p$$

and

$$\sup_{f \in W_\infty^m} ||f - s_{2n,m-1}(f)||_p = ||\varphi_{n,m}||_p.$$

Korneichuk N.P. [13,17,25].

**6.** For $n, m \in N$

$$E(W_2^m, S_{2n,m-1})_2 = \sup_{f \in W_2^m} ||f - s_{2n,m-1}(f)||_2 = \frac{1}{n^m}.$$

Subbotin U.N. [5].

**7.** A set $F \in L_1$ is called rearrangementally invariant if from the fact that $f \in F$ and $\pi(g) = \pi(f)$ it follows that $g \in F$. Let $K$ is arbitrary $CVD$-kernel and $F$ is arbitrary rearragementally invariant set. By $K * F$ we denote the set of all functions in the form

$$f = a\theta + K * \varphi, \ \ a \in R, \ \ \varphi \in F, \ \ \varphi \perp \theta,$$

where $\theta = 1$ if $K \perp 1$ and $\theta = 0$ in opposite case.

For any $CVD$-kernel $K$ and any rearragementally invariant set $F$ and for all $m = 1, 2, \ldots, \alpha, \beta > 0$

$$E(K * F, K * S_{2n,m})_{1;\alpha,\beta} = \sup_{\varphi \in F, \, \varphi \perp \theta} \int_0^{2\pi} \pi(K * \varphi_n(\cdot; \alpha, \beta); t) \pi(\varphi; t) dt.$$

Babenko V.F. [18].

In particular if $F = F_p$ is unit ball in the space $L_p$, $p \in [1, \infty]$ then

$$E(K * F, K * S_{2n,m})_{1;\alpha,\beta} = \inf_{\lambda} ||K * \varphi_n(\cdot; \alpha, \beta) - \lambda\theta||_{p'}.$$

**8.** Let $\omega(t)$ is convex from below modulus of continuity and $W^m H_\omega$, $m = 1, 2, \ldots$ is the class of all functions $f \in C$ such that

$$\forall x', x'' \ \ |f^{(m)}(x') - f^{(m)}(x'')| \leq \omega(|x' - x''|).$$

Then for any $m = 0, 1, 2, \ldots, n \in N$, $l \geq m$, $p = 1, \infty$

$$E(W^m H_\omega, S_{2n,l})_p = ||f_{n,m}(\omega)||_p.$$

Korneichuk N.P. ($l = m$) [7,10], Ligun A.A. ($l > m$) [6].

**9.** If $K$ is arbitrary $CVD$-kernel and $\omega(t)$ is convex from below modulus of continuity then for any $m = 0, 1, \ldots$ and $n \in N$

$$E(K * H_\omega, K * S_{2n,m})_\infty = ||K * f_{n,0}(\omega)||_\infty.$$

Babenko V.F. [18].

**10.** For any convex modulus of continuity $\omega(t)$ for all $n, m =\in N$ the following correlations

$$\sup_{f \in W^m H_\omega} ||f - s_{2n,m}(f)||_1 = E(W^m H_\omega, S_{2n,m})_1 = ||f_{n,m}(\omega)||_1$$

take place.

Korneichuk N.P. [21].

# Chapter 7

# Extremal properties of perfect splines

In the previous chapter perfect splines were introduced because in the series of cases extremal problems of splines with fixed knots as well as precise estimations of error of spline-interpolation are realized exactly at the standard perfect splines $\varphi_{n,m}(t)$ and $\varphi_{n,m}(\alpha,\beta;t)$.

In this chapter intrinsic extremal properties of collections of perfect splines $\Gamma_{2n,m}$ and $\Gamma_{2n,m}(\alpha,\beta)$ and some their generalizations are investigated. These collections of splines with free knots (with restriction on the number of knots and strict restriction on the values of $m$-th derivative) are not linear manifolds and it is natural that investigation of their properties requires new approach. One would observe that while in extremal problems of the sixth chapter standard splines realize the least upper bound, then in the problems of the seventh chapter these splines realize the greatest lower bound.

## 7.1 The uniform norms of perfect splines

Remained that $\Gamma_{2n,m}$ denotes the set of $2\pi$-periodic perfect splines of order $m$, $m$-th derivatives of which have the values $+1$ and $-1$ and change sign on the period at most $2n$ times.

The generalization of the set $\Gamma_{2n,m}$ on nonsymmetric case present us the set $\Gamma_{2n,m}(\alpha,\beta)$, $(\alpha,\beta>0)$ of the splines $\gamma(t)\in L_\infty^m$ for which $\gamma^{(m)}(t)$ has only values $\alpha$ and $-\beta$ and also changes sign on period at most $2n$ times.

The standard splines in $\Gamma_{2n,m}$ and $\Gamma_{2n,m}(\alpha,\beta)$ are correspondingly the functions $\varphi_{n,m}(t)$ and $\varphi_{n,m}(\alpha,\beta;t)$.

As before we put

$$E_0(f)_p = \inf_{\lambda \in R} \|f - \lambda\|_p.$$

**Theorem 7.1.1.** *For $\alpha, \beta > 0$, $n, m \in N$ the next correlations hold true*

$$\inf_{\gamma \in \Gamma_{2n.m}} \|\gamma\|_\infty = \|\varphi_{n,m}\|_\infty; \tag{7.1}$$

$$\inf_{\gamma \in \Gamma_{2n.m}(\alpha,\beta)} \|\gamma\|_\infty = E_0(\varphi_{n,m}(\alpha, y\beta))_\infty. \tag{7.2}$$

**Proof.** As $E_0(\varphi_{n,m})_\infty = \|\varphi_{n,m}\|_\infty$ so (7.1) is particular case of (7.2) (for $\alpha = \beta = 1$) and hence it is sufficiently to prove the correlation (7.2).

If we assume that (7.2) is not true then there exists perfect spline $\gamma \in \Gamma_{2n,m}(\alpha, \beta)$ for which

$$\|\gamma\|_\infty < E_0(\varphi_{n,m}(\alpha, \beta)_\infty)$$

and so for some $\lambda \in R$

$$\min_t \varphi_{n,m}(\alpha, \beta; t) < \gamma(t) - \lambda < \max_t \varphi_{n,m}(\alpha, \beta; t).$$

But then for any $\tau$ the difference

$$\delta_\tau(t) = \varphi_{n,m}(\alpha, \beta; t) - \gamma(t + \tau) + \lambda$$

has at least $2n$ sign changes on the period $[0, 2pi)$. In view of Rolle's theorem the derivative $\delta_\tau^{(m)}(t)$ has also at least $2n$ sign changes on the period.

On the other hand the perfect spline at most $2n$ times changes sign on period and

$$mes\{t : t \in (0, 2\pi), \gamma^{(m)}(t) = \alpha\} = \frac{2\pi\beta}{\alpha + \beta}.$$

Therefore there exists an interval $(a, b)$ such that

$$b - a \geq \frac{2\pi\beta}{n(\alpha + \beta)}$$

and $\gamma^{(m)}(t) = \alpha$ for all $t \in (a, b)$. As $\varphi_{n,m}(\alpha, \beta; t) = \alpha$ for $0 < t < 2\pi\beta/(n(\alpha + \beta))$ then for $\tau = a$ the difference $\delta_a^{(m)}(t)$ is identically equal to zero on the interval $(0, 2\pi\beta/(n(\alpha + \beta)))$ and hence changes sign on period at least $2n - 2$ times but that is contrary to previous fact.

Using this idea we can prove the next more general statement.

**Theorem 7.1.2.** *Let $\alpha, \beta > 0$, $m \in N$ and a function $f \in L_\infty^m$ is such that the set*

$$\{t : t \in R,\ f^{(m)}(t) = \alpha \| f^{(m)} \|_{\infty;\alpha^{-1},\beta^{-1}}\}$$

*contains an interval with length no less than $2\pi\beta/(n(\alpha+\beta))$ or the set*

$$\{t : t \in R,\ f^{(m)}(t) = -\beta \| f^{(m)} \|_{\infty;\alpha^{-1},\beta^{-1}}\}$$

*contains an interval with length no less than $2\pi\alpha/(n(\alpha+\beta))$. Then*

$$E_0(f)_\infty \geq E_0(\varphi_{n,m}(\alpha,\beta))_\infty \| f^{(m)} \|_{\infty;\alpha^{-1},\beta^{-1}}. \tag{7.3}$$

*In particular for any function $f \in L_\infty^m$ such that the set*

$$\{t : t \in R,\ f^{(m)}(t) = \pm \| f^{(m)} \|_\infty\}$$

*contains an interval with length no less than $\pi/n$ the next inequality is valid*

$$E_0(f)_\infty \geq \| \varphi_{n,m} \|_\infty \| f^{(m)} \|_\infty. \tag{7.4}$$

**Proof.** Put

$$g(t) = f(t)/\| f^{(m)} \|_{\infty;\alpha^{-1},\beta^{-1}}.$$

Due to conditions of our theorem, for any $t$

$$-\beta \leq g(t) \leq \alpha$$

and the set $\{t : g^{(m)}(t) = \alpha\}$ contains an interval with length not less than $2\pi\beta/(n(\alpha+\beta))$ or the set $\{t : g^{(m)}(t) = -\beta\}$ contains an interval with length no less than $2\pi\alpha/(n(\alpha+\beta))$. Now it is sufficiently to prove that

$$E_0(g)_\infty \geq E_0(\varphi_{n,m}(\alpha,\beta))_\infty.$$

If that is not true then for some $\lambda \in R$ and for any $\tau$ the difference

$$\delta_\tau(t) = \varphi_{n,m}(\alpha,\beta;t) - g(t+\tau) - \lambda$$

and so the function $\delta_\tau^{(m)}(t)$ has on period $[0, 2\pi)$ at least $2n$ changes of sign. But for suitable $\tau$ the function $\delta_\tau^{(m)}(t)$ vanishes on some interval where $\varphi_{n,0}(\alpha,\beta;t) = \alpha$ or on some interval where $\varphi_{n,0}(\alpha,\beta;t) = -\beta$. Hence for such $\tau$ this function has at most $2n-2$ changes of sign on the period.

In fact any perfect spine $\gamma \in \Gamma_{2n,m}(\alpha,\beta)$ satisfies the conditions of theorem 7.1.2. So the inequality (7.2) is the particular case of inequality (7.3).

Theorem 7.2.1 may be generalized on the case of convolution with $CVD$-kernels. Remind that with the kernel $K$ we relate the number $\theta = \theta(K) = 1$ if $K \perp 1$ and $\theta = 0$ in contrary case.

**Theorem 7.1.3.** *Let $n \in N$, $\alpha, \beta > 0$ and $K \in CVD$. Let also $\phi \bot \theta(K)$ is such that the set*

$$\{t : t \in R,\ \phi(t) = \alpha ||\phi||_{\infty;\alpha^{-1},\beta^{-1}}\}$$

*contains an interval with length no less than $2\pi\beta/(n(\alpha+\beta))$ or the set*

$$\{t : t \in R,\ \phi(t) = -\beta ||\phi||_{\infty;\alpha^{-1},\beta^{-1}}\}$$

*contains an interval with length no less than $2\pi\alpha/(n(\alpha+\beta))$. Then when $\theta = 0$ at least one inequality will take place*

$$||(K * \phi)_{\pm}||_{\infty} \geq ||(K * \varphi_{n,0})_{\pm}||_{\infty} ||\phi||_{\infty;\alpha^{-1},\beta^{-1}}$$

*and if $\alpha = \beta = 1$ then*

$$||K * \phi||_{\infty} \geq ||K * \varphi_{n,0}||_{\infty} ||\phi||_{\infty}.$$

*For $\theta = 1$ the next inequality is valid*

$$E_0(K * \phi)_{\infty} \geq E_0(K * \varphi_{n,0})_{\infty} ||\phi||_{\infty;\alpha^{-1},\beta^{-1}}.$$

In proof of theorem 7.1.2 we used only fact that $f = a + B_m * f^{(m)}$, where $B_m(t)$ is Bernully's kernel which is CVD-kernel. Therefore the proof of theorem 7.1.3 we don't give.

**Corollary 7.1.4.** *Let $n \in N$, $\alpha, \beta > 0$ and $K \in CVD$ and $\gamma \in \Gamma_{2n,0}(\alpha, \beta)$ and $\gamma \bot \theta(K)$. Then for $\theta = 0$ at least one inequality will take place*

$$||(K * \gamma)_{\pm}||_{\infty} \geq ||(K * \varphi_{n,0}(\alpha, \beta))_{\pm}||_{\infty}$$

*and if $\alpha = \beta = 1$ then*

$$||K * \gamma||_{\infty} \geq ||K * \varphi_{n,0}||_{\infty}.$$

*For $\theta = 1$*

$$E_0(K * \gamma)_{\infty} \geq E_0(K * \varphi_{n,0}(\alpha, \beta))_{\infty}.$$

Present now one extremal property of functions which are trigononetrically conjugate to perfect splines $\gamma \in \Gamma_{2n,1}$.

**Theorem 7.1.7.** *For $n \in N$ and any perfect spline $\gamma \in \Gamma_{2n,1}$ the exact inequality*

$$||\tilde{\gamma}||_{\infty} \geq ||\tilde{\varphi}_{n,1}||_{\infty}$$

*holds true.*

**Proof.** From the proof of the theorem 1.6.1 it is easy to see that if $\gamma \in \Gamma_{2n,1}$ then $\gamma'(t)$ has at most $2n$ changes of sign on the period $[0, 2\pi)$. Therefore the function $\tilde{\gamma}(t)$ has on the period at most $2n$ segments of monotony and hence

$$||\tilde{\gamma}||_\infty \geq \frac{1}{4n} V_0^{2\pi}(\tilde{\gamma}) = \frac{1}{4n} ||\tilde{\gamma}'||_1. \tag{7.5}$$

It is clear that

$$||\tilde{\gamma}'||_1 = 2 \int_0^{2\pi} \frac{1}{2} |(\gamma'(t)\tilde{+}1)| dt =$$

$$2 \int_0^{2\pi} r\left(\frac{1}{2}|(\gamma'(\cdot)\tilde{+}1)|, t\right) dt == 2 \int_0^{2\pi} r(|\tilde{\chi}_E|, t) dt, \tag{7.6}$$

where $\chi_E$ is the characteristic function of the set $E = \{t : t \in R,\ \gamma'(t) = 1\}$. As $mes(E \cap [0, 2\pi]) = \pi$ so due to theorem 1.6.1

$$\int_0^{2\pi} r(|\tilde{\chi}_E|, t) dt = \frac{2}{\pi} \int_0^{2\pi} arcsh \cot \frac{t}{4} dt = ||\tilde{\varphi}_{n,0}||_1 = 4n ||\tilde{\varphi}_{n,1}||_\infty. \tag{7.7}$$

Comparing the relations (7.5)-(7.7) we obtain the statement of our theorem.

## 7.2 The rearrangements of perfect splines

In this section we will prove a series extremal properties of periodic perfect splines which are express in the form of integral inequalities for rearrangements and also a series important corollaries from them.

Remind that notation $f \prec g$ means that for all $t \in [0, 2\pi]$

$$\int_0^t \pi(f, u) du \leq \int_0^t \pi(g, u) du$$

The main result of this section is contained in the next statement.

**Theorem 7.2.1.** *Let $n, m \in N$, $\alpha, \beta > 0$ and $\lambda \in R$. For any perfect spline $\gamma \in \Gamma_{2n,m}(\alpha, \beta)$ such that $\gamma \perp 1$ the next exact inequality*

$$(\varphi_{n,m}(\alpha, \beta) - \lambda)_\pm \prec (\gamma - \lambda)_\pm. \tag{7.8}$$

*is valid.*

*In particular if $\gamma \in \Gamma_{2n,m}$ and $\gamma \perp 1$ then*

$$(\varphi_{n,m} - \lambda)_\pm \prec (\gamma - \lambda)_\pm.$$

In view of theorem 1.5.1 in order to prove the theorem 7.2.1 it sufficies to prove next proposition.

**Theoren 7.2.2.** *Let $n, m \in N$, $\alpha, \beta > 0$ and $\delta, \theta > 0$. Then*

$$\min_{\gamma \in \Gamma_{2n,m}(\alpha,\beta)} E_0(\gamma)_{1;\theta\delta} = E_0(\varphi_{n,m}(\alpha,\beta))_{1;\theta,\delta}.$$

We need some auxiliary statements.

**Lemma 7.2.3.** *Let function $g \in L_i$ is such that $g \perp 1$ and for any $a \in R$*

$$mes\{x : \ x \in [0, 2\pi], \ g(x) = a\} = 0.$$

*Then*

$$E_0(g)_{1;\alpha,\beta} = \min_{\lambda} \left( \frac{\alpha+\beta}{2} \int_0^{2\pi} |g(t) - \lambda| dt + 2\pi\lambda \frac{\beta-\alpha}{2} \right). \tag{7.9}$$

Really for any $\lambda \in R$

$$\alpha \, sgn(g(t) - \lambda)_+ - \beta \, sgn(g(t) - \lambda)_- = \frac{\alpha+\beta}{2} sgn(g(t) - \lambda) - \frac{\beta-\alpha}{2}. \tag{7.10}$$

So

$$||g - \lambda||_{1;\alpha,\beta} = \int_0^{2\pi} (\alpha(g - \lambda)_+ + \beta(g - \lambda)_-) \, dt =$$

$$\int_0^{2\pi} (g(t) - \lambda) \left( \frac{\alpha+\beta}{2} sgn(g(t) - \lambda) - \frac{\beta-\alpha}{2} \right) dt =$$

$$\frac{\alpha+\beta}{2} \int_0^{2\pi} |g(t) - \lambda| dt + 2\pi\lambda \frac{\beta-\alpha}{2},$$

and the assartion of lemma becomes obvious.

It follows from the lemma 7.2.3 that criterion of best $(\alpha, \beta)$- approximation element by constant in the space $L_1$ (see the theorem 1.1.9) we can formulate in the form of

**Lemma 7.2.4.** *Let function $f \in L_1$ is such that for any $a \in R$*

$$mes\{x : \ x \in [0, 2\pi], \ f(x) = a\} = 0.$$

*Then a constant $\lambda_0 \in R$ is a constant of best $(\alpha, \beta)$-approximation in the space $L_1$ for a function $f$ if and only if*

$$\int_0^{2\pi} sgn(f(t) - \lambda_0) dt = 2\pi \frac{\beta-\alpha}{\beta+\alpha}.$$

Remind (see section 6.1) that if

$$x_1 < x_2 < \ldots < x_{2k} < 2\pi \leq x_1 + 2\pi, \quad k \leq n, \tag{7.11}$$

are nodes of a spline $\gamma \in \Gamma_{2n,m}(\alpha,\beta)$ such that $\gamma^{(m)}(t) = \alpha$ for $t \in (x_1, x_2)$ then

$$\sum_{i=1}^{2k} (-1)^i x_i = \frac{2\pi\beta}{\alpha+\beta}, \tag{7.12}$$

and for suitable $\lambda \in R$

$$\gamma(t) = \gamma(x_1, x_2, \ldots, x_{2k}; \lambda, t) = (\alpha+\beta)\sum_{i=1}^{2k}(-1)^{i+m}\beta_{m+1}(x_i - t) - \lambda. \tag{7.13}$$

Let

$$F(x_1, x_2, \ldots, x_{2k}; \lambda) = \int_0^{2\pi} |\gamma(t|dt).$$

Since for any $x_i$, $i = 1, 2, \ldots, 2k$ and any $\lambda$ the function $\gamma(t)$ has only finite number of zeros on period, the next proposition is valid.

**Lemma 7.2.5.** *The function $F(x_1, x_2, \ldots, x_{2k}; \lambda)$ has continuous partial derivatives of the first order and*

$$\frac{\partial F}{\partial \lambda} = -\int_0^{2\pi} sgn\,\gamma(t)dt,$$

$$\frac{\partial F}{\partial x_i} = (-1)^{i+m}(\alpha+\beta)\int_0^{2\pi} \beta_m(x_i - t) sgn\,\gamma(t)dt, \quad i = 1, 2, \ldots, 2k.$$

**The proof of theorem 7.2.2.** Let $\alpha, \beta, \theta, \delta$ and $n$ be fixed. Consider the next extremal problem

$$E_0(g)_{1,\theta,\delta} \to inf, \quad \gamma \in \Gamma_{2n,m}(\alpha,\beta). \tag{7.14}$$

Since the set $\Gamma_{2n,m}(\alpha,\beta)$ is locally compact in the topology of uniform convergency and $E_0(g)_{1,\theta,\delta}$ continuously depends on $\gamma$ the solution of problem (7.14) exists. Assume that the spline resolving the problem (7.14) possesses $2k$, $k \leq n$ nodes. In view of (7.9) and (7.10) and the lemma 7.2.5

the nodes (7.11) and the constant $\lambda$ of the best $(\theta,\delta)$-approximation of this spline $\gamma$ are also solution for the next extremal problem

$$\frac{\theta+\delta}{2}\int_0^{2\pi}|\gamma(t)|dt+2\pi\lambda\frac{\delta-\theta}{2}\to inf,$$

$$\sum_{i=1}^{2k}(-1)^i x_i=\frac{2\pi\beta}{\alpha+\beta}, \tag{7.15}$$

where the function $\gamma(t)$ is defined by the equality (7.12). In view of lemma 7.2.5 for studyng problem (7.15) we can apply Lagrange's multipliers method, which yields the following necessary conditions of extremum to be satisfied by the solutions $\lambda$ and $x_i$ for this problem:

$$(\theta+\delta)\int_0^{2\pi} sgn\,\gamma(t)dt+2\pi(\delta-\theta)=0; \tag{7.16}$$

$$(-1)^m\frac{\delta+\theta}{2}\int_0^{2\pi}\beta_m(x_i-t)sgn\,\gamma(t)dt=a,\quad i=1,2,\ldots,2k; \tag{7.17}$$

$$\sum_{i=1}^{2k}(-1)^i x_i=\frac{2\pi\beta}{\alpha+\beta}.$$

Let

$$g_0(t)=\theta\, sgn\,\gamma_+(t)-\delta\, sgn\,\gamma_-(t). \tag{7.18}$$

Since

$$g_0(t)=\frac{\theta+\delta}{2}sgn\,\gamma(t)+\frac{\theta-\delta}{2},$$

then in view of (7.16) $g_0\perp 1$.

Denote by $g_m(t)$ $m$-th periodic integral of the function $g_0(t)$ having mean value $(-1)^{m+1}a$ on period, i.e put

$$g_m(t)=(-1)^{m+1}a+(B_m*g_0)(t)=(-1)^{m+1}a+\int_0^{2\pi}B_m(t-u)g_0(u)du.$$

Moreover, let as before, $\mu(f)$ is the number of sign changes of function $f$ on period.

Since $\gamma(t)$ is a perfect spline from $\Gamma_{2k,m}(\alpha,\beta)$ then due to Rolle's theorem we can conclude that

$$\mu(g_m^{(m)})=\mu(\gamma)\le 2k,$$

hence $g_m \in \Gamma_{2k,m}(\theta, \delta)$. Besides that the equality (7.17) now we can write in the form

$$g_m(x_i) = 0, \quad i = 1, 2, \ldots, 2k. \tag{7.19}$$

Since $g_m(t)$ is a perfect spline having at most $2k$ nodes on a period, then $g_m(t)$ has at most $2k$ sign changes on period. ¿From this and from (7.19) it follows that

$$sgn\, g_m(t) = \varepsilon\, sgn\, \gamma^{(m)}(t), \quad |\varepsilon| = 1. \tag{7.20}$$

We will show that (up to shift of argument)

$$g_m(t) + \lambda = \varphi_{n,m}(\alpha, \beta; t).$$

Set

$$h_y(t) = \gamma(t) - \gamma(t + y)$$

and

$$H_y(t) = g_m(t) - g_m(t + y).$$

In general the functions $h_y(t)$ and $H_y(t)$ can have both isolated zeros and zero-segments, we shall call the the zero-segments also zeros.

**Lemma 7.2.6.** *For any $y \in R$*

$$\mu(H_y) \geq \mu(h_y^{(m)}).$$

**Proof.** For definitness we will assume that in (7.20) $\varepsilon = 1$. If $h_y^{(m)} > 0$, then taking into account that $\gamma^{(m)}(t)$ takes only values $\alpha$ and $-\beta$ we see that if $\gamma^{(m)}$ is continuous at a point $t$ then $\gamma^{(m)}(t) = \alpha$ and $\gamma^{(m)}(t + y) = -\beta$. Then in view of (7.20) $g_m(t) > 0$ and $g_m(t + y) < 0$ and consequently, $H_y(t) > 0$. Analogously if $h_y^{(m)}(t) < 0$ and $h_y^{(m)}(t)$ is continuous at a point $t$, then $H_y(t) < 0$.

By $\nu(f)$ we shall denote the number of zeros of the function $f$ on period $[0, 2\pi)$ that are counted according to the following rule: the simple isolated seros of $f$ are counted once, while the multiple zeros and sero-segments are counted twice.

**Lemma 7.2.7.** *Let $m = 2, 3, \ldots$ and $y \in R$. If $H_y(t) \not\equiv 0$ then*

$$\mu(H_y) \geq \mu(h_y^{(m)}) = \nu(H_y) = 2k.$$

**Proof.** Let $\nu(H_y) = 2s$. Then by Rolle's theorem on the period $[0, 2\pi)$ there exist $2s$ different zeros of the function $H_y'(t)$. But between zeros (including zero-segments) of function $H_y'(t)$ the function $H_y''(t)$ alternates its sign at least once. Therefore

$$\mu(H_y^{(m)}) \geq \mu(H_y'') \geq 2s = \nu(H_y). \tag{7.21}$$

Consequently there exist at least $2s$ points

$$t_1 < t_2 < \ldots < t_{2s} \leq t_1 + 2\pi,$$

at which the function $H_y^{(m)}(t)$ is continuous and alternates its sign passing from point $t_i$ to point $t_{i+1}$.

¿From the equality (7.18) it follows that the function $h_y(t)$ also alternates its sign between these points, so that

$$\mu(h_y) \geq 2s = \nu(H_y).$$

¿From this by Rolle's theorem, from (7.21) and from lemma 7.2.6 we obtain

$$\mu(h_y^{(m)}) \geq \mu(h_y) \geq \nu(H_y) \geq \mu(h_y) \geq \mu(H_y) \geq \mu(H_y^{(m)}).$$

**Lemma 7.2.8.** *Let there are on the segments $[a, b]$ and $[c, d]$ continuously differentiable functions $f(t)$ and $\psi(t)$ respectively such that*

$$f'(a) = f'(b) = \psi'(c) = \psi'(d) = 0.$$

*If*

*1) $f(t)$ and $\psi(t)$ are strictly increasing and*

$$\psi(c) \leq f(a) \leq \psi(d) \leq f(b) \tag{7.22}$$

*or*

$$f(a) \leq \psi(c) \leq f(b) \leq \psi(d) \tag{7.23}$$

*and*

*2) if $f(t)$ and $\psi(t)$ are strictly decreasing and*

$$\psi(d) \leq f(b) \leq \psi(c) \leq f(a)$$

*or*

$$f(b) \leq \psi(d) \leq f(a) \leq \psi(c)$$

*then there exist points $\xi \in [a, b]$ and $\eta \in [c, d]$ such that*

$$f(\xi) = \psi(\eta), \quad f'(\xi) = \psi'(\eta).$$

**Proof.** Let for definitness $f(t)$ and $\psi(t)$ are strictly increasing and inequalities (7.22) are valid. In the case when one from these inequalities becomes an equality, the statement of lemma is obvious. Let all inequalities in (7.22) are strict. At the interval $(f(a), \psi(d))$ $f^{-1}(y)$ and $\psi^{-1}(y)$ exist and are continuously differentiable. In addition $(f^{-1}(y))'$ is positive on this interval, takes on finite value at the point $\psi(d)$ and tends to plus infinity when $y$ tends to $f(a)$ and $(\psi^{-1}(y))'$ is positive, takes on finite value at the point $f(a)$ and tends to plus infinity when $y$ tends to $\psi(d)$. Therefore there exists a point $y_*$ such that

$$(\psi^{-1})'(y_*) = (f^{-1})'(y_*).$$

Let $\xi = f^{-1}(y_*)$ and $\eta = \psi^{-1}(y_*)$. Then $f(\xi) = \psi(\eta) = y_*$ and

$$f'(\xi) = \frac{1}{(f^{-1})'(y_*)} = \frac{1}{(\psi^{-1})'(y_*)} = \psi'(\eta)$$

**Continue the theorem 7.2.2 proof.** Let $T$ be the minimal period of the function $g_m(t)$ and let $a_1$ be the point of the smallest local maximum of $g_m(t)$. We'll prove that $g_m(t)$ possesses exactly two zeros on the interval $[a_1, a_1 + T)$.

Suppose the contrary i.e. that $g_m(t)$ has at least four zeros on the interval $[a_1, a_1 + T)$. But then there is at least one local maximum of $g_m(t)$ on this interval.

Let $a_2$ be the point of local maximum of $g_m(t)$ which lies nearest to $a_1$ from the right and $a_3$ is the local maximum point of $g_m(t)$ nearest to $a_1 + T$ from the left. Terefore, let $b_1$ be the local minimum point of $g_m(t)$ nearest from the right to $a_1$ and $b_2$ is the local minimum point nearest from the left to $a_1 + T$.

If $g_m(b_1) \geq g_m(b_2)$ then set $a = b_1$, $b = a_2$, $c = b_2$ and $d = a_1 + T$ and if $g_m(b_1) < g_m(b_2)$ then set $a = a_1$, $b = b_1$, $c = a_3$ and $d = b_2$. Denote by $f(t)$ and $\psi(t)$ the restrictions of the function $g_m(t)$ to the segments $[a, b]$ and $[c, d]$. It is clear that the functions $f(t)$ and $\psi(t)$ satisfy the conditions of lemma 7.2.8 and so that there exist points $\xi < \eta$, $\eta - \xi < T$ such that

$$g_m(\xi) = f(\xi) = \psi(\eta) = g_m(\eta)$$

and

$$g'_m(\xi) = f'(\xi) = \psi'(\eta) = g'_m(\eta),$$

i.e. such that the function

$$H_{\eta-\xi}(t) = g_m(t) - g_m(t + \eta - \xi)$$

has at the point $\xi$ a multiply zero and we get contradiction to the lemma 7.2.7, if $H_y(t) \not\equiv 0$. This means that $H_{\eta-\xi}(t) \not\equiv 0$ i.e that the number $\eta - \xi$ is a period of the function $g_m(t)$. But it is impossible because the number $T$ is a minimal period and $T > \eta - \xi$.

Thus we prove that on a minimal period of the function $g_m(t)$ there are exactly two zeros. Since zeros of $g_m(t)$ and $g_m^{(m)(t)}$ are distributed $T$-periodically so that

$$\gamma(t) = \varphi_{k,m}(\alpha, \beta; t + \tau) - \lambda.$$

We have proved that the perfect spline

$$\gamma(t) = \varphi_{k,m}(\alpha, \beta; t + \tau) - \lambda$$

solves problem (7.15). As for problem (7.14) then for $n > k$

$$E_0(\varphi_{n,m}(\alpha, \beta)_{1;\theta,\delta} = (k/n)^m E_0(\varphi_{n,m}(\alpha, \beta)_{1;\theta,\delta} < E_0(\varphi_{n,m}(\alpha, \beta)_{1;\theta,\delta}$$

and we see that the perfect spline

$$\gamma(t) = \varphi_{n,m}(\alpha, \beta; t + \tau) - \lambda.$$

solves it.

Theorem 7.2.2 as well as theorem 7.2.1 are proved.

¿From the theorem 7.2.1, taking into account the general facts presented in sections 1.3 and 1.5 we can deduce a series of useful corollaries. Thus for instance from the theorem 7.2.1 and from the proposition 1.5.1 it follows at once

**Theorem 7.2.9.** *Let $n, m \in N$, $\alpha, \beta > 0$ and $\lambda \in R$. For any perfect spline $\gamma \in \Gamma_{2n,m}(\alpha, \beta)$, $\gamma \perp 1$ the following exact inequality*

$$|\varphi_{n,m}(\alpha, \beta) - \lambda| \prec |\gamma - \lambda|. \tag{7.24}$$

*is valid. In particular, if $\gamma \in \Gamma_{2n,m}$ and $\gamma \perp 1$ then*

$$|\varphi_{n,m} - \lambda| \prec |\gamma - \lambda|. \tag{7.25}$$

In its turn from theorems 7.2.9 and 3.1.1 it follows

**Theorem 7.2.10.** *Let $n, m \in N$; $\alpha, \beta > 0$ and $\lambda \in R$. For any N-function $\Phi$ and any perfect spline $\gamma \in \Gamma_{2n,m}(\alpha,\beta)$ such that $\gamma \perp 1$ the next inequality is valid*

$$\int_0^{2\pi} \Phi(|\varphi_{n,m}(\alpha,\beta;t) - \lambda|)dt \leq \int_0^{2\pi} \Phi(|\gamma(t) - \lambda|)dt \tag{7.26}$$

*and in particular, for all $p \in [1;\infty]$*

$$\|\varphi_{n,m}(\alpha,\beta) - \lambda\|_p \leq \|\gamma - \lambda\|_p. \tag{7.27}$$

It follows from inequality (7.24) for $\lambda = 0$ that if $\gamma \in \Gamma_{2n,m}(\alpha,\beta)$ and $\gamma \perp 1$ then

$$\max_t \gamma(t) = r(\gamma_+, 0) \geq r((\varphi_{n,m}(\alpha,\beta))_+, 0) = \max_t \varphi_{n,m}(\alpha,\beta;t)$$

and

$$\min_t \gamma(t) = -r(\gamma_-, 0) \leq -r((\varphi_{n,m}(\alpha,\beta))_-, 0) = \min_t \varphi_{n,m}(\alpha,\beta;t)$$

i.e, if $\gamma \in \Gamma_{2n,m}(\alpha,\beta)$ and $\gamma \perp 1$ then for all $t$

$$\min_t \gamma(t) \leq \varphi_{n,m}(\alpha,\beta;t) \leq \max_t \gamma(t).$$

¿From here and from theorem 7.2.10 the next theorem follows.

**Theorem 7.2.11.** *Let $n, m \in N$; $\alpha, \beta > 0$ and $\lambda \in R$. For any N-function $\Phi$ and any perfect spline $\gamma \in \Gamma_{2n,m}(\alpha,\beta)$*

$$\int_0^{2\pi} \Phi(|\gamma(t)|)dt \geq \inf_\lambda \int_0^{2\pi} \Phi(|\varphi_{n,m}(\alpha,\beta;t) - \lambda|)dt, \tag{7.28}$$

*and, in particular, for any $p \in [1,\infty]$*

$$\|\gamma\|_p \geq E_0(\varphi_{n,m}(\alpha,\beta))_p. \tag{7.29}$$

*For any nonnegative on the all real axis perfect spline $\gamma \in \Gamma_{2n,m}(\alpha,\beta)$*

$$\int_0^{2\pi} \Phi(|\gamma(t)|)dt \geq \int_0^{2\pi} \Phi(|\varphi_{n,m}(\alpha,\beta;t) - \min_t \varphi_{n,m}(\alpha,\beta;t)|)dt, \tag{7.30}$$

*and, in particular, for any $p \in [1,\infty]$*

$$\|\gamma\|_p \geq \|\varphi_{n,m}(\alpha,\beta;t) - \min_t \varphi_{n,m}(\alpha,\beta;t)\|_p. \tag{7.31}$$

For $\alpha = \beta = 1$ we obtain from the theorem 7.2.10 that the following theorem is valid

**Theorem 7.2.12.** *Let $n, m \in N$ and $\Phi$ be an arbitrary $N$-function. Then for any perfect spline $\gamma \in \Gamma_{n,m}$ the next inequality is valid*

$$\int_0^{2\pi} \Phi(|\gamma(t)|)dt \geq \int_0^{2\pi} \Phi(|\varphi_{n,m}|)dt, \tag{7.32}$$

*and, in particular, for any $p \in [1, \infty]$*

$$||\gamma||_p \geq E_0(\varphi_{n,m})_p. \tag{7.33}$$

*For any nonnegative on the all real axis perfect spline $\gamma \in \Gamma_{2n,m}(\alpha, \beta)$*

$$\int_0^{2\pi} \Phi(|\gamma(t)|)dt \geq \int_0^{2\pi} \Phi(|\varphi_{n,m} + \min_u \varphi_{n,m}(u)|)dt, \tag{7.34}$$

*and, in particular, for any $p \in [1, \infty]$*

$$||\gamma||_p \geq ||\varphi_{n,m}(\alpha, \beta) + \min_t \varphi_{n,m}(\alpha, \beta; t)||_p. \tag{7.35}$$

*All inequalities (7.24)-(7.25) are best possible since $\varphi_{n,m} \in \Gamma_{2n,m}$ and $\varphi_{n,m}(\alpha, \beta) \in \Gamma_{2n,m}(\alpha, \beta)$.*

Finally we note that by theorem 1.7.8 for $\gamma \in \Gamma_{2n,m}$

$$\frac{||\gamma^{m)}||_1}{||\varphi_{1,1}||_1 V_0^{2\pi}(\gamma^{(m)})} \leq$$

$$\left(\left(\frac{\mu(\gamma')}{2}\right)^{1-1/p} \left(\frac{4||\gamma||_p}{||\varphi_{1,m}||_p V_0^{2\pi}(\gamma^{(m)})}\right)\right)^{1/(m+1/p)}. \tag{7.36}$$

But if $\gamma \in \Gamma_{2n,m}$ and $\mu(\gamma^{(m)}) = 2k,\ k < n$ then $||\gamma^{(m)}||_1 = 2\pi$ and $\mu(\gamma') \leq 2k$. From this and from (7.26) it follows that

$$\frac{1}{k} \leq \left(k^{1-1/p} \frac{4||\gamma||_p}{4k||\varphi_{1,m}||_p}\right)^{m+1/p},$$

or that is the same

$$||\gamma||_p \geq k^{-m}||\varphi_{1,m}||_p = ||\varphi_{k,m}||_p > ||\varphi_{n,m}||_p,$$

i.e. inequality (7.33) can be deduced from the theorem 1.7.8.

## 7.3 Generalized perfect splines

Let a CVD-kernel $K$ is given and positive numbers $\alpha, \beta$ and a number $\sigma \in (-\beta, \alpha)$ are given also. By $\Gamma^{\sigma}_{2n,0}(\alpha, \beta)$ we denote the set of functions $\gamma \in \Gamma_{2n,0}(\alpha, \beta)$ which have mean value on the period $[0, 2\pi]$ equal to $\sigma$.

Functions of the form $\gamma * K$ where $\gamma \in \Gamma^{\sigma}_{2n,0}(\alpha, \beta)$ we will call the generalized perfect $(\alpha, \beta)$-splines.

Denote by $\varphi^{\sigma}(\alpha, \beta, t)$ the function of the period $2\pi/n$, having value $\alpha$ on the interval $[0, 2\pi(\sigma + \beta)/(n(\alpha + \beta))$ and value $-\beta$ on the interval $[2\pi(\sigma - \beta)/(n(\alpha + \beta))$.

It is clear that $\varphi^{\sigma}(\alpha, \beta, t)$ is a unique (up to the shift of argument) $2\pi/n$-periodic function from $\Gamma^{\sigma}_{2n,0}(\alpha, \beta)$. Clearly also

$$\varphi^{0}_{n}(\alpha, \beta; t) = \varphi_{n,0}(\alpha, \beta; t).$$

The next theorem gives us a generalization of the theorem 7.2.1.

**Theorem 7.3.1.** *Let $K \in CVD$, $n \in N$; $\alpha, \beta > 0$; $\sigma \in (-\beta, \alpha)$ is arbitrary number if $\theta(K) = 0$ and $\sigma = 0$ if $\theta(K) = 1$. Then for any $\gamma \in \Gamma^{\sigma}_{2n,0}(\alpha, \beta)$ and $\lambda \in R$*

$$(K * \varphi^{\sigma}(\alpha, \beta, t) - \lambda)_{\pm} \prec (K * \gamma - \lambda)_{\pm} \tag{7.37}$$

It is clear that for $K(t) = B_m(t)$ the inequality (7.37) becomes an (7.9).

**Proof** of the theorem 7.3.1 is analogous in a large extend to the proof of theorem 7.2.1 but we will give this proof in detail to illustrate the means which one can use in the similar situations for the passage from the convolution with Bernully kernels to the convolution with arbitrary kernels.

In view of the theorem 1.5.1 to prove (7.37) it is sufficiently to prove that for all $\theta, \delta > 0$

$$E_0(K * \varphi^{\sigma}(\alpha, \beta, t))_{1;\theta,\delta} \leq E_0(K * \gamma)_{1;\theta,\delta}. \tag{7.38}$$

In the proof (7.38) we will assume (as in the proof of the theorem 6.7.1) that the kernel $K$ is analytic on the real axis.

By the same way as the relations (7.26) and (7.27) we establish that if $\gamma \in \Gamma^{\sigma}_{2n,0}(\alpha, \beta)$,

$$x_1 < x_2 < \ldots < x_{2k} < x_1 + 2\pi$$

are its nodes and $\gamma(t) = \alpha$ for $t \in (x_1, x_2)$ then

$$\sum_{j=1}^{2k} (-1)^j x_j = \frac{2\pi(\beta+\sigma)}{\alpha+\beta} \tag{7.39}$$

and

$$\gamma(t) = \sigma + (\alpha+\beta) \sum_{j=1}^{2k} (-1)^j B_1(x_j - t). \tag{7.40}$$

Then if $f = K * \gamma$ then

$$f(t) = \sigma \int_0^{2\pi} K(t)dt + (\alpha+\beta) \sum_{j=1}^{2k} (-1)^j \int_0^{2\pi} B_1(x_j - u)K(t-u)du. \tag{7.41}$$

Consider the extremal problem

$$E_0(f)_{1;\theta,\delta} \to \inf, \quad f \in K * \Gamma_{2n,0}^{\sigma}(\alpha,\beta). \tag{7.42}$$

Since for any $K \in L_1$ the set $K * \Gamma_{2n,0}^{\sigma}(\alpha,\beta)$ is locally compact in the space $C$ then there exists a spline $\gamma_* \in \Gamma_{2n,0}^{\sigma}(\alpha,\beta)$ such that the function $f_* = K * \gamma_*$ realizes the infimum in the problem (7.42). Suppose that $\gamma_*$ has on a period $2k < 2n$ points of discontinuity

$$x_1^* < x_2^* < \ldots < x_{2k}^*$$

(without loss of generality we can assume that $\gamma_*(t) = \alpha$ for $t \in (x_1^*, x_2^*)$). Then taking into account the lemma 7.2.7 and relations (7.39) and (7.41) we conclude that points $x_1^*, x_2^*, \ldots, x_{2k}^*$ and the constant $\lambda_*$ of the best $(\theta,\delta)$-approximation for $f_*$ realize also infimum in the following problem

$$F(x_1, x_2, \ldots, x_{2k}; \lambda) \to inf; \tag{7.43}$$

$$\sum_{j=1}^{2k} (-1)^j x_j = \frac{2\pi(\beta+\sigma)'}{\alpha+\beta}$$

$$0 < x_1 < x_2 < \ldots < x_{2k} < 2\pi,$$

where

$$F(x_1, x_2, \ldots, x_{2k}; \lambda) = \pi\lambda(\beta-\alpha)+$$

$$\frac{\theta+\delta}{2} \int_0^{2\pi} |(\alpha+\beta) \sum_{j=1}^{2k} (-1)^j \int_0^{2\pi} B_1(x_j - u)K(t-u)du - \lambda|dt. \tag{7.44}$$

The function $F$ is continuously differentiable in some neighbourhood of the point $(x_1^*, x_2^*, \ldots, x_{2k}^*; \lambda^*)$ in the sense that the partial derivatives $\partial F/\partial x_j$, $j = 1, 2, \ldots, 2k$; $\partial F/\partial \lambda$ exist and are continuous and by Lagrange's multipliers method we obtain the following necessary conditions of extremum for the problem (7.43):

$$\frac{(\alpha+\beta)(\theta+\delta)}{2} \int_0^{2\pi} sgn\big((\alpha+\beta)\sum_{j=1}^{2k}(-1)^j \int_0^{2\pi} B_1(x_j - u)K(t-u)du$$

$$-\lambda\big)\left(K(t-x_l) - \frac{1}{2\pi}\int_0^{2\pi} K(u)du\right) dt + \xi = 0, \qquad (7.45)$$

$$l = 1, 2, \ldots, 2k$$

and

$$\frac{\theta+\delta}{2} \int_0^{2\pi} sgn\left((\alpha+\beta)\sum_{j=1}^{2k}(-1)^j \int_0^{2\pi} B_1(x_j-u)K(t-u)du - \lambda\right) dt -$$

$$\pi(\beta - \alpha) = 0. \qquad (7.46)$$

Taking into account (7.41) and lemma 7.2.8 we see that the condition (7.46) means the orthogonality to constants for the function

$$g(t) = \theta sgn\,(f(t) - \lambda)_+ - \delta\, sgn\,(f(t) - \lambda)_-.$$

If now

$$G(t) = K(-\cdot) * g(t) - K(-\cdot) * g(x_1),$$

then the condition (7.45) means that for $l = 1, 2, \ldots, 2k$

$$G(x_l) = 0. \qquad (7.47)$$

Taking into account that $K \in CVD$, we obtain that $\mu(g) \leq 2k$ and since $K(-\cdot) \in CVD$ we get $\mu(G) \leq 2k$. Comparing this fact with (7.47) it is easy to verify that $G(t)$ vanishes on $[o, 2\pi)$ only at the points $x_j^*$ and alternates sign at every of this points. In the case $\theta = 1$ we can reson as follows. For any trigonometric polynomial $\tau(t)$, having sufficiently small $C$-norm

$$\mu(G - \tau) = \mu(g) = \mu(f_* - \lambda^*) \leq \mu(\gamma_*) = 2k.$$

On the other hand, if at some of the points $x_j^*$ the function $G(t)$ does not alternate sign or besides points $x_j^*$ the function $G(t)$ has another zeros in

$[0, 2\pi)$, then we can choose $\tau(t)$ such that $\mu(G - \tau) > 2k$ contrary to the precedent fact. The case $\theta = 0$ is only slightly more difficult.

Thus

$$sgn\, G(t) = \pm sgn\, \gamma_*(t). \tag{7.48}$$

Now our goal is to prove that (7.48) is possible (up to translation of argument) only if

$$f_*(t) = K * \varphi_k^\sigma(\alpha, \beta).$$

If $y \in R$ we put

$$h_{y,0} = \gamma_*(t) - \gamma_*(t + y),$$

$$h_y(t) = f_*(t) - f_*(t + y),$$

$$H_{y,0}(t) = g(t) - g(t + y),$$

$$H_y(t) = G(t) - G(t + y).$$

**Lemma 7.3.2.** *For any $y \in R$*

$$\mu(H_y) \geq \mu(h_{y,0}), \quad \mu(H_{y,0}) \leq \mu(h_y). \tag{7.49}$$

**Proof.** If on period there exist $2d$ points

$$t_1 < t_2 < \ldots < t_{2d},$$

such that in their neihgbourhood $h_{y,0}$ has nonzero valuea with alternating signs under the passage from one neighbourhood to another, then in view of (7.48) $H_y(y)$ also has nonzero value at this points with alternating sign. Consequently $\mu(H_y) \geq \mu(h_{y,0})$. Analogously taking into account the fact that by definition of $g(t)$

$$sgn\, g(t) = sgn\, (f_*(t) - \lambda)$$

we can prove the second inequality (7.49).

We need in follows the next technical lemma.

**Lemma 7.3.3.** *Let $K$ be the kernel different from constant, and is analitic on the all real axis and let segments $\Delta_1$ and $\Delta_2$ lie on the same interval of monotony for $K$. Then there exist functions $\omega_1, \omega_2 \in C$ such that*

*1) the support of the function $\omega_i$, $i = 1,2$ is contained in the set*

$$\cup_{l \in Z}(\Delta_i + 2l\pi);$$

$$2)\ ,\ \int_{\Delta_i} \omega_i(t)\, dt = 0, \quad i = 1,2;$$

*3) $(-1)^i \omega_i(t)$ alternates sign on the segment $\Delta_i$ exactly once from negative to positive;*

*4) there exists $\xi \in R$, $\xi > 0$ such that*

$$\xi \int_{\Delta_1} K(t)\omega_1(t)dt + \int_{\Delta_2} K(t)\omega_2(t)dt = 0 \tag{7.50}$$

*and*

$$\xi \int_{\Delta_1} K'(t)\omega_1(t)dt + \int_{\Delta_2} K'(t)\omega_2(t)dt \neq 0.$$

**Proof.** The existence of the functions $\omega_1$ and $\omega_2$ having the properties 1)-3) is obvious. Moreover the integrals

$$\int_{\Delta_i} K'(t)\omega_i(t)dt \quad i = 1,2$$

are different from zero and have opposite signs. So for any $\omega_1$ and $\omega_2$ it is easy to satisfy the condition (7.50). Fixing some function $\omega_2$ and assuming that if (7.50) then also the equality

$$\xi \int_{\Delta_1} K'(t)\omega_1(t)dt + \int_{\Delta_2} K'(t)\omega_2(t)dt = 0,$$

is valid, then we obtain that for any $\omega_1$ such that $\omega_1$ is continuous on $\Delta_1$ , alternates sign exactly once and has on $\Delta_1$ mean value which is equal to zero and the next equality will be valid

$$\frac{\int_{\Delta_1} K'(t)\omega_1(t)dt}{\int_{\Delta_1} K(t)\omega_1(t)dt} = \frac{\int_{\Delta_2} K'(t)\omega_2(t)dt}{\int_{\Delta_2} K(t)\omega_2(t)dt}.$$

It is possible only if on $\Delta_1$ the kernel $K$ satisfes to the equation

$$K'(t) + \eta K(t) \equiv const. \tag{7.51}$$

Since among analytic solutions of the equation (7.51) have the period $2\pi$ then only constants we arrive at the contrary with the fact that $K \not\equiv const$.

**Lemma 7.3.4.** *For any $y$ the function $H_y(t)$ either vanishes identically or has simple isolated zeros.*

**Proof.** Let $H_y(t) \not\equiv 0$. Since $H_y(t)$ is analytic on real axis, this function is also to have only isolated zeros.

Since $K \in CVD$ we have $\mu(h_y) \leq \mu(h_{y,0})$, that together with the first of inequalities (7.49) gives us $\mu(H_y) \geq \mu(h_y)$. On the other hand taking into account the relation $K(-\cdot) \in CVD$ and the second inequality (7.49), we obtain

$$\mu(H_y) \leq \mu(H_{y,0}) \leq \mu(h_y).$$

Thus

$$\mu(H_y) = \mu(H_{y,0}). \tag{7.52}$$

Suppose that $H_y(t)$ has multiply zeros $t_0$, i.e.

$$H_y(t_0) = H_y'(t_0) = 0.$$

Choose some interval where $H_{y,0}(t)$ is nonzero and which lies on the segment of monotony of the function $K(t - t_0)$. On this segment we choose two segments $\Delta_1$ and $\Delta_2$ having empty intersection. By lemma 7.3.2 there exist continuous functions $\omega_1$ and $\omega_2$ which possess concerning $K(t - t_0)$ the properties 1)-4) from lemma 7.3.2.

Put $\omega_\varepsilon = \varepsilon(\xi\omega_1 + \omega_2)$, where $\varepsilon \in R, \varepsilon \neq 0$. By the property 4) it will be $(K(-\cdot) * \omega_\varepsilon)(t_0) = 0$ and $(K(-\cdot) * \omega_\varepsilon)'(t_0) \neq 0$. A number $\varepsilon$ we can choose such that $\mu(H_{y,0} + \omega_\varepsilon) = \mu(H_{y,0})$, in a neighbourhood of any point of sign change of $H_y$ the function $H_y + K(-\cdot) * \omega_\varepsilon$ also has sign change and in neighbourhood of the point $t_0$ the function $H_y + K(-\cdot) * \omega_\varepsilon$ has at least two sign changes. Then we obtain (taking into account (7.52))

$$\mu(K(-\cdot) * (H_{y,o} + \omega_\varepsilon)) = \mu(H_y + K(-\cdot) * \omega_\varepsilon) \geq$$

$$\geq \mu(H_y) + 2 = \mu(H_{y,0}) + 2 = \mu(H_{y,0} + \omega_\varepsilon) + 2,$$

contrary to the fact that $K(-\cdot) \in CVD$.

Now we can word by word repeat the corresponding part of the proof of theorem 7.2.1 (with replacing $g_m$ by $G$ and reference on the lemma 7.2.7 by reference on the lemma 7.3.3) we establish that $G$ has a period $2\pi/k$ , but it is possible only when $f_*(t) = K * \varphi_k^\sigma(\alpha, \beta; \cdot + \tau)$. To complete the proof of the theorem it remains to take into account that for $n > k$

$$E_0(K * \varphi_n^\sigma(\alpha, \beta))_{1;\theta,\delta} < E_0(K * \varphi_k^\sigma(\alpha, \beta))_{1;\theta,\delta}.$$

Like theorem 7.2.1,the theorem 7.3.1 has a series of useful corollaries generalizing the theorem 7.2.8-7.2.11. For example the next theorems are valid.

**Theorem 7.3.5.** *Under the conditions of the theorem 7.3.1*

$$|K * \varphi_n^\sigma(\alpha,\beta) - \lambda| \prec |K * \gamma - \lambda|, \quad \lambda \in R,$$

*and consequently for any $N$-function $\Phi$*

$$\int_0^{2\pi} \Phi(|K * \varphi_n^\sigma(\alpha,\beta;t)) - \lambda|)\,dt \le \int_0^{2\pi} \Phi(|K * \gamma(t) - \lambda|)\,dt.$$

**Theorem 7.3.6.** *Under the conditions of the theorem 7.3.1 for any $N$-function $\Phi$*

$$\int_0^{2\pi} \Phi\left(|K * \varphi_n^\sigma(\alpha,\beta;t) - \min_u K * \varphi_n^\sigma(\alpha,\beta;u)|\right) dt \le$$

$$\int_0^{2\pi} \Phi\left(|K * \gamma(t) - \min_u K * \gamma(u)|\right) dt.$$

We omit proof of theorem 7.3.3 and 7.3.4 which are based on the theorem 7.3.1 since it has been shown above as using the theorem 7.2.1 allows us to prove theorems 7.2.8 and 7.2.11.

## 7.4 Existence theorem for perfect splines

Here we present some statements of existence theorem type for perfect splines. These statements are useful for investigation of various extremal problems and in particular in the problems of estimation of widths of various classes differentiable functions.

Results of this section are based essentially on the next Borsuk's theorem.

**Theorem 7.4.1.** *Let $X$ be linear normed space rith dimension $n+1$ and let*

$$S_n = \{x : x \in X_{n+1}, \ ||x|| = R\}$$

*be a sphere in $X_{n+1}$. If operapor (vector field) $P(x)$ which maps $S^n$ into $n$-dimensional space $Y_n$ is continuous and odd (i.e. $P(-x) = -P(x)$ for all $x$) then there exists a point $x_0 \in S^n$ such that*

$$P(x_0) = 0.$$

Proof of Borsuk's theorem one can find for example in (Borsuk [1]).

**Theorem 7.4.2.** *Let $F_j$, $j = 1, 2, \ldots, 2n$ are continuous and odd functions on $C^{m-1}$ such that*

$$\sum_{j=1}^{2n} F_j^2(1) \neq 0. \tag{7.53}$$

*Then for any $m = 1, 2, \ldots$ there exists perfect spline $\gamma_* \in \Gamma_{2n,m}$ such that*

$$F_j(\gamma_*) = 0, \quad j = 1, 2, \ldots, 2n. \tag{7.54}$$

**Proof.** In view of (7.53) we can assume that $F_1(1) \neq 0$. Let $S^{2n+1}$ be a sphere with a center in origin and of radius $\sqrt{2\pi}$ in $R_{2n+1}$. For a vector

$$\xi = (\xi_1, \xi_2, \ldots, \xi_{2n+1}) \in S^{2n+1}$$

put

$$\eta_i = \sum_{j=1}^{i} \xi_i^2, \quad i = 1, 2, \ldots, 2n+1$$

and define on $[0, 2\pi)$ the function

$$\gamma_0(\xi, t) = sgn\, \xi_i, \quad t \in [\eta_{i-1}, \eta_i), \quad i = 1, 2, \ldots, 2n+1. \tag{7.55}$$

Next put

$$\gamma_k^0(\xi, t) = \int_0^t \gamma_{k-1}^0(\xi, u)du -$$

$$-\frac{1}{2\pi} \int_0^{2\pi} \int_0^t \gamma_{k-1}^0(\xi, u)dudt, \quad k = 1, 2, \ldots, m$$

and

$$\gamma_m(\xi, t) = \gamma_m^0(\xi, t) - \frac{F_1(\gamma_m^0(\xi))}{F_1(1)} \tag{7.56}$$

$$\gamma_m(\xi, t + 2\pi) = \gamma_m(\xi, t).$$

On the sphere $S^{2n+1}$ we define the vector field

$$P(\xi) = (P_1\xi, P_2(\xi), \ldots, P_{2n}(\xi)),$$

where

$$P_1(\xi) = \int_0^{2\pi} \gamma_m^{(m)}(\xi, t)dt = \int_0^{2\pi} \gamma_0(\xi, t)dt, \tag{7.57}$$

$$P_j(\xi) = F_j(\gamma_m(\xi)), \quad j = 2, 3, \ldots, 2n.$$

This field $P(\xi)$ is continuous on $S^{2n+1}$ and for $\xi \in S^{2n+1}$ it will be $P(-\xi) = -P(\xi)$. So due to theorem 7.4.1 there exists a vector $\xi_0 \in S^{2n+1}$ such that $P(\xi_0) = 0$.

But the equality $P_1(\xi_0) = 0$ means that

$$\int_0^{2\pi} \gamma_0(\xi_0, t)dt = 0$$

and consequently

$$\gamma_m(\xi_0) \in \Gamma_{2n,m}.$$

Puting $\gamma_*(t) = \gamma_n(\xi_0, t)$, we note that the equality (7.54) for $j = 1$ follows from (7.56) and for $j = 2, 3, \ldots, 2n$ it follows from (7.57) and from the equality $P(\xi_0) = 0$.

Analogously we can prove the next theorem

**Theorem 7.4.3.** *Under the conditions of theorem 7.4.2 for any kernel $K \in L_1$ there exists spline $\gamma_0 \in \Gamma_{2n,m}$, $\gamma_0 \bot 1$ such that*

$$F_j(K * \gamma_0) = 0, \quad j = 1, 2, \ldots, 2n.$$

**Theorem 7.4.4.** *Let $\Re$ be a 2n-dimensional subspace of the space $L_1$ which contains the constant. There exists a perfect spline $\gamma_*(t) \in \Gamma_{2n,m}$ such that $\gamma_*^{(m)} \bot \Re$, i.e.*

$$\int_0^{2\pi} \gamma_*^{(m)} g(t)dt = 0, \quad \forall\, g(t) \in \Re.$$

*and for given $p \in [1, \infty]$*

$$\inf_{\lambda} \|\gamma_* - \lambda\|_p = \|\gamma_*\|_p.$$

**Proof.** As in proof of the theorem 7.4.2 we define on the sphere $S^{2n+1} \subset R_{2n+1}$ the function $\gamma_0(\xi, t)$ by means of the equality (7.55) and let $e_1(t) \equiv 1, e_2(t), \ldots, e_{2n}(t)$ is a basis of the space $\Re$. On the sphere $S^{2n+1}$ we define the vector field

$$P_j(\xi) = \int_0^{2\pi} e_j \gamma_0(\xi, t)dt, \quad j = 1, 2, \ldots, 2n,$$

and it is obvious that $P_j$ will be continuous and odd. By Borsuk's theorem there exists a vector $\xi_0 \in S^{2n+1}$ such that $P_j(\xi_0) = 0$, $j = 1, 2, \ldots, 2n$ and in particular

$$P_1(\xi_0) = \int_0^{2\pi} \gamma_0(\xi_0, t)dt = 0.$$

Now we define $\gamma_*(t)$ as $m$-th periodic integral of the function $\gamma_0(\xi_0)$ (we assume that $\gamma_0(\xi_0)$ is continued from period $2\pi$ on the real axis).

**Theorem 7.4.5.** *Let $\Phi$ be a strictly increasing and continuously differentiable N-function. Let $L_\Phi$ be a linear space of real $2\pi$-periodic functions $f(t)$ such that*

$$\int_0^{2\pi} \Phi(|f(t)|dt < \infty,$$

*and let $\Re$ be $2n$-dimensional subspace of $L_\Phi$. Then there exists a perfect spline $\gamma \in \Gamma_{2n,m}$ such that*

$$\inf_{\phi \in \Re} \int_0^{2\pi} \Phi(|\gamma(t) - \phi(t)|)dt = \int_0^{2\pi} \Phi(|\gamma(t)|)dt. \qquad (7.58)$$

*In particular for any $p \in [1, \infty]$ there exists a perfect spline $\gamma \in \Gamma_{2n,m}$ such that*

$$E(\gamma, \Re)_p = ||\gamma||_p. \qquad (7.59)$$

**Proof.** Let $e_1 \equiv 1, e_2, \ldots, e_{2n}$ be a basis of the subspace $\Re$ and for $f \in C$

$$F_j(f) = \int_0^{2\pi} \Phi'(|f(t)|)e_j(t) sgn\, f(t)dt, \quad j = 1, 2, \ldots, 2n.$$

It is clear that $F_j$ are odd and continuous on $C$ functions and consequently in view of theorem 7.4.2 there exists a perfect spline $\gamma(t)$ such that $\gamma \in \Gamma_{2n,m}$ and

$$F_j(\gamma) = \int_0^{2\pi} \Phi'(|\gamma(t)|)e_j(t) sgn\, \gamma(t)dt, \quad j = 1, 2, \ldots, 2n.$$

But then the equality (7.58) immediately follows from theorem 1.1.10 and equality (7.59) and (7.58) for $\Phi(t) = t^p$, $p \in [1, \infty)$ if we take into account the possibility of passing to the limit then $p \to \infty$.

Note that from theorem 7.4.2 it follows the next generalization (in the case $\alpha = \beta = 1$) of the theorem 6.1.7 on zeros of the perfect splines.

**Corollary 7.4.6.** *Let $t_k \in [0.2\pi)$, $k = 1, 2, \ldots, 2n$ are arbitrary points and $\nu_k$, $0 \le \nu_k \le m - 1$ are arbitrary integer numbers moreover at least one of them is equal to zero. Then there exists a perfect spline $\gamma \in \Gamma_{2n,m}$ such that*

$$\gamma^{(\nu_k)}(t_k) = 0, \quad k =, 2, \ldots, 2n$$

To complete the present section we 'll formulate a result about $\omega$-perfect spline which proof one can find in the articles of Korneichuk N.P.[22,24], Motornyi V.P.,Ruban V.I.[1], Ligun A.A.[8].

Let arbitrary modulus of continuity $\omega(t)$ is given. The function $\gamma \in C^m$ will be called $\omega$-perfect spline of order $m$ with $2n$ nodes if there exist points (nodes)

$$t_1 < t_2 < \ldots < t_{2n} \le t_1 + 2\pi,$$

such that for $t \in (t_{i-1}, t_i)$

$$\gamma^{(m)}(t) = \frac{1}{2}\varepsilon_i \min\{\omega(2|t - t_{i-1}|), \omega(2|t_i - t|)\}, \quad (|\varepsilon_i| = 1).$$

The set of all $\omega$-perfect splines of order $m$ with $2n$ nodes we shall denote by $\Gamma^{\omega}_{2n,m}$.

**Theorema 7.4.7.** *Let $\omega$ is convex from below modulus of continuity $n \in N$ and $m = 1, 2, \ldots$. Then for any perfect spline $\gamma \in \Gamma^{\omega}_{2n,m}$ the next exact on $\Gamma^{\omega}_{2n,m}$ inequality is valid*

$$||\gamma||_p \ge ||f_{n,m}(\omega)||_p \quad (p \in [1, \infty]).$$

# Chapter 8

# Extremal properties and widths

## 8.1 Widths of odd order

On definition by Kolmogorov N-width of a central-symmetric sef @ in a linear normed space $X$ we denote next value

$$d_N(@, X) = \inf\{\sup_{x\in @} \inf_{u\in @_n} \|x - u\| : @_n \subset X, dim\, @_N = N\}.$$

As $X$ we will consider the spaces $C$ and $L_1$ of $2\pi$-periodic functions. As @ we will consider the class of convolutions of CVD-kernel $K(t)$ with unit ball from $C$ or $L_1$. As above we put

$$F_p = \{f(t) : f \in L_p, \|f\|_p \leq 1\}$$

above estimate of $d_{2n-1}(K * F_p, L_p)$ for $p = \infty$ and $p = 1$ we obtain calculating the best approximation of the class $K * F_p$ by subspace $T_{2n-1}$ of trigonometric polynomials of order $n - 1$.

**Theorem 8.1.1.** *Let $K(t)$ be an arbitrary CVD-kernel and $n \in N$. Them*

$$E(K, T_{2n-1})_1 = \|K * \varphi_{n,0}\|_\infty, \tag{8.1}$$

*where $\varphi_{n,0}(t) = sgn \sin nt$.*

**Proof.** Put

$$A_h(t) = \frac{1}{2\pi} \sum_{m=-\infty}^{\infty} \frac{e^{imt}}{ch\, mh} \; (h > 0)$$

The function $A_h(t)$ is analytic on the all real axis CVD-kernel If $K(t)$ is CVD-kernel then $K_h := K * A_h$ is analitic on the all real axis CVD-kernel. For any $\delta \in (0, \pi)$

$$\lim_{h \to 0} \int_{-\delta}^{\delta} A_h(t)dt = 1$$

so

$$\lim_{h \to 0} \|K - K_h\|_1 = 0,$$

and consequently it is sufficiently to prove the theorem 8.1.1 in the case when $K(t)$ is analytic on the all real axis CVD-kernel.

Calculating zeros of the functions $f(t)$ we will calculate any zero at whith $f(t)$ changes sign, and twice we will calculate any zero at which $f(t)$ has not change of sign.The number of zeros of the function $f$, calculating according to this rule on period $[0, 2\pi)$ we denote by $z(f)$.

**Lemma 8.1.2.** *For any polynomial $\tau(t) \in T_{2n-1}$ and for any CVD-kernel $z(K - \tau) \leq 2n$.*

**Proof.** If the statement of the lemma is not true then there exists a polynomial $\tau_*(t) \in T_{2n-1}$ such that $z(K - T_*) > 2n$. Then in view of Rolle's theorem $\nu(K' - \tau_*') \leq 2n + 2$ and consequently for sufficiently small $\varepsilon > 0$

$$\mu(K'(\cdot) - \tau_*'(\cdot - \varepsilon)) \geq 2n + 2.$$

If

$$f_{k,h} = \frac{1}{(2h)^k} (\delta_{k,h} * f)$$

is $K$-th Steklov's function with the step $h$ for the function $f$, then for small $h > 0$

$$\mu((K'(\cdot) - \tau_*'(\cdot - \varepsilon))_{k,h}) \geq 2n + 2 \tag{8.2}$$

Let now

$$\tau_*'(\cdot - \varepsilon) = a\theta + K * \tau_\varepsilon, \;\; , \;\; \tau_\varepsilon \in T_{2n-1},$$

where $\theta = 1$ if $K \perp 1$ and $\theta = 0$ in the opposite case. Then the difference $(K(\cdot) - \tau_*(\cdot - \varepsilon))_{k,h}$ we can represent in the form $K * (\delta_{k,h}' - \tau_\varepsilon)$. Now from the fact that $K$ is CVD-kernel anf from (8.2) it follows that $\mu(\delta_{k,h}' - \tau_\varepsilon) \geq 2n+2$ and in particular

$$\mu(\delta_{3,h}' - \tau_\varepsilon) \geq 2n + 2 \tag{8.3}$$

But $\delta'_{3,h}(t)$ is odd, continuous and piece-wise linear function with the nodes at the points $\pm h$ and $\pm 3h$,and $\delta'_{3,h}(t)$ is equal to $-(4h^2)^{-1}$ at the point $h$ and $\delta'_{3,h}(t)$ is equal to 0 at the point $3h$. If $\tau_\varepsilon(t) \not\equiv 0$ then we choose $\varepsilon > 0$ such that $\tau_\varepsilon(0) \neq 0$. Then for small $h > 0$ the function $\delta'_{3,h}(t) - \tau_\varepsilon(t)$ changes sign 2 times on the interval $(-3h, 3h)$ and $\delta'_{3,h}(t) - \tau_\varepsilon(t) = \tau_\varepsilon(t)$ for $3h \leq |t| \leq \pi$, i.e. in any case for small $h > 0$ there will be $\mu(\delta'_{3,h} - \tau_\varepsilon) \leq 2n$ contrary to (8.3).

Lemma is proved.

**Continue the proof of the theorem 8.1.1.** As

$$\varphi_{n,0}(t) = 2\sum_{j=1}^{2n}(-1)^{k+1}B_1\left(t - \frac{j\pi}{n}\right)$$

so

$$(K * \varphi_{n,0})'(t) = \sum_{j=1}^{2n}(-1)^j K\left(t - \frac{j\pi}{n}\right).$$

Thus if $\xi$ is a point of extremum of the function $K * \varphi_{n,0}$ then

$$\sum_{j=1}^{2n}(-1)^j K\left(\xi - \frac{j\pi}{n}\right) = 0.$$

Besides that if

$$\varphi_n(f,t) = \sum_{j=1}^{2n}(-1)^j f\left(t - \frac{j\pi}{n}\right) = 0,$$

then $\varphi_n(f,t)$ is odd $2\pi/n$-periodic function. Consequently for all $\tau \in T_{2n-1}$ $\varphi_\tau$ is odd $2\pi/n$-periodic polynomial from $T_{n+1}$ i.e.

$$\varphi_n(\tau,t) \equiv 0 \quad \forall\, \tau \in T_{2n-1}. \tag{8.4}$$

Let

$$\tau(K,t) = \frac{1}{2n}\sum_{k=1}^{2n}\frac{\sin(n-1/2)(t-x_k)}{\sin(t-x_k)/2}\,\frac{\cos n(t-\xi)}{2n}\varphi_n(K,\xi) =$$

$$\frac{1}{2}\,\frac{\sin(n-1/2)(t-x_k)}{\sin(t-x_k)/2}$$

interpolates the function $K(t)$ at the points $x_\nu$, $\nu =, 2, \ldots, 2n$ and belongs to the set $T_{2n-1}$. Taking into account the lemma 8.1.2 we conclude that

$$sgn(K(t) - \tau(K,t)) = \pm sgn \sin n(t - x_1).$$

But then

$$\int_0^{2\pi} \tau(t) sgn(K(t) - \tau(K,t))dt = 0 \quad \forall \tau \in T_{2n-1}$$

and in view of theorem 1.1.9

$$E(K, T_{2n-1})_1 = |\int_0^{2\pi} K(t) sgn \sin n(t - x_1)dt| =$$

$$|(K * \varphi_{n,0}(x_1))| = ||K * \varphi_{n,0}||_\infty.$$

**Theorem 8.1.3.** Let $K(t)$ be an arbitrary $CVD$-kernel and $n \in N$. Then for $p = 1, \infty$

$$E(K * F_p, T_{2n-1})_p = ||K * \varphi_{n,0}||_1. \tag{8.5}$$

**Proof.** By definition $K * F_p$ is the set of all functions of the form

$$f(t) = a\theta + (K * \varphi)(t), \ \ ||\varphi||_p \leq 1.$$

Put for $f \in K * F_p$

$$U_n(f,t) = a\mu + (\tau(K) * \varphi)(t),$$

where $\tau(t)$ is a polynomial of best $L_1$-approximation of the kernel $K$. It is easy to see that $U_n(f) \in T_{2n-1}$.

Since

$$f - U_n(f) = K * \varphi - \tau(K) * \varphi = (K - \tau(K)) * \varphi,$$

then using the generalized Minkowski inequality (see, for example, Korneichuk [11, p .299]) we obtain that

$$||f - U_n(f)||_p \leq ||\varphi||_p ||K - \tau(K)||_1 \leq ||K - \tau(K)||_1 =$$

$$E(K, T_{2n-1})_1 = ||K * \varphi_{n,0}||_1$$

and, consequently, for all $f \in K * F_p$

$$E(f, T_{2n-1})_p \leq ||f - U_n(f)||_p \leq ||K * \varphi_{n,0}||_1.$$

Thus

$$E(K * F_p, T_{2n-1})_p \le \sup_{f \in K*F_p} \|f - U_n(f)\|_p \le \|K * \varphi_{n,0}\|_1.$$

The inequality of opposite sense for $p = 1, \infty$ is valid because

$$E(K * \varphi_{n,0}, T_{2n-1})_p = E(K * *\psi_h, T_{2n-1})_p - \varepsilon(h) = \|K * \varphi_{n,0}\|_p - \varepsilon(h),$$

where $\psi_h(t)$ is a derivative of the Steklov's function for $(1/(4n))\varphi_{n,0}(t)$ and $\varepsilon(h) \to 0$ for $h \to 0$.

The next corollary immediately follows from the theorem 8.1.3 if we put $K(t) = B_m(t)$.

**Corollary 8.1.4.** *For all* $n, m \in N$ *and* $p = 1, \infty$

$$E(W_p^m, T_{2n-1})_p = \|B_m * \varphi_{n,0}\|_\infty = \|\varphi_{n,m}\|_\infty.$$

**Theorem 8.1.5.** *For all* $n, m \in N$ *and* $p = 1, \infty$

$$d_{2n-1}(W_p^m, L_p) = \|\varphi_{n,m}\|_\infty. \tag{8.6}$$

**Proof.** In view of corollary 8.1.4 for $p = 1, \infty$

$$d_{2n-1}(W_p^m, L_p) \le \|\varphi_{n,m}\|_\infty. \tag{8.7}$$

On the other hand if

$$U_{2n,m} = \{s(t) : s \in S_{2n,m}, \|s\|_\infty \le \|\varphi_{n,m}\|_\infty\}$$

is $2n$-dimensional ball of the radius $\|\varphi_{n,m}\|_\infty$ in the subspace $S_{2n,m}$ of the space $C$ then in view of inequality (6.11) $U_{2n,m} \subset W_\infty^m$. Using theorem 7.4.1 of the widths of ball we obtain

$$d_(W_\infty^m, L_\infty) \ge d_{2n-1}(U_{2n,m}, L_\infty) \ge \|\varphi_{n,m}\|_\infty.$$

Comparing this inequality and (8.7) we prove (8.6) for $p = \infty$.

Analogously using in the estimation from below (6.54) instead of (6.11) we obtain the inequality

$$d_{2n-1}(W_1^m, L_1) \ge \|\varphi_{n,m}\|_\infty$$

which together with (8.7) gives us (8.6) for $p = 1$.

As for the class $K * F_p$ with $CVD$-kernel $K$, then for $p = \infty$ $2n$-dimensional ball

$$U_{2n} = \{f(t) : f \in K * s,\ s \in S_{2n,0},\ \|f\|_\infty \le \|K * \varphi_{n,0}\|_\infty\}$$

in view of theorem 6.5.4 is contained in $K * F_\infty$. This fact due to the theorem of width of ball gives for $d_{2n-1}(K * F_\infty, L_\infty)$ the estimation from below. Taking into account theorem 8.1.3 we obtain the following statement.

**Theorem 8.1.6.** *Let $n \in N$ and $K \in CVD$. Then*

$$d_{2n-1}(K * F_p, L_\infty) = \|K * \varphi_{n,0}\|_\infty.$$

Thus the subspace $T_{2n-1}$ is an extremal subspace in the sense of Kolmogorov width for the class $W_p^m$ and $K * F_p$ for $p = 1, \infty$.

## 8.2 Perfect splines and widths

In this section we using the extremal properties of the perfect splines will obtain the exact values of the width in a new situations.

Remaind that Kolmogorov width of the class @ in the space $X$ is the value

$$d_N(@, X) = \inf_{a,\, \Re_N} E(@, a + \Re_N)_X,$$

where infinum is taken over all $N$-dimensional subspace $\Re_N$ of the space $X$ and over all elements $a \in X$.

If @ is the central-symmetric set (i.e. $f \in @ \to f \in @$) then in definition we can take $a = 0$ and we obtain the definition of width which was already used.

Present first some results on the best approximations which give us above estomations of width of corresponding classes.

A set $F \in L_1$ will be called rearrangementally invariant if from $f \in F$ and $\pi(g, t) = \pi(f, t)$ for all $t \in [0, 2\pi]$ it follows that $g \in F$.

Let $W^m F$ is the class of functions $f(t)$ from $L_1^m$ such that $f^{(m)} \in F$.

**Theorem 8.2.1.** *If $F \subset L_1$ is arbitrary rearrangementally invariant set then for any $n, m \in N$*

$$E(W^m F, T_{2n-1})_1 = \sup_{\phi \in F,\ \phi \perp 1} \int_0^{2\pi} \pi(\varphi_{n,m}, t)\pi(\phi, t)dt.$$

**Proof.** In view of duality theorem of the best $L_1$-approximation by subspace (see theorem 1.2.4) we have

$$E(W^m F, T_{2n-1})_1 = \sup_{f \in W^m F} E(f, T_{2n-1})_1 =$$

$$\sup_{f \in W^m F} \sup_{\|\psi\|_\infty \leq 1,\, \psi \perp T_{2n-1}} \int_0^{2\pi} f(t)\psi(t)dt.$$

Integrating $m$ times by parts in the last integral and taking into account the proposition 1.3.3 we obtain

$$E(W^m F, T_{2n-1})_1 = \sup_{\phi \in F,\, phi \perp 1} \sup_{\|\psi\|_\infty \leq 1,\, \psi \perp T_{2n-1}} (-1)^m \int_0^{2\pi} \phi(t)\psi(t)dt \leq$$

$$\sup_{\phi \in F,\, phi \perp 1} \sup_{\|\psi\|_\infty,\, \psi \perp T_{2n-1}} (-1)^m \int_0^{2\pi} \pi(\phi, t)\pi(\psi, t)dt. \tag{8.8}$$

In the following we shall need

**Lemma 8.2.2.** *Let $n, m \in N$, $\psi \in W_\infty^m$ and $\psi \perp T_{2n-1}$. Then for any $\lambda \in R$*

$$(\psi - \lambda)_\pm \prec (\varphi_{n,m} - \lambda)_\pm. \tag{8.9}$$

**Proof.** We will use the theorem 1.4.6. Put $f = g_1$ where $g_1(t)$ is a periodic antiderivative for $\psi(t)$ having zero mean value on a period and let $g = \varphi_{n,m+1}$. We will show that functions $f$ and $g$ satisfy the conditions of the theorem 1.4.6. To show this we make sure that $\|f\|_\infty \leq \|g\|_\infty$ and $\|f'\|_\infty \leq \|g'\|_\infty$ (other conditions will be satisfied automatically in view of Rolle's theorem).

For $i = 0, 1$ we have

$$\|f^{(i)}\|_\infty = \sup_t \int_0^{2\pi} B_{m+i}(t-u)\psi^{(m)}(u)du \leq$$

$$\sup_t \sup_{\|h\|_{infty},\, h \perp T_{2n-1}} \int_0^{2\pi} B_{m+i}(t-u)h(u)du$$

and taking into account the theorem 1.2.4 and corollary 8.1.4 we obtain

$$\|f^{(i)}\|_\infty = \sup_t E(B_{m+i}(t-\cdot), T_{2n-1})_1 =$$

$$E_(B_{m+i}, T_{2n-1})_1 = \|\varphi_{n,m+i}\|_\infty = \|g^{(i)}\|_\infty.$$

Now in view of theorem 1.4.6 for any $\lambda \in R$

$$(f' - \lambda)_\pm \prec (g' - \lambda)_\pm.$$

It is obvious that this relation coincides with (8.9).

**Continue the proof of theorem 8.2.1.** Taking into account the relation (8.8), lemma 8.2.2 and proposition 1.3.14 we obtain

$$E(W^m F, T_{2n-1})_1 \leq$$

$$\sup_{\phi \in F, phi \perp 1} \sup_{\|\psi\|_\infty \leq 1, \psi \perp T_{2n-1}} (-1)^m \int_0^{2\pi} \pi(\phi, t)\pi(\psi, t)dt \leq$$

$$\sup_{\|\psi\|_\infty \leq q, \psi \perp T_{2n-1}} \int_0^{2\pi} \pi(\varphi_{n,m}, t)\pi(\psi, t)dt.$$

It will follow from the results on widths that this estimate is best possible.

**Theorem 8.3.2.** *If $F \subset L_1$ is arbitrary rearrangementally invariant set, then for any $n, m \in N$*

$$d_{2n-1}(W^m F, L_1) = d_{2n}(W^m F, L_1) = E(W^m F, T_{2n-1})_1 =$$

$$\sup_{\phi \in F, \phi \perp 1} \int_0^{2\pi} \pi(\varphi_{n,m}, t)\pi(\phi, t)dt.$$

**Proof.** Theorem 8.2.1 gives for the widths $d_n(W^m F, L_1)$ the above estimations since

$$d_{2n}(W^m F, L_1) \leq d_{2n-1}(W^m F, L_1) \leq E(W^m F, T_{2n-1})_1.$$

It remains to prove that

$$d_{2n}(W^m F, L_1) \geq \sup_{\phi \in F, \phi \perp 1} \int_0^{2\pi} \pi(\varphi_{n,m}, t)\pi(\phi, t)dt := B. \tag{8.10}$$

Fix $a \in L_1$ and arbitrary $2n$-dimensional subspace $\Re = \Re_{2n} \in L_1$ which contains constants and prove that

$$E(W^m F, a + \Re)_1 \geq B. \tag{8.11}$$

Considering subspaces which contain constant we do not lose generality since if $\Re$ does not contain constant, then

$$E(W^m F, a+\Re)_1 = +\infty.$$

In view of the theorem 1.2.4, for any function $f \in L_1$ we have

$$E(f, a+\Re)_1 = \sup_{\|g\|_\infty \le 1} \left( \int_0^{2\pi} f(t)g(t)dt - \sup_{u \in a+\Re} \int_0^{2\pi} u(t)g(t)dt \right) =$$

$$\sup_{\|g\|_\infty \le 1} \left( \int_0^{2\pi} f(t)g(t)dt - \sup_{u \in \Re} \int_0^{2\pi} (a(t)+u(t))g(t)dt \right) =$$

$$\sup_{\|g\|_\infty \le 1} \left( \int_0^{2\pi} (f(t)-a(t))g(t)dt - \sup_{u \in \Re} \int_0^{2\pi} u(t)g(t)dt \right).$$

We can calculate the first supremum taking into account only function $g \perp \Re$ since in the opposite case

$$\sup_{u \in \Re} \int_0^{2\pi} u(t)g(t)dt = +\infty.$$

Therefore

$$E(f, a+\Re)_1 = \sup_{\|g\|_\infty \le 1,\, g \perp \Re} \int_0^{2\pi} (f(t)-a(t))g(t)dt. \tag{8.12}$$

In view of theorem 7.4.4 there exists a function $\gamma \in \Gamma_{2n,m}$ such that $\gamma^{(m)} \perp \Re$ and $\gamma \perp 1$. Therefore taking into account (8.12) we obtain

$$E(W^m F, a+\Re)_1 := \sup_{f \in W^m F} E(f, a+\Re)_1 \ge$$

$$\max\Big( \sup_{f \in W^m F} \int_0^{2\pi} f(t)\gamma^{(m)}(t)dt - \int_0^{2\pi} a(t)\gamma^{(t)}(t)dt,$$

$$\sup_{f \in W^m F} \int_0^{2\pi} f(t)\gamma^{(m)}(t)dt + \int_0^{2\pi} a(t)\gamma^{(m)}(t)dt\Big)$$

Consider the value

$$A_\pm = \sup_{f \in W^m F} \int_0^{2\pi} f(t)(\pm\gamma^{(m)}(t))dt$$

Integrating by parts $m$ times and taking into account the proposition 1.4.3 we obtain

$$A_{\pm} = \sup_{\phi \in F,\, \phi \perp 1} (-1)^m \int_0^{2\pi} \phi(t)(\pm\gamma(t))dt =$$

$$\sup_{\phi \in F,\, \phi \perp 1} \int_0^{2\pi} \pi(\phi, t)\pi(\pm(-1)^m \gamma, t)dt.$$

Since $\pm(-1)^m\gamma \in \Gamma_{2n,m}$ and $\gamma \perp 1$, then in view of theorem 7.2.1 for any $\lambda \in R$

$$(\varphi_{n,m} - \lambda)_{\pm} \prec (\pm(-1)^m\gamma - \lambda)_{\pm}. \tag{8.13}$$

Taking into account (8.10) and theorem 1.3.4 we obtain that $A_{\pm} \geq B$ and thus

$$E(W^m F, a + \Re)_1 \geq$$

$$\max\left(B + \int_0^{2\pi} a(t)\gamma^{(m)}(t)dt, B - \int_0^{2\pi} a(t)\gamma^{(m)}(t)dt\right) \geq$$

$$B = \sup_{\phi \in F,\, \phi \perp 1} \int_0^{2\pi} \pi(\varphi_{n,m}, t)\pi(\phi, t)dt,$$

which is that to be proved.

¿From theorem 8.2.3 it follows

**Corollary 8.2.4.** *Let $n, m \in N$, $p \in [1, \infty]$ and $p' = p/(p-1)$. Then*

$$d_{2n-1}(W_p^m, L_1) = d_{2n}(W_p^m, L_1) = E(W_p^m, T_{2n-1})_1 = ||\varphi_{n,m}||_{p'}.$$

**Proof.** If

$$F = F_p = \{f(t) : f \in L_p,\ ||f||_p \leq 1\}$$

then $W^m F = W_p^m$. The set $F_p$ is rearrangementally invariant and in view of proposition 1.3.4 and theorem 1.2.4 for any function $g \in L_{p'}$

$$\sup_{\phi \in F_p,\, \phi \perp 1} \int_0^{2\pi} \pi(g, t)\pi(\phi, t)dt = E_1(g)_{p'}.$$

And since $E_1(\varphi_{n,m})_{p'} = ||\varphi_{n,m}||_{p'}$ then from theorem 8.2.3 we obtain the statement of corollary 8.2.4.

It is also true the next more general statement.

**Corollary 8.2.5.** *Let $n, m \in N$ and let $\Phi(t)$ be arbitrary $N$-function and $\Psi(t)$ is $N$-function conjugated to $\Phi(t)$ (see sec 1.1). Then*

$$d_{2n-1}(W_{\Phi}^{m}, L_1) = d_{2n}(W_{\Phi}^{m}, L_1) = \|\varphi_{n,m}\|_{L(\Psi)}.$$

Using the duality theorem for the best $(\alpha,\beta)$-approximation (theorem 1.2.6) and choosing

$$F = F_{p;1/\alpha,1/\beta} = \{f : f \in L_p, \|f\|_{p;1/\alpha,1/\beta}\},$$

we obtain from theorem 8.2.3

**Corollary 8.2.6.** *If $n, m \in N$, $p \in [1, \infty]$ and $0 < \alpha, \beta \leq \infty$ then*

$$d_{2n-1}(W^{m}_{p;1/\alpha,1/\beta}, L_1) = d_{2n}(W^{m}_{p;1/\alpha,1/\beta}, L_1) = E_1(\varphi_{n,m}(\alpha,\beta))_{\infty}.$$

The next statement also follows from the theorem 9.2.3.

**Corollary 8.2.7.** *Let $n, m \in N$, $f \in L_1$ and*

$$F(f) = \{g(t) : g \in L_1, \pi(g,t) = \pi(f,t)\, \forall t\}.$$

*Then*

$$d_{2n-1}(W^m F(f), L_1) = d_{2n}(W^m F(f), L_1) = \int_0^{2\pi} \pi(f,t)\pi(\varphi_{n,m},t)dt.$$

Using theorem 8.2.1 to a large extent analogously to the theorem 8.1.1 we can prove the next more general theorem

**Theorem 8.2.8.** *If $K(t)$ is arbitrary CVD-kernel and $F \subset L_1$ is arbitrary rearrangementally invariant set then for any $n \in N$*

$$E(K * F, T_{2n-1})_1 = \sup_{\phi \in F,\, \phi \perp \theta} \int_0^{2\pi} \pi(K * \varphi_{n,0}, t)\pi(\phi, t)dt,$$

*where $\theta = 1$ if $K \perp 1$ and $\theta = 0$ in opposite case.*

In view of this theorem

$$d_{2n}(K * F, L_1) \leq d_{2n-1}(K * F, L_1) \leq$$

$$\sup_{\phi \in F,\, \phi \perp \mu} \int_0^{2\pi} \pi(K * \varphi_{n,0}, t)\pi(\phi, t)dt.$$

On the other hand, repeating almost word to word the proof of theorem 8.2.3 (certainly instead of theorem 7.2.1 it is necessary to use theorem 7.3.1) it is easy to prove that

$$d_{2n-1}(K * F, L_1) \geq d_{2n-1+\mu}(K * F, L_1) \geq$$

$$\sup_{\phi \in F,\, \phi \perp \mu} \int_0^{2\pi} \pi(K * \varphi_{n,0}, t)\pi(\phi, t)dt.$$

Comparing these relations we see that next theorem holds true.

**Theorem 8.2.9.** *If $K(t)$ is arbitrary CVD-kernel and $F$ is arbitrary rearrangementally invariant set, then for any $n \in N$*

$$d_{2n-1}(K * F, L_1) = d_{2n-1+\mu}(K * F, L_1) =$$

$$\sup_{\phi \in F,\, \phi \perp \mu} \int_0^{2\pi} \pi(K * \varphi_{n,0}, t)\pi(\phi, t)dt.$$

Using the extremal properties of perfect splines which were established above we can obtain the exact values of widths of the classes $W_\infty^m$ in the spaces $L_p$ and in more general spaces.

**Theorem 8.2.10.** *For any $n, m \in N$ and any Orlich space $L_\Phi$ the following equalities are valid*

$$d_{2n}(W_\infty^m, L_\Phi) = E(W_\infty^m, S_{n,m-1})_{L_\Phi} = ||\varphi_{n,m}||_{L_\Phi}.$$

*In particular for all $p \in [1, \infty]$*

$$d_{2n}(W_\infty^m)_p = E(W_\infty^m, S_{n,m-1})_p = ||\varphi_{n,m}||_p.$$

**Proof.** Using the corollary 6.1.10 we can write

$$d_{2n}(W_\infty^m, L_\Phi) \geq E(W_\infty^m, S_{n,m-1})_{L_\Phi} \geq$$

$$\sup_{f \in W_\infty^m} ||f - s_{2n,m-1}(f)||_{L_\Phi} = ||\varphi_{n,m}||_{L_\Phi}.$$

On the other hand in view of theorem 7.4.5 for any $2n$-dimensional subspace $\Re \in L_\Phi$ there exists a perfect spline $\gamma \in \Gamma_{2n,m} \subset W_\infty^m$ such that

$$E(\gamma, \Re)_{L_\Phi} = ||\gamma||_{L_\Phi}.$$

So

$$d_{2n}(W_\infty^m, L_\Phi) \geq \inf_{\gamma \in \Gamma_{2n,m}} ||\gamma||_{L_\Phi}.$$

It remains to use the theorem 7.2.12.

The normed space $Y \subset L_1$ is called symmetric if from the facts $f \in Y$ and $|g| \prec |f|$ it folllows that $g \in Y$ and $||g||_Y \leq ||f||_Y$.

In the proof of theorem 8.2.10 it was proved fact that if $Y$ is arbitrary symmetric space then for any $n, m \in N$

$$d_{2n}(W_\infty^m) = E(W_\infty^m, S_{2n,m-1})_Y = ||\varphi_{n,m}||_\infty.$$

**Remark.** In theorem 8.1.2 subspace $T_{2n-1}$ is an extremal subspace for the widths $d_{2n-1}(W^m F, L_1)$ and $d_{2n}(W^m F, L_1)$. Using the proposition 6.1.2, corollary 6.1.10 and duality theorem 1.2.4 we can prove that subspace $S_{2n,m}$ is also an extremal subspace for the width $d_{2n}(W^m F, L_1)$.

Note also that the next statement is valid.

**Theorem 8.2.11.** *If $K(t)$ is arbitrary CVD-kernel and $Y$ is arbitrary symmetric space then for any $n \in N$*

$$d_{2n}(K * F_\infty, Y) = ||K * \varphi_{n,0}||_Y.$$

Now using the theorem 7.1.5 we obtain estimations of widths for classes of conjugate functions.

Remind that $\tilde{W}_p^m$ ($n \in N$, $p \in [1, \infty]$) is the space of functions $a + \tilde{f}(t)$ where $a \in R$ and $\tilde{f}$ is a function conjugated with the function $f \in W_p^m$.

**Theorem 8.2.12.** *For $n \in N$ and $p = 1, \infty$ the next relations are valid*

$$d_{2n-1}(\tilde{W}_p^1)_p = d_{2n}(\tilde{W}_p^1)_p = ||\tilde{\varphi}_{n,1}||_\infty. \tag{8.14}$$

**Proof.** Results of Favard-Ahiezer-Krein (see Ahiezer [6], p. 242)

$$E(\tilde{W}_p^m, T_{2n-1})_p = \|\tilde{\varphi}_{n,m}\|_\infty$$

give us the above estimation for all $m \in N$

$$d_{2n-1}(\tilde{W}_p^m)_p \leq E(\tilde{W}_p^m, T_{2n-1})_p = \|\tilde{\varphi}_{n,m}\|_\infty. \tag{8.15}$$

Estimations from below we can obtain only for $m = 1$.

Consider first the case $p = \infty$. In view of the theorem 7.4.4 for any $2n$-dimensional subspace $\Re \subset L_\infty$ which contains constants there exists a perfect spline $\gamma \in \Gamma_{2n,m}$ such that

$$E(\tilde{\gamma}, \Re)_\infty = E_1(\tilde{\gamma})_\infty.$$

So for any such subspace

$$E(\tilde{W}_\infty^m)_\infty \geq \inf_{\gamma \in \Gamma_{2n,m}} E_1(\tilde{\gamma})_\infty,$$

and consequently

$$d_{2n}(\tilde{W}_\infty^m)_\infty \geq \inf_{\gamma \in \Gamma_{2n,m}} E_1(\tilde{\gamma})_\infty. \tag{8.16}$$

Comparing (8.16) and theorem 7.1.5 we obtain for $m = 1$

$$d_{2n}(\tilde{W}_\infty^1)_\infty \geq \|\tilde{\varphi}_{n,1}\|_\infty. \tag{8.17}$$

Let now $p = 1$. Using theorem 1.2.4 we obtain that for any $2n$-dimensional subspace $\Re \in L_1$q which contains constant

$$E(\tilde{W}_1^m, T_{2n-1})_1 = \sup_{f \in W_1^m} \sup_{\|g\|_\infty \leq 1,\, g \perp \Re} \int_0^{2\pi} \tilde{f}(t) g(t) dt =$$

$$\sup_{f \in W_1^m} \sup_{g \in W_\infty^m,\, g^{(m)} \perp \Re} \int_0^{2\pi} f^{(m)}(t) \tilde{g}(t) dt =$$

$$\sup_{\|\phi\|_\infty \leq 1,\, \phi \perp 1} \sup_{g \in W_\infty^m,\, g^{(m)} \perp \Re} \int_0^{2\pi} \phi(t) \tilde{g}(t) dt = \sup_{g \in W_\infty^m,\, g^{(m)} \perp \Re} E_1(\tilde{g})_\infty.$$

Since in view of theorem 7.4.4 there exists a spline $\gamma \in \Gamma_{2n,m} \subset W_\infty^m$ such that $\gamma^{(m)} \perp \Re$ then

$$E(\tilde{W}_1^m, \Re)_1 \geq \inf_{\gamma \in \Gamma_{2n,m}} E_1(\tilde{\gamma})_\infty$$

and consequently

$$d_{2n}(\tilde{W}_1^m, L_1) \geq \inf_{\gamma \in \Gamma_{2n,m}} E_1(\tilde{\gamma})_\infty .$$

Using the theorem 7.1.5 we obtain

$$d_{2n}(\tilde{W}_1^m, L_1) \geq \|\tilde{\varphi}_{n,1}\|_\infty . \tag{8.18}$$

Comparing the estimations (8.15), (8.17) and (8.18) we obtain the statement of the theorem.

## 8.3 Spline and optimal recovery

Problems of approximation which we considered in sections 8.1 and 8.2 from theoretical point of view can be interpretated as problem of recovery of functions an apriory and discerete information having an applied shade. In connection with this new point of view it is possible to assess at their true value the significance of the splines which extremal properties are very important in solving of these problems.

In a normed functional space $X$ the set $M_N = \{\mu_1, \mu_2, \ldots, \mu_N\}$ of functionals $\mu_i$ defined on $X$ we can consider as some coding method which associates with the function $f \in X$ a point in $R_N$ i.e. vector

$$T(f, M_N) = \{\mu_1(f), \mu_2(f), \ldots, \mu_N(f)\} \tag{8.19}$$

which contains $N$ units of information on $f(t)$. Any mapping $\Phi : R \to X$ we can consider as a method of recovery of a function $f$ on vector (8.19). With this vector we associate the function $g(f; M_N, \Phi, t)$.

As error of recovery it is naturally to take the value

$$\|f - g(f; M_N, \Phi)\|_X \tag{8.20}$$

and it is obvious that the problem of estimation of this error will be correct if we have some apriory information on function $f$ which at least assures the boundness of the norm (8.20) on the set of functions $f$ having the same vector (8.19) apriory information defines in $X$ some class functions @ and value

$$\rho(@; M_N, \Phi)_X = \inf_{f \in @} \|f - g(f; M_N, \Phi)\|_X$$

is an error of recovery on a class @ by method $\Phi$ on information $T(f, M_N)$ and method $\Phi_*$ such that

$$\rho(@; M_N)_X = \inf_{\Phi : R_N \to X} \rho(@; M_N, \Phi)_X = \rho(@; M_N, \Phi_*)_X$$

is the best method of recovery.

If $\Im_N$ is some class of vectors $M_N$ in the form of (8.19) then we put

$$\rho(@; \Im_N)_X = \inf_{M_N} \rho(@; M_N)_X.$$

Information $T(f, M_N^*)$, $M_N^* \in \Im_N$ such that

$$\rho(@; M_N^*)_X = \rho(@; \Im_N)_X$$

will be called $\Im_N$-optimal for the class @ and metric $X$ and the function $g(f; M_N^*, \Phi_*, t)$ will be called $\Im_N$-optimal method of recovery of functions from @.

Consider some concrete situations. Let $A_{N,m}$, $N, m = 1, 2, \ldots$ be a class of sets $M_N = \{\mu_1, \mu_2, \mu_2, \ldots, \mu_N\}$ where $\mu_k$ are arbitrary linear continuous on $C^{(m-1)}$ functionals.

**Theorem 8.3.1.** *A spline $s_{2n,m-1}$ from $S_{2n,m}$ which interpolates a function $f \in C^{m-1}$ at the points*

$$\tau_{m,k} = \frac{k\pi}{n} + (1 - (-1)^m)) \frac{\pi}{4n}, \quad k = 1, 2, \ldots, 2n, \tag{8.21}$$

*gives us $A_{2n,m}$-optimal methods of recovery on the class $W_\infty^m$ in the metric $L_p$, $p \in [1, \infty]$.*

**Proof.** Let for fixed $m$ $s_k(t)$, $k = 1, 2, \ldots, n$ be a spline from $S_{2n,m}$ uniquely determined (in view of theorem 6.8.1) by the conditions

$$s_k(\tau_{m,i}) = \begin{cases} 1, & i = k, \\ 0, & i \neq k. \end{cases}$$

The spline $s_{2n,m-1}(f, t)$ has unique representation in the form

$$s_{2n,m-1}(f, t) = \sum_{k=1}^{2n} f(\tau_{m,k}) s_k(t),$$

so $s_{2n,m-1}(f, t)$ is a method of recovery of $f(t)$ on information

$$\{f(\tau_{m,1}), f(\tau_{m,2}), \ldots, f(\tau_{m,2n})\}.$$

By the corollary 6.1.10

$$\rho(W_\infty^m, A_{2n,m}) \leq \sup_{f \in W_\infty^m} \|f - s_{2n,m}(f, t)\|_p = \|\varphi_{n,m}\|_p. \tag{8.22}$$

On the other hand, let $M_{2n}$ be arbitrary set of functionals from $A_{2n,m}$. If $T(1, M_{2n}) = 0$ then for any method $\Phi$

$$\sup_{f\in W^m_\infty} \|f - g(f; M_{2n}, \Phi)\|_p \geq$$

$$\sup_\lambda \|\lambda - g(\lambda; M_{2n}, \Phi)\|_p = \sup_\lambda \|\lambda - g(0; M_{2n}, \Phi)\|_p = \infty.$$

So calculating the exact lower bound of the error of recovery over all $M_{2n} \in A_{2n,m}$ we can assume that $T(1, M_{2n}) \neq 0$.

But in view of theorem 7.4.2 there exists a perfect spline $\gamma_* \in \Gamma_{2n,m} \subset W^m_\infty$ such that $T(\gamma_*, M_{2n}) = 0$. This fact allows us to write for any methods of $\Phi$:

$$\sup\{\|\|\|f - g(f; M_{2n}, \Phi)\|_p : f \in W^m_\infty\} \geq$$

$$\max\{\|\gamma_* - g(\gamma_*; M_N, \Phi)\|_p; \|\gamma_* + g(\gamma_*; M_N, \Phi)\|_p\} =$$

$$\max\{\|\gamma_* - g(0; M_N, \Phi)\|_p; \|\gamma_* + g(0*; M_N, \Phi)\|_p\} \geq$$

$$\|\gamma_*\|_p \geq \inf_{\gamma\in\Gamma_{2n,m}} \|\gamma\|_p,$$

and since it is true for any $M_{2n} \in A_{2n,m}$ we taking into account the theorem 7.2.5 obtain the estimation

$$\rho(W^m_\infty; A_{2n,m})_{L_p} \geq \|\varphi_{n,m}\|_p,$$

which together with (8.22) gives us the statement of the theorem.

Using the method of estimating of width from the proof of theorem 8.2.3 and taking into account the relation (see Korneichuk [24])

$$\sup_{f\in W^m_p} \|f - s_{2n,m-1}(f,t)\|_1 = \|\varphi_{n,m}\|_{p'},\ 1/p + 1/p' = 1$$

we can obtain the following statement

**Theorem 8.3.2.** *For $n, m \in N$ the spline $s_{2n,m-1}(f,t)$ gives us $A_{2n,m}$-optimal method of recovery of functions from the class $W^m_p$, $p \in [1,\infty]$ in the metric $L_1$.*

Let

$$x_1 < x_2 < \ldots, x_n \leq 2\pi$$

and $A^1_n$ be a class of sets $M_{2n}$ of functionals of the form

$$\mu_k(f) = f(x_k),\ \ \mu_{n+k}(f) = f'(x_k),\quad k = 1, 2, \ldots, n.$$

In the end of section 6.1 we introduced the spline $s_{2n,m,*}(f,t)$ from $S_{2n,m}$ which interpolates a function $f \in C^1$ and its derivative at the points

$$\tau_{m,k} = \frac{2k\pi}{n} + (1-(-1)^m))\frac{\pi}{2n}, \quad k = 1,2,\ldots,n.$$

¿From corollary 6.1.11 and theorem 7.2.5 it follows that the next statement is true.

**Theorem 8.3.3.** *For $m,n = 1,2,...$ the spline $s_{2n,m-1}(f,t)$ gives us $A_n$-optimal method of recovery on the set $W_\infty^m$ in the metric $L_p$, $p \in [1,\infty]$.*

Let

$$x_1 < x_2 < \ldots, x_n \leq 2\pi$$

$\rho_\nu$, $\nu = 1,2,\ldots,k$ be integer, $1 \leq \rho_\nu \leq m$ and

$$\sum_{\nu=1}^{k} \rho_\nu = 2n$$

and let

$$\Psi = \{\psi_{\nu,j}(t)\}_{\nu=1,j=0}^{k,\rho_\nu-1}$$

be some set of functions from $C$. The function

$$L(f;\Psi,t) = \sum_{\nu=1}^{k}\sum_{j=0}^{\rho_\nu-1} f^{(j)}(x_\nu)\psi_{\nu,j}(t) \tag{8.23}$$

we will consider as recovery method for function $f \in C^{(m-1)}$ on information

$$\{f^{(j)}(x_j)\}_{\nu=1,j=0}^{k,\rho_\nu-1}. \tag{8.24}$$

In section 6.1 we introduced the spline

$$s(f;\Delta_k,\rho) \in S_{m-1}(\Delta_{2n})$$

defined by the equality

$$s^{(j)}(f;\Delta_k,\rho,x_\nu) = f^{(j)}(x_\nu), \;\; \nu = 1,2,\ldots,k, \;\; j = 0,1,\ldots,\rho_\nu - 1$$

and also a perfect spline $\gamma_m(\Delta_k,\rho) \in \Gamma_{2n,m}$, having $\rho_\nu$-multiple zeros at points $x_\nu$, $\nu = 1,2,\ldots,k$.

**Proposition 8.3.4.** *Spline $s(f;\Delta_k,\rho,t)$ gives us the best method of recovery of functions from class $W_\infty^m$ on information (8.24) and among all methods of the form (8.23) in the spase $X$, having monotonic norm.*

Really, for any method of the form (8.23)

$$\sup_{f\in W_\infty^m} \|f - L(f;\Psi)\|_X \geq \|\gamma_m(\Delta_k,\rho) - L(\gamma_m(\Delta_k,\rho);\Psi)\|_X = \|\gamma_m(\Delta_k,\rho)\|_X.$$

Above estimatiom follows from theorem 6.1.8.

**Return to the theorem 8.3.1.** If we consider the problem of optimal recovery not over all set $A_{2n,m}$ of according method, but on fixed vector of information $\{f(\tau_{m,k})\}_{k=1}^{2n}$ where points $\tau_{m,k}$ are defined by (8.21) then we can intensify by them the statement of the theorem 8.3.1. Fix a point $t_* \neq t_{m,k}$ and consider the sum

$$\alpha(f,c) = \sum_{k=1}^{2n} f(\tau_{m,k})$$

as a method of recovery of the value $f(t_*)$ on information $\{f(\tau_{m,k})\}$. It is requierd to choose a vector of coefficients $c = \{c_1, c_2, \ldots, c_{2n}\}$ such that the error of recovery

$$|f(t_*) - \alpha(f,c)|$$

has minimal value .

Let $W_{\infty,0}^m$ be the set of functions $f \in W_\infty^m$ such that $f(\tau_{m,k}) = 0$, $k = 1, 2, \ldots, 2n$.

Supposing the contraty and using Rolle's theorem (see, for example, the proof of theorem 6.1.8) it is easy to prove that for $f \in W_{\infty,0}^m$ the inequality

$$|f(t)| \leq |\varphi_{n,m}(t)|$$

holds true at any points $t$. But then for any vector of coefficients $c$

$$\sup_{f\in W_\infty^m} |f(t_*) - \alpha(f,c)| \geq \sup_{f\in W_{\infty,0}^m} |f(t_*)|.$$

In view of the corollary 6.1.10

$$|f(t_*) - s_{2n,m-1}(f,t_*)| \leq |\varphi_{n,m}(t_*)|$$

and we obtain the following statement.

**Proposition 8.3.5** *Interpolating spline $f_{2n,m}(f,t)$ gives us the optimal method of recovery of functions from $W_\infty^m$ at every point $t$ on information $f(\tau_{m,,k})$.*

Note that components $c_k$ of optimal vector $c$ are defined by equality $c_k = s_k(t_*)$ where $s_k(t)$ are fundamental splines in $S_{2n,m-1}$.

## 8.4 Optimal recovery on a finite interval

Results of this section we present schematically.

Let $W_p^m[a,b]$, $p \in [1,\infty]$ be the class of functions $f$ defined on interval $[a,b]$ and such that the derivative of $(m-1)$-th order is absolutely continuous on $[a,b]$ and $||f^{(m)}||_{L_p[a,b]} \leq 1$.

As in periodic case the width of classes $W_p^m[a,b]$ will be expressed as norm of some perfect splines of order $m$.

As in periodic case we call the perfect spline of order $m$ on the segment $[a,b]$ the function $\gamma \in W_\infty^m[a,b]$ such that $\gamma^{(m)}(t)$ is piece-wise constant and has only values $\pm 1$. If $\gamma^{(m)}(t)$ changes sign on $(a,b)$ at most $N-1$ times then the set of such perfect splines we denote by $\Gamma_{N,m}[a,b]$.

In the set $\Gamma_{N,m}[a,b]$ for any $q \in [1,\infty]$ there exists a spline

$$\psi_{N,m}(t) = \psi_{N,m}(q,t)$$

having the minimal $L_q$-norm, i.e. (we will write $||\cdot||_q$ instead of $||\cdot||_{L_q[a,b]}$ in this section)

$$||\psi_{N,m}||_q = \inf_{\gamma \in \Gamma_{N,m}[a,b]} ||\gamma||_q.$$

The perfect spline $\psi_{N,m}(t)$ has inside $[a,b]$ exactly $N-1$ nodes and exactly $N+m-1$ different zeros.

If

$$\Delta_{N^*[a,b]} = \{a = x_0 < x_1 < \ldots < x_N = b\}$$

is a partition formed by the nodes of the spline $\psi_{N,m}(q,t)$ and let also $S_{m-1}(\Delta_N^*[a,b])$ is corresponding subspace of splines of order $m-1$.

Let moreover $s(f,t)$ is a spline from $S_{m-1}(\Delta_N^*[a,b])$ which interpolates the $f(t)$ at zeros of $\psi_{N,m}(q,t)$ then

$$\sup_{f \in W_\infty^m[a,b]} ||f - s(f)||_q = ||\psi_{N,m}(q)||_q, \quad q \in [1,\infty].$$

¿From this taking into account that

$$dim\, S(\Delta_N^*[a,b]) = N + m - 1,$$

we obtain the above estimation

$$d_{N-m-1}(W_\infty^m[a,b], L_q[a,b]) \leq ||\psi_{N,m}(q)||_q, \quad q \in [1,\infty].$$

On the other hand using theorem 7.4.2 we can prove that for the best $L_q$-approximation of the set $\Gamma_{N,m}[a,b]$ by the subspace $\Re$ of dimension $N+m-1$, which contains the set $P_{m-1}$ of algebraic polynomials of degree $m-1$ the next inequality is valid

$$E(\Gamma_{N,m}[a,b], \Re)_q \geq \inf_{\gamma \in \Gamma_{N,m}[a,b]} ||\gamma||_q.$$

Since $P_{m-1} \subset W_\infty^m[a,b]$ then estimating from below the $N+m-1$-th width we can take into account only subspace $\Re$ which contains $P_{m-1}$. We obtain the next statement.

**Proposition 8.4.1.** *For all $N = 2,3,\ldots$, $m = 1,2,\ldots$ and $q \in [1,\infty]$*

$$d_{N-m-1}(W_\infty^m[a,b], L_q[a,b]) = ||\psi_{N,m}(q)||_q$$

*and width is realised by splines $s(f,t) \in S_{m-1}(\Delta_N^*[a,b])$ which interpolate a function $f \in W_\infty^m[a,b]$ at zeros of perfect spline $\psi_{N,m}(q,t)$.*

Denote by $\Gamma_{N,m}^0[a,b]$ the set of perfect on $[a,b]$ splines $\gamma(t)$ of order $m$, having on $(a,b)$ at most $N+m-1$ nodes and satisfying the boundary conditions

$$\gamma^{(\nu)}(a) = \gamma^{(\nu)}(b) = 0, \quad \nu = 0,1,\ldots,m-1.$$

The spline $\psi_{N,m}^0(q,t)$ from $\Gamma_{N,m}^0[a,b]$ having minimal $L_q$-norm, has on $(a,b)$ exact $N+m-1$ nodes an $N-1$ zeros.

Let $\Delta_N^0[a,b]$ be a partition formed by zeros of the spline $\psi_{N,m}^0(q,t)$ and $\sigma(f,t)$ is a spline from $S_{m-1}(\Delta_N^0[a,b])$ which interpolates a function $f(t)$ at the nodes of $\psi_{N,m}^0(q,t)$. Then (see Korneichuk [12] sec. 5.3)

$$\sup_{f \in W_p^m[a,b]} ||f - \sigma(f)||_1 = ||\psi_{N,m}^0(p')||_{p'}$$

and consequently

$$d_{N+m-1}(W_p^m[a,b], L_1[a,b]) \leq ||\psi_{N,m}^0(p')||_{p'}.$$

The estimation from below one can obtain using Borsuk's theorem. Thus the next proposition holds true.

**Proposition 8.4.2.** *For all $N = 2, 3, \ldots, m = 1, 2, \ldots$ and $p \in [1, \infty]$*

$$d_{N+m-1}(W_p^m[a,b], L_1[a,b]) = \|\psi^0_{N,m}(p')\|_{p'}.$$

*Moreover the subspace $S_{m-1}(\Delta^0_N[a,b])$ is an extremal subspace for the width $d_{N+m-1}(W_p^m[a,b], L_1[a,b])$.*

## 8.5 Supplementary results

**1.** For all $r = 1, 2, \ldots, m \geq r$ and $p \in [1, \infty]$

$$E^{\pm}(W_p^m, T_{2n-1})_1 = E^{\pm}(W_p^m, S_{2n,m})_1 = \inf_{\lambda} \|B_r - \lambda\|_{p'}.$$

Doronin V.G.,Ligun A.A. [2].

**2.** If $f \in L_1^r$; $\alpha, \beta > 0$, $n, r = 1, 2, \ldots, m \geq r$ and $\Re$ is $T_{2n-1}$ or $S_{2n,m}$ then

$$E(f, \Re)_{1;\alpha,\beta} \leq \int_0^{2\pi} \pi(\varphi_{n,r}(\alpha, \beta, t)), \pi(f^{(r)}, t) dt.$$

Inequality is exact.

If $F$ is arbitrary $\pi$-invariant set then

$$E(W^r F, \Re)_{1;\alpha,\beta} \leq \sup_{g \in F,\, g \perp 1} \int_0^{2\pi} \pi(\varphi_{n,r}(\alpha, \beta, t)), \pi(g, t) dt.$$

Babenko V.F.[19].

In the paper of V.F.Babenko [18] these results are extended on the classes of convolutions with $CVD$-kernel.

**3.** Let $\tilde{W}^r_\infty$ be class of functions $\tilde{f}$ which trigonometrically conjugated with functions $f \in W^r_\infty$. If $r, n, m$ and $\Re$ such that if **2.**, then

$$E(\tilde{W}^r_\infty, \Re)_1 = \|\tilde{\varphi}_{n,r}\|_1.$$

Babenko V.F.[19],[13].

**4.** If $s(f,t)$ is a spline from $S_{2n,r}$ which interpolates the function $f(t)$ at zeros of $\varphi_{n,r+1}(t)$ then

$$\sup_{f \in W^{r+1}_\infty} \|\tilde{f} - \tilde{s}(f,t)\|_1 = \|\tilde{\varphi}_{n,r+1}\|_1.$$

Babenko V.F.[13].

**5.** Let $\omega(t)$ be convex upwards modulus of continuity, $W^r H_\omega$ be the class of function $f \in C^r$ such that

$$\omega(f^{(r)}, t) \leq \omega(t), \quad t \geq 0$$

where

$$\omega(f, t) = \sup_{|t_1 - t_2| \leq t} |f(t_1) - f(t_2)|$$

is modulus of continuity of the function $f(t)$. Then

$$d_{2n-1}(W^r H_\omega, C) = d_{2n}(W^r H_\omega, C) = ||f_{n,r}(\omega)||_C.$$

Subspaces $T_{2n-1}$ and $S_{2n,r}$ are extremal subspaces.

Korneichuk N.P.[5], Ruban V.I.[3].

**6.** Under the conditions of 5.

$$d_{2n-1}(W^r H_\omega, L_1) = d_{2n}(W^r H_\omega, L_1) = ||f_{n,r}(\omega)||_1.$$

Subspaces $T_{2n-1}$ and $S_{2n,r}$ are extremal subspaces.

Motornyi V.P., Ruban V.I. [1], Ruban V.I.[5].

**7.** For any convex upwards modulus of continuity $\omega(t)$, for all $n = 1, 2, \ldots$ and any $p \in [1, \infty]$

$$d_{2n}(H_\omega, L_p) = ||f_{n,0}(\omega)||_p.$$

Korneichuk N.P. [19].

**8.** For any convex upwards modulus of continuity $\omega(t)$, for all $n = 1, 2, \ldots$ and any $p \in [1, \infty]$

$$d_{2n}(W^1 H_\omega, L_p) = ||f_{n,1}(\omega)||_p.$$

Korneichuk N.P. [19].

# Chapter 9

# Extremal properties of monosplines

In fifthies S.M.Nikolski formulated and for some cases solved the problem of the best quadrature formula for classes of functions. It was elucidated that this problem on the class of functions with restricted in the space $L_p$ $m$-th derivative was reduced to the problem of minimization in the adjoint metric of the special form of the norm of piecewise polynomial functions which later were called monosplines.

Of course, among splines with free knots monosplines and perfect splines alone are of the great interest to investigators. However, no doubt that study of monosplines was to a considerable extent stimulated by the problems of the best quadratures, as well as investigation of perfect splines was stimulated by the problem of widths.

The deepest results connected with the elucidation of extremal properties of monosplines have been obtained in the papers on optimization of quadratures. Though investigation in this direction have been conducted quite intensively, the problem turned out to be difficult and precise results have been appearing gradually, step by step, with big intervals.

In this chapter results are adduced in the most general form in which they have been obtained lately.

## 9.1 The best approximation by constant

Let natural numbers $m, k$ and $n$ are given. By $@_{n,m}$ denote the set of all $2\pi$-periodic functions of the form

$$M(t) = -\sum_{j=1}^{m} a_j B_m(x_j - t), \tag{9.1}$$

where

$$x_1 < x_2 < \ldots < x_k < x_1 + 2\pi, \quad k \le n, \quad a_j \in R,$$

$$\sum_{j=1}^{k} a_j = 2\pi \tag{9.2}$$

and $B_m(t)$ is Bernully's kernel.

Function from $@_{n,m}$ will be called monospline of order $m$ (and of minimal defect) with the nodes $x_1, x_2, \ldots, x_k$ and coefficients $a_1, a_2, \ldots, a_k$.

In the case $m \ge 2$ monospline $M \in @_{n,m}$ has continuous on all real axis derivative $M^{(m-2)}(t)$ and its derivative $M^{(m-1)}(t)$ generally is discontinuous at the nodes. From (9.1) we see that

$$M^{(m-1)}(t) = (-1)^m \sum_{j=1}^{k} a_j B_1(x_j - t).$$

¿From (9.2) and from the equality

$$B_1(t) = \frac{\pi - t}{2}, \quad t \in (0, 2\pi)$$

it follows that on every interval $(x_j, x_{j+1})$, $x_{k+1} = x_1 + 2\pi$

$$M^{(m)}(t) = (-1)^{m+1} \frac{1}{2\pi} \sum_{j=1}^{k} a_j = (-1)^{m+1}.$$

Therefore on every interval $(x_j, x_{j+1})$

$$M^{(m-1)}(t) = (-1)^{m+1} c_j$$

where $c_j$ is some constant and so $M^{(m-1)}(t)$ has on a period at most $2k \le 2n$ sign shanges. Using Rolle's theorem we obtain the next statement.

**Theorem 9.1.1.** *For any monospline $N \in @_{n,m}$ and any constant $a \in R$*

$$\mu(M - a) \leq 2n.$$

As above $\mu(f)$ is the number of sign changes of the function $f$ on a period.

Introduce the standard monospline in $@_{n,m}$

$$M_{n,m}(t) = -\frac{2\pi}{n}\sum_{j=1}^{n} B_m\left(\frac{2j\pi}{n} - t\right). \tag{9.3}$$

**Theorem 9.1.2.** *For all $n = 2, 3, \dots$ and $m = 1, 2, \dots$*

$$\inf_{f \in @_{n,m}} E_0(f)_\infty = E_0(M_{n,m})_\infty.$$

**Proof.** For $m = 1$ the statement of theorem is obvious. Let $m \geq 2$. Suppose that for some monospline $M \in @_{n,m}$

$$E_0(M)_\infty < E_0(M_{n,m})_\infty.$$

Then for some $\lambda$ for all $t \in R$

$$\min_u M_{n,m}(u) < M(t) - \lambda < \max_u M_{n,m}(u),$$

and hence for all $\alpha$ the difference

$$\delta_\alpha = M_{n,m}(t) - (M(t + \alpha) - \lambda)$$

will have on a period at least $2n$ sign changes.

Choose $\alpha = \alpha_*$ such that one of the nodes of monospline $M(t + \alpha_*)$ coincides with one of the nodes $M_{n,m}(t)$ (i.e. with the point of the form $2\pi/n$). Consider the difference $\delta_{\alpha_*}(t)$. At the points $t$ which are not the nodes of $M_{n,m}(t)$ and $M_{n.m}(t + \alpha_*)$ it will be $\delta_{\alpha_*}^{(m)}(t) = 0$. Therefore $\delta_{\alpha_*}^{(m-1)}(t)$ is piece-wise constant function which can alternates sign only at the nodes of monosplines $M_{n,m}(t)$ and $M_{n,m}(t + \alpha_*)$. But since one of the nodes of monospline $M_{n,m}(t + \alpha_*)$ coincides with some node of $M_{n,m}(t)$ then function $\delta_{\alpha_*}^{(m-1)}(t)$ has at most $2n - 1$ points of discontinuity (and, consequently, points of sign changes) on a period. Then by Rolle's theorem the difference $\delta_{\alpha_*}(t)$ will have at most $2n-1$ sign changes contrary to the previous fact.

Theorem is proved.

The idea of the proof of theorem 9.1.2 allows us to prove a series statements of such type.

Let $K$ be a $CVD$-kernel and as in chapter 6 with $K$ we will connect the number $\theta(K)=0$ or 1. Denote by $@_n(K)$ the set of function $M(t)$ of the form

$$M(t)=\int_0^{2\pi}K(u)du-\sum_{j=1}^{k}a_jK(x_j-t), \tag{9.4}$$

where

$$x_1<x_2<\ldots<x_k<x_1+2\pi=x_{k+1},\quad k\le n,\quad a_j\in R.$$

If $\sigma\in R$ is given then by $@_{n,\sigma}(K)$ we will denote the set of all functions $M(t)$ of the form (9.4) such that

$$\sum_{j=1}^{k}a_j=2\pi\sigma.$$

Let also

$$M_{n,\sigma}(t)=M_{n,\sigma}(K,t)=\int_0^{\pi}K(u)du-\frac{2\pi\sigma}{n}\sum_{j=1}^{n}K\left(\frac{2\pi j}{n}-t\right). \tag{9.5}$$

Functions from $@_{n,\sigma}(K)$ will be called generalized monosplines or, if it do not lead to misunderstanding, simply monosplines.

The next statement is a generalization of the theorem 9.1.2.

**Theorem 9.1.3.** *Let $K(t)$ be continuous $CVD$-kernel and $n\in N$. If $\theta(K)=1$ then*

$$\inf_{M\in @_{n,1}(K)}E_0(M)_\infty=E_0(M_{n,1}(K,\cdot))_\infty. \tag{9.6}$$

*If $\theta(K)=0$ then*

$$\inf_{M\in @_{n,1}(K)}\|M\|_\infty=\inf_{\sigma}\|M_{n,\sigma}(K,\cdot)\|_\infty.$$

It is easy to prove theorem 9.1.3 following to the scheme of the theorem 9.1.2 but we will obtain this theorem as a simple consequence of more general statement.

Denote by $@_{n,\sigma}^h(K)$, $h \in (0, 2\pi)$ the set of all functions of the form

$$M_h(t) = \frac{1}{2h} \int_{t-h}^{t+h} M(u)du,$$

where $M \in @_{n,\sigma}(K)$ (i.e. $M(t)$ has the form (9.3)) and is such that

$$x_1 + h < x_2 - h < x_2 + h < x_3 - h < \ldots < x_k + h < x_1 + 2\pi - h$$

and put

$$@_n^h(K) = \sup_{\sigma \in R} @_{n,\sigma}^h(K).$$

**Theorem 9.1.4.** *Let $K \in CVD$, $n \in N$ and $h \in (0, \pi/n)$. If $\theta(K) = 1$ then ($M_{n,\sigma}(K) = M_{n,\sigma}$)*

$$\inf_{M \in @_{n,1}(K)} E_0(K)_\infty = E_0((M_{n,1})_h)_\infty.$$

*If $\theta(K) = 0$ then*

$$\inf_{M \in @_{n,1}(K)} ||K||_\infty = \inf_\sigma ||(M_{n,\sigma})_h||_\infty.$$

**Proof.** Let $\theta(K) = 1$. Assume that for some $M \in @_{n,1}^h(K)$

$$E_0(M)_\infty < E_0((M_{n,1})_h)_\infty.$$

As in the proof of the theorem 9.1.2 taking into account that $(M_{n,1})_h(t)$ is $2\pi/n$-periodical we obtain that in this case there exists a number $\lambda \in R$ such that for any $\alpha \in R$ for the difference

$$\delta_\alpha(t) = (M_{n,1})_h(t) - M(t + \alpha) + \lambda$$

it will be

$$\mu(\delta_\alpha) \geq 2n. \tag{9.7}$$

Let $\chi_A(t)$ be the characteristic function of a set $A \subset [0, 2\pi)$ which is continued with period $2\pi$ on the all real axis and

$$H(t) = H(M,t) = 1 - \frac{1}{2h} \sum_{j=1}^{k} a_j \chi_{[x_j - h, x_j + h]}(t) := 1 - H_1(M,t), \tag{9.8}$$

where $x_j$ are nodes and $a_j$ are coefficients of the monospline $M(t)$. Then as it shows the simple calculation we can represent $M(t)$ in the form

$$M(t) = \int_0^{2\pi} K(u-t)H(M;u)du. \tag{9.9}$$

Since together with $K(t)$ the kernel $K(-t)$ is a $CVD$-kernel, taking into account (9.7) it is easy to make shure that for any $\alpha \in R$ the difference

$$\delta'_\alpha(t) = H((M_{n,1})_h, t) - H(M, t+\alpha)$$

will have at least $2n$ sign changes on a period, i.e.

$$\mu(\delta'(\alpha)) \geq 2n. \tag{9.10}$$

Let $i$ be such that $a_i \leq 2\pi/n$ (existence of such $i$ is obvious). Put $\alpha = \alpha_* = 2\pi/n - x_i$ and show that

$$\mu(\delta'_{\alpha_*}) \leq 2n-2. \tag{9.11}$$

As soon as (9.11) will be proved, we obtain the contrary with the (9.10) and the theorem in the case $\theta(K) = 1$ will be proved.

Let $\mu(\delta_{\alpha_*}) \geq 2n$. Then there are points

$$y_1 < y_2 < \ldots < y_{2n} < y_1 + 2\pi$$

such that $\delta'_{\alpha_*}(t)$ takes on nonzero values at this points with alternating signs.

Let for definitness that the difference $\delta'_{\alpha_*}(t)$ takes on negative values at the points $y_2, y_4, \ldots, y_{2n}$. Since in one interval of positivity of the function $h(M, t+\alpha)$ the difference $\delta'_{\alpha_*}(t)$ has constant sign, the points $y_2, \ldots, y_{2n}$ will be situated on the different intervals of positivity of the function $H(M, t+\alpha)$ and hence one of these points will be lie in the interval $(2\pi/n - h, 2\pi/n + h)$. But on all this interval the function $\delta'_{\alpha_*}(t)$ by construction is nonnegative, thus (9.11) holds true.

Let now $\theta(K) = 0$.

**Lemma 9.1.5.** *The greatest lower bound*

$$\inf\{\|(M_{n,\sigma})_h\|_\infty : \sigma \in R\}$$

*is attained for $\sigma = \sigma_*$ such that*

$$\max_t (M_{n,\sigma_*})_h(t) = -\min_t (M_{n,\sigma_*})_h(t).$$

**Proof.** Represent $(M_{n,\sigma})_h(t)$ in the form

$$(M_{n,\sigma})_h(t) = \int_0^{2\pi} K(u)du - \sigma \int_0^{2\pi} K(t-u)H_1((M_{n,1})_h, u)du.$$

It is known that $CVD$-kernel $K(t)$ with $\theta(K) = 0$ has constant sign. Let for definitness $K(t)$ is nonnegative. Then the function

$$\psi((t) = \int_0^{2\pi} K(u-t)H_1((M_{n,1})_h, u)du$$

is strictly positive. If for same $\sigma_0 > 0$

$$\max_t \left( \int_0^{2\pi} K(u)du - \sigma_0 \psi(t) \right) > -\min_t \left( \int_0^{2\pi} K(u)du - \sigma_0 \psi(t) \right), \quad (9.12)$$

then for $\sigma < \sigma_0$ and sufficiently closed to $\sigma_0$ it will be

$$||(M_{n,\sigma})_h||_\infty < ||(M_{n,\sigma_0})_h||_\infty. \quad (9.13)$$

If instead of (9.12) the inequality of opposite sense holds true then (9.13) will be valid for all $\sigma > \sigma_0$ and sufficiently closed to $\sigma_0$. Since the existence of $\sigma_*$ and its positivity are obvious, lemma is proved.

Suppose now for same $M \in @_n^h(K)$

$$||M||_\infty < ||(M_{n,\sigma_*})_h||_\infty. \quad (9.14)$$

Then taking into account the lemma 9.1.5 and the fact that the function $(M_{n,\sigma_*})(t)$ has period $2\pi/n$ we conclude that the difference

$$\delta_\alpha(t) = (M_{n,\sigma_*})_h(t) - M(t+\alpha)$$

has for any $\alpha \in R$ at least $2n$ sign changes on a period. But

$$\delta_\alpha(t) = \int_0^{2\pi} K(u-t) \left[ H((M_{n,\sigma_*})_h, u) - H(M, u+\alpha) \right] du$$

and since $K(-\cdot) \in CVD$. Then for the derivative

$$\delta'_\alpha(t) = H((M_{n,\sigma_*})_h, t) - H(M, t+\alpha)$$

we will have $\mu(\delta'_\alpha) \geq 2n$ for any $\alpha \in R$.

At the same time as in the case $\theta(K) = 1$ we can choose $\alpha = \alpha_*$ such that the function $\delta'_{\alpha_*}(t)$ will have on a period at most $2n-2$ sign shanges.

Thus the inequality (9.14) is impossible and the theorem 9.1.4 is completely proved.

Now the theorem 9.1.3 we can prove in such a way.

Let, for example, for some $M \in @_{n,1}(K)$

$$E_0(M)_\infty < E_0(M_{n,1}(K))_\infty .$$

Then for all sufficiently small $h > 0$ it will be

$$E_0(M_h)_\infty < E_0((M_{n,1}(K))_h)_\infty .$$

that is contraty with the first part of the theorem 9.1.4.

## 9.2 Monosplines of minimal defect

Using the extremal properties of perfect splines (chapter 7) we will prove inequalities for rearrangements of monosplines. We will use notations connected with monosplines which we introduced in present section.

**Theorem 9.2.1.** *Let $n = 2, 3, \dots$ and $m = 1, 2. \dots$. For any monospline $M \in @_{n,m}$ and for any $\lambda \in R$*

$$(M_{n,m} - \lambda)_\pm \prec (M - \lambda)_\pm . \tag{9.15}$$

**Proof.** As in the proof of the theorem 7.2.1 and 7.3.1 to prove inequality (9.15) it suffices to show that for $M \in @_{n,m}$ and for any $\eta, \theta > 0$

$$E_0(M_{n,m})_{1;\eta,\theta} \le E_0(M)_{1;\eta,\theta} . \tag{9.16}$$

By definition a monospline $M(t)$ from $@_{n,m}$ has the form

$$M(t) = -\sum_{j=1}^{k} \alpha_j B_m(x_j - t),$$

where

$$\sum_{j=1}^{k} \alpha_j = 2\pi$$

and

$$x_1 < x_2 < \dots < x_k < x_1 + \pi = x_{k+1}, \quad k \le n.$$

By theorem 6.1.7 there is nonnegative perfect spline $\gamma_0 \in \Gamma_{2n,m}$ such that

$$\gamma_0(x_j) = \gamma_0'(x_j) = 0, \quad j = 1\ldots., k.$$

For the spline $\gamma_0(t)$ the next equation will be true

$$E_0^+(\gamma_0)_1 = \int_0^{2\pi} \gamma_0(t)td. \tag{9.17}$$

By the duality theorem for the best $(\alpha, \beta)$-approximations (see theorem 1.2.6) the next relation holds true

$$E_0(M)_{1;\eta,\theta} = \sup_{\|g\|_{\infty;\eta^{-1}\leq 12,\,\theta^{-1},\, g\perp 1}} \int_0^{2\pi} M(t)g(t)dt,$$

and since for the perfect spline $\gamma_0$

$$\|\gamma_0^{(m)}\|_{\infty;\eta^{-1},\theta^{-1}} \leq 1, \quad \gamma_0^{(m)} \perp 1,$$

then

$$E_0(M)_{1;\eta,\theta} \geq \int_0^{2\pi} M(t)\gamma_0^{(m)}(y)dt = -\int_0^{2\pi} \sum_{j=1}^{k} \alpha_j B_m(x_j - t)\gamma_0^{(m)}(t)ft =$$

$$-\sum_{j=1}^{k} \alpha_j \left(\gamma_0(x_j) - \frac{1}{2\pi}\int_0^{2\pi} \gamma_0(t)dt\right) = \int_0^{2\pi} \gamma_0(t)dt.$$

Thus in view of (9.17)

$$E_0(M)_{1;\eta,\theta} \geq E^+(\gamma_0)_1. \tag{9.18}$$

On the other hand

$$E_0(M_{n,m})_{1;\eta,\theta} = -\frac{2\pi}{n}\sum_{j=1}^{n}\int_0^{2\pi} B_m\left(\frac{2j\pi}{n} - u\right) sgn_{\eta,\theta}(M_{n,m}(u) - \lambda)du \tag{9.19}$$

where $\lambda$ is the constant of the best $(\eta, \theta)$-approximation for $M_{n,m}(t)$ in the space $L_1$. But the function

$$F(t) := \int_0^{2\pi} B_m(t-u)sgn_{\eta,\theta}(M_{n,m}(u) - \lambda)du$$

differs from $\varphi_{n,m}(\eta, \theta; t)$ only by shift of argument and due to (9.19)

$$E_0(M_{n,m})_{1;\eta,\theta} = -\frac{2\pi}{n}\sum_{j=1}^{n} F\left(\frac{2\pi j}{n}\right) = -2\pi F\left(\frac{2\pi j}{n}\right) \leq$$

$$-2\pi \min_t \varphi_{n,m}(\eta, \theta; t) = E_0^+(\varphi_{n,m}(\eta, \theta))_1. \tag{9.20}$$

Finally, by theorem 7.2.2

$$E_0^+(\varphi_{n,m}(\eta, \theta))_1 \leq E_0^+(\gamma_0)_1. \tag{9.21}$$

Comparing the inequalities (9.18), (9.20) and (9.21), we complete the proof of the inequality (2.16) and also the theorem 9.2.1.

By means of theorem 1.3.11 we deduce from the theorem 9.2.1

**Theorem 9.2.2.** *For any N-function $\Phi$ and for all $n = 2, 3, \dots$, $m = 1, 2, \dots$*

$$\inf_{M \in @_{n,m}} \inf_{\lambda \in R} \int_0^{2\pi} \Phi(|M(t) - \lambda|)dt = \inf_{\lambda \in R} \int_0^{2\pi} (|M_{n,m}(t) - \lambda|)dt \tag{9.22}$$

*and in particular*

$$\inf_{M \in @_{n,m}} R_0(M)_p = E_0(M_{n,m})_p, \quad 1 \leq p \leq \infty. \tag{9.23}$$

¿From the theorem 9.2.1 it follows

**Corollary 9.2.3.** *For all $n = 2, 3, \dots$, $m = 1, 2 \dots$, $p \in [1, \infty]$ and $\alpha, \beta > 0$*

$$\inf_{M \in @_{n,m}} E_0(M)_{p;\alpha,\beta} = E_0(M_{n,m})_{p;\alpha,\beta}. \tag{9.24}$$

Analogously but using the theorem 7.3.1 instead of the theorem 7.2.1 we obtain the next statement.

**Theorem 9.2.4.** *Let $K \in CVD$, $\sigma \in R$ is arbitrary, if $\theta(K) = 0$ and $\sigma = 1$, if $\theta(K) = 1$. For all $M \in @_{n,\sigma}(K)$, $n = 2, 3, \dots$ and $\lambda \in R$*

$$(M_{n,\sigma}(K, \cdot) - \lambda)_\pm \prec (M(\cdot) - \lambda)_\pm.$$

As in the proof of the theorem 7.3.1 we without loss of the generality can assume that the kernel $K(t)$ is analytic on the real axis.

**Lemma 9.2.5.** *Let CVD-kernel $K$ is analtic on the real axis; $\eta, \theta > 0$,*

$$x_1 < x_2 < \ldots < x_n < x_1 + 2\pi.$$

*Then there exists a function $\gamma \in \Gamma_{2n,0}(\eta,\theta)$, $\gamma \perp 1$ such that a generalized perfect spline $K*\gamma(t)$ is equal at the points $x_1, x_2, \ldots, x_n$ to* $\min\{(K*\gamma)(t) : t\}$.

To prove the lemma we fixing $x_1, x_2, \ldots, x_n$ define a set $\Re$ of functions $g(t)$ of the form

$$g(t) = \sum_{j=1}^{n} a_j K(x_j - t) + \sum_{j=1}^{n} b_j K'(x_j - t), \tag{9.25}$$

where $a_j, b_j \in R$, $j = 1, 2, \ldots, n$ and

$$\sum_{j=1}^{n} a_j = -1.$$

For this set of functions we study an extremal problem

$$E_0(g)_{1,\eta,\theta} \to inf, \quad g \in \Re. \tag{9.26}$$

Suppose that the infinum in (9.26) is realized by a function $g_* \in \Re$ (existence of $g_*$ is obvious). Applying Lagrange's principle (see the proof of the theorem 7.3.1) we see that if $\lambda$ is the constant of the best $(\eta,\theta)$-approximation for $g_*(t)$ in the space $L_1$ and

$$\gamma(t) = sgn_{\eta,\theta}(g_*(t) - \lambda),$$

then $\gamma \perp 1$ and for some number $\xi$

$$\int_0^{2\pi} K(x_j - t)\gamma(t)dt = \xi, \quad j = 1, 2, \ldots, n$$

and

$$\int_0^{2\pi} K'(x_j - t)\gamma(t)dt = 0, \quad j = 1, 2, \ldots, n$$

Since $K \in CVD$ it is easy to see that $\gamma \in \Gamma_{2n,0}(\eta,\theta)$ (in particular $\mu(\gamma) \le 2n$) and either $\xi = \min\{K * \gamma(t) : t\}$ or $\xi = \max\{K * \gamma(t) : t\}$. However we have

$$0 < E_0(g_*)_{1;\eta,\theta} = \int_0^{2\pi} g_*(t)\gamma(t)dt =$$

$$\sum_{j=1}^{n} a_j^* \int_0^{2\pi} K(x_j - t)\gamma(t)dt = -\xi$$

(here $a_j^*$ are corresponding coefficients of the function $g_*(t)$ in representation (9.25)). Thus $\xi < 0$ and therefore

$$(K * \gamma)(x_j) = \min_t K * \gamma(t), \quad j = 1, 2, \ldots, n.$$

Lemma is proved.

By scheme of the proof of the inequality (9.16), using the lemma 9.2.5 instead of the theorem 6.1.7 we obtain that for any $M \in @_{\eta,\theta}(K)$

$$E_0(M_{n,\sigma}(K,\cdot))_{1;\eta.\theta} \le E_0(M(\cdot))_{1;\eta.\theta}$$

and in view of the theorem 1.5.1 it completes the proof of theorem 9.2.4.

**Corollary 9.2.6.** *Under the conditions of theorem 9.2.4, for any N-function $\Phi(t)$ the next statements are valid. If $\theta(K) = 1$ then*

$$\inf_{M \in @_{n,\sigma}} \inf_{\lambda \in R} \int_0^{2\pi} \Phi(|M(t) - \lambda|)dt = \inf_{\lambda \in R} \int_0^{2\pi} \Phi(|M_{n,1}(K;t) - \lambda|)dt. \tag{9.27}$$

*In particular, for any $p \in [1, \infty]$ and any $\alpha, beta > 0$ If $\theta(K) = 0$ then*

$$\inf_{M \in @_{n,\sigma}} E_0(M)_{p;\alpha,\beta} = E_0(M_{n,\sigma}(K))_{p;\alpha,\beta}.$$

**Corollary 9.2.7.** *Let $n = 2, 3, \ldots$, $K$ is $CVD$-kernel and $\Phi$ is arbetrary N-function. Then for $\theta(K) = 1$*

$$\inf_{M \in @_n(K)} \inf_{\lambda \in R} \int_0^{2\pi} \Phi(|M(t) - \lambda|)dt =$$

$$\inf_{\lambda \in R} \int_0^{2\pi} \Phi(|M_{n,1}(K;t) - \lambda|)dt.$$

*For $\theta(K) = 0$*

$$\inf_{M \in @_n(K)} \int_0^{2\pi} \Phi(|M(t)|)dt = \inf_{\sigma \in R} \int_0^{2\pi} \Phi(|M_{n,\sigma}(K;t)|)dt.$$

*In particular for $\theta(K) = 1$*

$$\inf_{M\in @_n(K)} \inf_{\lambda} \|M_{n,1} - \lambda\|_p = \inf_{\sigma\in R} \int_0^{2\pi} \Phi(|M_{n,\sigma}(K;t)|)dt.$$

*anf for $\theta(K) = 0$ and $p \in [1,\infty]$*

$$\inf_{M\in @_n(K)} \|M_{n,1}\|_p = \inf_{\sigma\in R} \|M_{n,\sigma}(K;\cdot)\|_p.$$

## 9.3 Monosplines of variable defect

In this section we will study a set of monosplines which is wider then $@_{n,m}$. Let $n = 2,3,\ldots$ and $m = 1,2\ldots$. Denote by $@^*_{n,m}$ the set of functions (monosplines) of the form

$$M(t) = -\sum_{j=1}^{k} \sum_{i\in A_j} a_{i,j} B_{m-i}(x_j - t), \tag{9.28}$$

where

$$x_1 < x_2 < \ldots < x_k < x_1 + 2\pi, \quad k \le n,$$

$$A_j \subset \{0,1,\ldots,m-1\}, \quad j = 1,2,\ldots,k$$

and

$$\sum_{j=1}^{k} |A_j| \le n$$

($|A_j|$ is a number of elements of the set $A_j$), $a_{i.j} \in R$ and

$$\sum_{1\le j\le k,\, 0\in A_j} a_{0,j} = 2\pi.$$

In particular case when $A_j = \{0\}$, $j = 1,2,\ldots,k$ we return to the set $@_{n,m}$ so standard monospline $M_{n,m}(t)$ belongs to the set $@^*_{n,m}$.

The main result of this section is the following statement intensifying the theorem 9.2.1.

**Theorem 9.3.1.** *Let $n = 2,3,\ldots$, $m = 1,2,\ldots$. For any monospline $M \in @^*_{n,m}$ and for any $\lambda \in R$ the best possible inequality is valid*

$$(M_{n,m} - \lambda)_{\pm} \prec (M - \lambda)_{\pm}. \tag{9.29}$$

To prove this theorem it is sufficiently to repeat the proof of the theorem 9.2.1, using instead of the theorem 6.1.7 the next theorem 9.3.2. Proof of theorem 9.3.2 is the main contents of this section.

**Theorem 9.3.2.** *Let $\eta, \theta > 0$ be fixed and let points $x_1, x_2, \ldots, x_k$ and sets $A_1, A_2, \ldots, A_k$ such that any set $A_j$, $j = 1,2,\ldots,n$ contains exactly one element $i_j$. To simplify the exposition, we will also assume that $A_j \subset \{0,1,\ldots,m-2\}$. For any $i = 0,1,\ldots,m-2$ denote by $n_i$ the number of indexes $j$ such that $i \in A_j$. Now for any $j$ put $A'_j = \{i_j, i_{j+1}\}$. Denote by $@^{**}_{n,m}$ the set of functions of the form*

$$M(t) = -\sum_{j=1}^{k} \sum_{i \in A'_j} a_{i,j} B_{m-i}(x_j - t), \tag{9.30}$$

*where numbers $a_{i,j} \in R$ are such that*

$$\sum_{j:0 \in A'_j} a_{0,j} = 2\pi.$$

It is easy to understand that to prove theorem 9.3.2 it suffices to prove the next three lemmas.

**Lemma 9.3.3.** *For any $M \in @^{**}_{n,m}$ and for any constant $\lambda$ the difference $M - \lambda$ is different from zero almost everywhere and*

$$\mu(M - \lambda) \le 2n.$$

**Lemma 9.3.4.** *There exists the perfect spline $\gamma_0 \in \Gamma_{2n,m}(\eta, \theta)$ such that*

$$\gamma_0^{(i)}(x_j) = 0, \quad i \in A_j, \quad j = 1,2,\ldots,n \tag{9.31}$$

*and having at some point a positive value.*

**Lemma 9.3.5.** *The spline $\gamma_0$ from the lemma 9.3.4 is nonnegative everywhere.*

**Proof of lemma 9.3.3.** Let $M \in @^{**}_{n,m}$, i.e. $M(t)$ is of the form (9.30). Interchanging the order of summation we obtain

$$M(t) = -\sum_{i=0}^{m-2} \sum_{1 \le j \le n, i \in A_j} (a_{i.j} B_{m-i}(x_j - t) + a_{i+1,j} B_{m-i-1}(x_j - t)).$$

Put for $i = 0, 1, \ldots, m-2$

$$m_i(t) = \sum_{1 \le j \le n, i \in A_j} (a_{i.j} B_{m-i}(x_j - t) + a_{i+1,j} B_{m-i-1}(x_j - t)) \quad (9.32)$$

and for $k = 0, 1, \ldots, m-2$

$$N_k(t) = \sum_{\nu=1}^{k} m_\nu(t). \quad (9.33)$$

It is obvious that if $M \in @^{**}_{n,m}$ then $M(t) = N_{m-2}(t)$.

Let $\delta_h(t)$ be $2\pi$-periodic continuation on the all real axis of the function $(1/(2h))\chi_{[-h,h]}$ and

$$\delta_{hh}(t) = (\delta_h * \delta_h)(t).$$

Then

$$f_{hh}(t) = (f * \delta_{hh})(t)$$

is the second Steclov function with the step $2h$ for function $f(t)$.

We will prove by induction that for any $k = 0, 1, \ldots, m$ and any $s = 0, 1, \ldots, m-k$ and for all sufficiently small $h > 0$ it will be

$$\mu((N_k)^{(s)}_{hh}) \le 2 \sum_{i=0}^{k} n_i. \quad (9.34)$$

¿From this we will obtain the statement of lemma 9.3.3.

Thus we will prove (9.34).

First of all we note that by (9.32) for all $k = 0, 1, \ldots, m-2$ the function $(N_k)^{(s)}_{hh}(t)$ is of the form

$$(N_k)^{(m-k)}_{hh}(t) =$$

$$a + (-1)^{m-k-1} \sum_{1 \le j \le n, k \in A_j} (a_{i.j} \delta_{hh}(x_j - t) + a_{i+1,j} \delta'_{hh}(x_j - t)). \quad (9.35)$$

Let $k = 0$. Then $N_0(t) = m_0(t)$ and using (9.35) it is easy to make that

$$\mu((N_0)_{hh}^{(m)}) \leq 2n_0.$$

For all $s < m$ the inequality (9.34) follows from the Rolle's theorem.

Suppose that (9.34) is valid for $k = k_0$ and for all $s = 0, 1, \ldots, m - k_0$.

Prove that then (9.34) is valid for $k = k_0 + 1$ and all $s = 0, 1, \ldots, m - k_0 - 1$.

If $m_{k_0+1}(t) \equiv 0$ then it is obvious. Let $m_{k_0+1}(t) \not\equiv 0$. Then in view of (9.33) and (9.35) we have

$$(N_{k_0+1})_{hh}^{(m-k_0-1)}(t) = (N_{k_0})_{hh}^{(m-k_0-1)}(t) + a+$$

$$(-1)^{m-k_0} \sum_{1 \leq j \leq n,\, k_0+1 \in A_j} (a_{i.j}\delta_{hh}(x_j - t) + a_{i+1,j}\delta'_{hh}(x_j - t)). \qquad (9.36)$$

If the point $x^*$ is not a knot of monospline $N_{k_0}(t)$ then there exists a neighbourhood $V_{x^*}$ of this point and there exists a constant $B$ such for all sufficiently small $h > 0$ it will be

$$|(N_{k_0})_{hh}^{(m-k_0-2)}(t)| \leq B, \;\; t \in V_{x^*}.$$

So for any fixed $a, b, c$ and sufficiently small $h > 0$ it will be

$$\mu\left((N_{k_0})_{hh}^{(m-k_0-1)}) + a + b\,\delta_{hh}(x^* - \cdot) + c\,\delta'_{hh}(x^* - \cdot)\right) \leq$$

$$\mu\left((N_{k_0})_{hh}^{(m-k_0-1)}\right) + 2.$$

¿From this and from (9.36) it follows that

$$\mu(N_{k_0+1})_{hh}^{(m-k_0-1)}) = \mu((N_{k_0})_{hh}^{(m-k_0-1)} + (m_{k_0+1})_{hh}^{(m-k_0-1)}) \leq$$

$$\mu((N_{k_0})_{hh}^{(m-k_0-1)}) + 2n_{k_0+1}.$$

Taking into account the induction hypothesis we obtain

$$\mu((N_{k_0+1})_{hh}^{(m-k_0-1)}) \leq 2\sum_{\nu=0}^{k_0} n_\nu.$$

¿From this and from Rolle's theorem it follows that (9.34) is valid for $k = k_0 + 1$.

When $k = m - 2$ we obtain the statement of lemma 9.3.3.

**Proof of lemma 9.3.4.** We will reason analogously to the proof of lemma 9.2.5. If points $x_1, x_2, \ldots, x_n$ and one-element sets $A_1, A_2, \ldots, A_n$ are fixed, consider the set $\Re$ of function $g(t)$ of the form

$$g(t) = \sum_{j=1}^{n} \left(a_{i_j,j} B_{m-i_j}(x_j - t) + a_{i_j+1,j+1} B_{m-i_j-1}(x_j - t)\right), \qquad (9.37)$$

where $\{i_j, i_j + 1\} = A_j'$ and

$$\sum_{i:i_j=0} a_{0,j} = -2\pi.$$

Consider the extremal problem

$$E_0(g)_{1;\eta,\theta} \to \inf, \quad g \in \Re. \qquad (9.38)$$

Let $g_* \in \Re$ be an extremal element in problem (9.38) (existence of $g^*(t)$ is obvious). Using Lagrange's principle, we obtain that if $\lambda_0$ is a constant of the best $(\eta, \theta)$-approximation for $g^*(t)$ in the metric of space $L_1$ and

$$\gamma(t) = agn_{\eta,\theta}(g^*(t) - \lambda_0),$$

then $\gamma \perp 1$ and for some constant $\xi$ the function

$$\gamma_0(t) = \int_0^{2\pi} B_m(t - u)\gamma(u)du - \xi$$

satisfies the conditions (9.31). Taking into account the lemma 9.3.3 we see that $\gamma \in \Gamma_{2n,0}(\eta, \theta)$. Thus $\gamma_0 \in \Gamma_{2n,m}(\eta, \theta)$.

Further we have

$$0 < E_0(g^*)_{1;\eta,\theta} = \int_0^{2\pi} g^*(t)\gamma(t)dt =$$

$$\sum_{j:0 \in A_j} a_{o,j} \int_0^{2\pi} B_m(x_j - t)\gamma(t)dt = -2\pi\xi,$$

here $a_{o,j}$ are coefficients in representation (9.37) for function $g^*(t)$. Thus $\xi < 0$ and, consequently, $\gamma_0(t)$ has positive values at some points.

Lemma 9.3.4 is proved.

**Proof of lemma 9.3.5.** First of all we note that if a function $f \in C^3[a,b]$ is such that $f(a) = f(b) = 0$ and for some $t_0 \in (a,b)$

$$f'(t_0) = f''(t_0), \quad f'''(t_0) \neq 0,$$

then there exists $t^* \in (a,b)$, $t^* \neq t_0$ such that $f'(t^*) = 0$.

Let $\gamma_0$ be the spline from lemma 9.3.4. To prove the lemma 9.3.5 it is sufficiently to show that it has constant sign. Suppose that it is not true, i.e. $\gamma_0$ alternates sign on a period at least once. Then $\gamma_0$ will have on a period at least $2n_0 + 1$ zeros calculating with their multiplicity (if every zero we calculate at most three times).

¿From this in view of the remark from the begining of the proof we obtain that the function $\gamma_0'$ will have on a period at least $2(n_0 + n_1) + 1$ zeros calculating with their multiplicity (if every multiple zero we calculate twice).

Estimating analogously the number of zeros of functions $\gamma_0''(t)$, $\gamma_0'''(t)$, $\ldots, \gamma^{(m-2)}(t)$, we obtain that $\gamma_0^{(m-2)}(t)$ has on a period at least $2(n_0 + n_1 + \ldots + n_{m-2}) + 1 = 2n + 1$ zeros if every multiple zero we calculate twice. So $\gamma_0^{(m-1)}(t)$ has at least $2n + 1$ different zeros on a period is impossible, that since $\gamma_0^{(m-1)}$ is a periodic polygonal line having at most $2n$ nodes.

Lemma 9.3.5 is proved.

¿From lemmas 9.3.3 and 9.3.4 it follows that theorem 9.3.2 holds true and consequently the theorem 9.3.1 holds true to.

To finish this section we note that using theorem 9.3.1 we can intensify the corollaries 9.2.2 and 9.2.3, replacing in the relations (9.22), (9.23) and (9.24) the set of monosplines $@_{n,m}$ by the set $@^*_{n,n}$.

In particular, the next statement is valid.

**Corollary 9.3.6.** *For all $n = 2, 3, \ldots$, $m = 1, 2, \ldots$ and for $\alpha, \beta > 0$ and $p \in [1, \infty]$*

$$\inf_{M \in @^*_{n,m}} E_0(M)_{p;\alpha,\beta} = E_0(M_{n,m})_{p;\alpha,\beta}.$$

## 9.4 Quadrarute formulas and monosplines

In this section with the help of extremal properties of monosplines, proved in previous sections, we shall present a series problems on the best quadrature formulas for certain classes of periodic differentiable functions.

Let @ is some class of sufficiently smooth periodical functions, $Q$ is certain set of functionals defined on the functions from @ and being linear combinations of values of functions and its derivatives at the same points from the period. Fixing the functional $q \in Q$ we obtain the quadrature formula

$$\int_0^{2\pi} f(t)dt \asymp q(f) \tag{9.39}$$

for approximate calculation the integrals of a function $f \in @$.

Formula of the form (9.39) in which $q \in Q$ will be called quadrature formulas of the type $Q$.

A classic organization of the problem on the best quadrarure formula belongs to S.M.Nikolski.

For central-symmetric classes @ this problem we may formulate in following manner. Put

$$R(f,q) = \int_0^{2\pi} f(t)dt - q(f) \tag{9.40}$$

and

$$R(@,q) = \sup_{f \in @} |R(f,q)|. \tag{9.41}$$

The value (9.41) is called an error of quadrature formula (9.40) on the class @.We are to obtain next value

$$R(@,Q) = \inf_{q \in Q} R(@,q) \tag{9.42}$$

and show a functional $q^* \in Q$ realizing in (9.42) lower bound. Then the quadrature formula

$$\int_0^{2\pi} f(t)dt \asymp q^*(f) \tag{9.43}$$

will get the least on the class @ error among all quadrature formulas of the type @ and in this sense it will be the best on the class @.

Denote by $Q_n$, $n = 2, 3, ...$ the set of all functionals of the form

$$q(f) = \sum_{i=1}^{n} a_i f(x_i), \tag{9.44}$$

where

$$x_1 < x_2 < \ldots < x_n < x_1 + 2\pi, \quad a_1, \ldots, a_n \in R.$$

Now we will study the problem on the best quadrature formula of the type $Q_n$ for classes $K * F_p$ ($K$ is $CVD$-kernel and $p \in [1, \infty]$) and in particular for the classes $W_p^l$.

Let for any $\sigma \in R$

$$q_{n,\sigma}(f) = \frac{2\pi\sigma}{n}\sum_{i=1}^{n} f\left(\frac{2\pi i}{n}\right). \tag{9.45}$$

Remind also that

$$M_{n,\sigma}(K,t) = \int_0^{2\pi} K(u)du - \frac{2\pi\sigma}{n}\sum_{i=1}^{n} K\left(\frac{2\pi i}{n} - t\right).$$

Note now that the next statement will take place.

**Theorem 9.4.1.** *Let the CVD-kernel $K$ and $p \in [1,\infty]$ are such that $K * F_p \subset C$ and $n = 2,3,\ldots$. Among all quadrature formulas of the type $Q_n$ the formula*

$$\int_0^{2\pi} f(t)dt \asymp q_{n,\sigma^*}(f), \tag{9.46}$$

*is the best for the class $K * F_p$, where $q_{n,\sigma^*}$ is defined by equality (9.45).*
*However in the case when $\theta = \theta(K) = 1$ we have that $\sigma^* = 1$ and*

$$R(K * F_p, Q_n) = E_0(M_{n,1}(K,\cdot))_{p'}, \;\; (1/p + 1/p' = 1)$$

*and in the case when $\theta = \theta(K) = 0$ the number $\sigma^*$ is such that*

$$\inf_{\sigma \in R} ||M_{n,\sigma}(K,\cdot)||_{p'} = ||M_{n,\sigma^*}(K,\cdot)||_{p'}$$

*and*

$$R(K * F_p, Q_n) = ||M_{n,\sigma^*}(K,\cdot)||_{p'}.$$

At the same time note that for $K = B_l$, $l = 1,2\ldots$ from this theorem the next statement follows

**Corollary 9.4.2.** *If $l \in N$ and $p \in [1,\infty]$ are such that $W_p^l \subset C$ then among all quadrature formulas of the type $Q_n$ the formula (9.45) is the best on the class $W_p^l$ for $\sigma^*$.*

**Proof of the theorem 9.4.1.** Remind that if $f \in K * F$, $F \in L_1$ then

$$f(t) = a\theta + \int_0^{2\pi} K(t-u)\phi(u)du, \tag{9.47}$$

where $a \in R$, $\phi \in F$ and $\phi \perp \theta$. Taking into account the equalities (9.41), (9.44) and (9.47) we obtain

$$R(f,q) = \int_0^{2\pi} \left( a\theta + \int_0^{2\pi} K(t-u)\phi(u)du \right) -$$

$$-\sum_{i=1}^{n} a_i \left( a\theta + \int_0^{2\pi} K(x_i - u)\phi(u)du \right) =$$

$$a\theta \left( 2\pi - \sum_{i=1} na_i \right) + \int_0^{2\pi} \phi(u) \left( \int_0^{2\pi} K(t-u)dt - \sum_{i=1}^{n} a_i K(x_i - u) \right) du.$$

Thus

$$R(f,q) = a\theta \left( 2\pi - \sum_{i=1} na_i \right) + \int_0^{2\pi} \phi(u)M(u)du, \tag{9.48}$$

where

$$M(u) = \int_0^{2\pi} K(t)dt - \sum_{i=1}^{n} a_i K(x_i - u) \tag{9.49}$$

is (generalized) monospline from $@_n(K)$.

In the case when $\theta = 0$ the outintegral member in the right part of (9.48) is equal to zero and we obtain

$$R(K * F_p, q) = \sup_{\|\phi\|_p \leq 1} \int_0^{2\pi} \phi(u)M(u)du = \|M\|_{p'}.$$

But then in view of corollary 9.2.7 for any monospline $M$ of the form (9.49)

$$\|M\|_{p'} \geq \|M_{n,\sigma^*}(K,\cdot)\|_{p'} = R(K * F_p, Q_{n,\sigma^*}).$$

In the case $\theta = 0$ theorem is proved.

If $\theta = 1$ and

$$\sum_{i=1}^{n} a_i \neq 2\pi$$

then

$$R(K * F_p, q) = \infty$$

so for the class $K * F_p$ the best quadrature formula of the type $Q_n$ we are to find only among formulas for which

$$\sum_{i=1}^{n} a_i = 2\pi.$$

In this case the outintegral member in (9.48) is equal to zero and

$$R(K * F_p, q) = \sup_{\|\phi\|_p \leq 1, \, \phi \perp 1} \int_0^{2\pi} \phi(u) M(u) du.$$

¿From this with the help of duality theorem (theorem 1.2.4) we obtain

$$R(K * F_p, q) = E_0(M)_{p'}.$$

Moreover due to corollary 9.2.7 we have

$$E_0(M)_{p'} \geq E_0(M_{n,1}(K, \cdot))_{p'} = R(K * F_p, q_{n,1}).$$

For $\theta = 1$ theorem is proved.

For the classes $W_p^l$ we can prove more strong statement.

Denote by $Q_{n,l}$, $n = 2, 3, \ldots$, $l = 1, 2, \ldots$ the set of functionals of the form

$$q(f) = \sum_{i=1}^{m} \sum_{j \in A_i} a_{i,j} f^{(j)}(x_i), \tag{9.50}$$

where $m \leq n$;

$$x_1 < x_2 < \ldots < x_n < x_1 + 2\pi; \;\; a_{i,j} \in R;$$

$$A_1, A_2, \ldots, A_m \subset \{0, 1, \ldots, l-1\}$$

and

$$\sum_{i=1}^{m} |A_i| \leq n.$$

(remind that $|A|$ is number of elements for the set $A$).

**Theorem 9.4.3.** *If $l \in N$ and $p \in [1, \infty]$ are such that $W_p^l \subset C^{(l-1)}$ then for any $n = 2, 3, \ldots$ among all quadrature formulas of the form $Q_{n,l}$ the formula of the form (9.45) is the best for the class $W_p^l$ for $\sigma^* = 1$.*

**Proof.** The best for the class $W_p^l$ quadrature formula of the type $Q_{n,l}$ we can find only among formulas for which in representation (9.49) of functional $q \in Q_{n,l}$

$$\sum_{1 \leq i \leq m, \, 0 \in A_1} a_{i,0} = 2\pi. \tag{9.51}$$

(in opposite case $R(W^l_p, q) = +\infty$). If (9.50) is valid then for $R(f,q)$ and $f \in W^l_p$ it is easy to qet the next representation

$$R(f,q) = \int_0^{2\pi} M(u) f^{(l)}(u) du, \tag{9.52}$$

where

$$M(u) = -\sum_{i=1}^{m} \sum_{j \in A_i} a_{i,j} B_l - j(x_i - u)$$

is monospline from the set $@^*_{n,l}$.

Now, taking into account the duality theorem (theorem 1.2.4) and corollary 9.3.6 we obtain

$$R(W^l_p, q) = E_0(M)_{p'} \le E_0(M_{n,l})_{p'},$$

where

$$M_{n,l}(t) = -\frac{2\pi}{n} \sum_{i=1}^{n} B_l \left( \frac{2i\pi}{n} - t \right)$$

and so

$$R(W^l_p, q_{n,1}) = E_0(M_{n,l})$$

the theorem is proved.

Now we shall extend previous results concerning with optimization of quadrature formulas on more general classes of functions $W^l F$ and $K * F$ where $F$ is arbitrary rearrangementally invariant set. These classes, in general, are not central-symmetric. Therefore it is naturally to modify the organization of problem on the best quadrature formula .

Let as before @ is some class of sufficiently smooth $2\pi$-periodical functions (it must not be central-symmetric) and $Q$ is a set of functionals defined on the functions from @ and giving us the quadrature formulas of the form (9.39). For $q \in Q$ we put

$$R^+(@, q) = \sup_{f \in @} R(f,q),$$

$$R^-(@, q) = \inf_{f \in @} R(f,q)$$

and

$$I(@, q) = \left[ R^-(@, q);\ R^+ @, q) \right].$$

We tell that the quadrature formula

$$\int_0^{2\pi} f(t) dt \asymp q^*(f)$$

$(q^* \in Q)$ is the best for the class @ among all formulas of the type $Q$ if for any another formula of the form (9.40) it will be

$$I(@, q^*) \subset I(@, q). \tag{9.53}$$

The theorem 9.4.3 is generalized by following theorem.

**Theorem 9.4.4.** *Let $l = 1, 2, \ldots$ and rearrangememtally invariant set $F \in L_1$ such that $W^l F \subset C^{(l-1)}$. The quadrature formula (9.45) for $\sigma^* = 1$ is the best for class $W^l F$ quadrature formula among all formulas of the type $Q_{n,l}$.*

**Proof.** The best for the class $W^l F$ quadrature formula of the type $Q_{n,l}$ we ought to find only among formulas which are exact on constants (in opposite case $I(W^l F, q) = (-\infty, \infty)$). But amomg such formulas the error $R(f, q)$ may be represented in the form (9.51). Using consistently the statements 1.3.4, 9.3.1 and 1.3.14 we obtain

$$R^+(W^l F, q) = \sup_{f^{(l)} \in F} \int_0^{2\pi} M(u) f^{(l)}(u) du =$$

$$\sup_{\phi \in F,\, \phi \perp 1} \int_0^{2\pi} \pi(M, u)\pi(\phi, u) du \geq$$

$$\sup_{\phi \in F,\, \phi \perp 1} \int_0^{2\pi} \pi(M_{n,l}, u)\pi(\phi, u) du = R^+(W^l F, q_{n,1}).$$

Analogously we show that

$$R^-(W^l F, q) \leq R^-(W^l F, q_{n,1}).$$

The theorem is proved.

For any fixed $\sigma \in R$ denote by $Q_n(\sigma)$ the set of all functionals $q \in Q_n$ for which

$$\sum_{i=1}^{n} a_i = 2\pi\sigma.$$

**Theorem 9.4.5.** *Let $K$ is $CVD$-kernel, $F$, $K * F \subset C$ is rearramgementally invariant set, $\sigma \in R$ is arbitrary and $Q = Q_{n,\sigma}$ if $\theta = 0$ and $\sigma = 1$ and $Q = Q_n$ if $\theta = 1$. Then the quadrature formula*

$$\int_0^{2\pi} f(u) du \asymp q_{n,\sigma}(f)$$

*is the best for the class $K * F$ among all quadrature formulas of the type $Q$.*

To prove this theorem as in the proof of theorem 9.4.1 for any $q \in Q_n(\sigma)$ and any $f \in K * F$ we obtain

$$R(f, q) = \int_0^{2\pi} \phi(u) M(u) du,$$

where $M$ is generalized monospline from the set $@_{n,\sigma}(K)$. Applying consistently the statements 1.3.4, 9.2.4 and 1.3.4 we verify that

$$R^+(K * F, q) \geq R^+(K * F, q_{n,\sigma})$$

and

$$R^-(K * F, q) \leq R^-(K * F, q_{n,\sigma}).$$

The theorem is proved.

The method used for the proof of theorem 9.4.1 allows us (with the help of theorem 9.4.1) to investigate one problem of optimization of "intervals" quadrature formulas.

Let $n = 2, 3, \ldots$ and $h \in (0, \pi/n)$ are given. Denote by $Q_n^h$ the set of functionals of the form

$$q(f) = \frac{1}{2h} \sum_{i=1}^{n} a_i \int_{x_i - h}^{x_i + h} f(t) dt,$$

where $a_i \in R$,

$$x_1 + h \leq x_2 - h < x_2 + h \leq \ldots \leq x_n - h < x_n + h \leq x_1 + 2\pi - h.$$

Put (for $\sigma \in R$)

$$q_{n,\sigma}^*(f) = \frac{2\pi\sigma}{n} \frac{1}{2h} \sum_{i=1}^{n} \int_{\frac{2i\pi}{n} - h}^{\frac{2i\pi}{n} + h} f(t) dt.$$

**Theorem 9.4.6.** *Let $K \in CVD$, $n = 2, 3, \ldots$ and $h \in (0, \pi/n)$. The quadrature formula*

$$\int_0^{2\pi} f(t) dt \asymp q_{n,\sigma^*}^*(f), \quad \sigma^* = 1$$

*is the best for the class $K * F_1$ among all formulas of the type $Q_n^h$. Moreover*

$$R(K * F_1, Q_n^h) = E_0((M_{n,1}(K, \cdot))_h)_\infty$$

*if* $\theta = \theta(K) = 1$ *(*$(M_{n,1}(K,\cdot))_h$ *is meaning on Steklov with the step* $2h$ *for the function* $M_{n,1}(K,\cdot)$ *). If* $\theta = 0$ *then* $\sigma^*$ *such that*

$$\inf_{\sigma \in R} \|(M_{n,\sigma}(K,\cdot))_h\|_\infty = \|(M_{n,\sigma^*}(K,\cdot))_h\|_\infty$$

*and*

$$R(K * F_1, Q_n^h) = \|(M_{n,\sigma^*}(K,\cdot))_h\|_\infty.$$

In conclusion of this section we consider still one problem on the best quadrature formula for class $W_p^l$ which decision is reduced to investigation of extremal properties of monosplines.

Let $n = 2, 3, \ldots, l = 1, 2, \ldots$ and $\rho = 0, 1, \ldots, l-1$.

Denote by $Q_{n,l}^\rho$ the set of all functionals of the form

$$q(f) = -\sum_{i=1}^{n} \sum_{j=0}^{\rho} a_{i,j} f^{(j)}(x_i), \tag{9.54}$$

where

$$x_1 < x_2 < \ldots < x_n < x_1 + 2\pi, \quad a_{i,j} \in R.$$

As before we will constrein the consideration of such functionals of the form (9.54) for which

$$\sum_{i=1}^{n} a_{i,0} = 2\pi. \tag{9.55}$$

But for such functionals $q$ the error $R(f,q)$ may be represented in the form

$$R(f,t) = \int_0^{2\pi} M(u) f^{(l)}(u) du,$$

where

$$M(u) = -\sum_{i=1}^{n} \sum_{j=0}^{\rho} a_{i,j} B_{l-j}(x_i - u). \tag{9.56}$$

The totality of functions of the form (9.56), for which $x_i$ and $a_{i,j}$ are as in (9.54) and (9.55), denote by $@_{n,l}^\rho$.

Functions from $@_{n,l}^\rho$ are called monosplines of order $l$ with defect $\rho + 1$, nodes $x_i$ and coefficients $a_{i,j}$.

Taking into account that for $t \in (0, 2\pi)$

$$B_1(t) = \frac{\pi - t}{2\pi},$$

it is easy to see that on each interval $(x_i, x_{i+1})$ the monospline $M \in @^{\rho}_{n,l}$ has the form

$$M(t) = \frac{(-1)^{l+1}}{l!} t^l + \sum_{\nu=0}^{l-1} c_{i,\nu} t^{\nu}. \tag{9.57}$$

For arbitrary $l$ and $\rho = l-1$ and $\rho = l-2$ consider the problem of minimization of the value $E_0(M)_p$, $(1 \le p \le \infty)$ for $M \in @^{\rho}_{n,l}$.

Denote by $p_{l,q}(a,h,t)$ the polynomial of the least deviation from zero in the space $L_q[a-h, a+h]$, $(1 \le q \le \infty)$ among all polynomials of the form

$$p_l(f) = t^l + \sum_{\nu=0}^{l-1} c_{\nu} t^{\nu}.$$

We will write $p_{l,q}$ instead of $p_{l,q}(0,1,t)$.

It is clear that

$$p_{l,q}(a,h,t) = h^l p_{l,q}\left(\frac{t-a}{h}\right)$$

and consequently

$$||p_{l,q}(a,h,\cdot)||^q_{L_q[a-h,a+h]} = h^{lq+1} N,$$

where

$$N = ||p_{l,q}||^q_{L_q[-1,1]}.$$

¿From the results of section 2.3 it follows also that

$$p_{l,q}(a,h,a+t) = (-1)^l P_{l,q}(a,h,a-t), \quad t \in (-h,h)$$

because polynomial $p_{l,q}$ is even for even $l$ and odd if $l$ is odd.

Define for $t \in (2i\pi/n, 2\pi(i+1)/n)$, $i = 1,2,\dots,n$ the monospline $M^{l-1}_{n,l} \in @^{l-1}_{n,l}$ by the next way

$$M^{l-1}_{n,l}(t) = \frac{(-1)^{i+1}}{l!} \left(\frac{\pi}{n}\right)^l p_{l,q}\left(\frac{nt-(2i+1)\pi}{\pi}\right).$$

The coefficients $a^*_{i,j}$ of last polynomial (that is easy to see from (9.56)) may be found by formulas

$$a^*_{i,j} = \frac{(-1)^j}{l!} \left(\frac{\pi}{n}\right)^{j+1} \left(p^{(l-j-1)}_{l,q}(1) - p^{(l-j-1)}_{l,q}(-1)\right)$$

and consequently for $j = 0,1,\dots,(r-1)/2$

$$a^*_{i,2j} = \frac{1}{l!} \left(\frac{\pi}{n}\right)^{2j+1} 2p^{(l-2j-1)}_{l,q}(1)$$

and $a^*_{i,2j+1} = 0$.

Thus for all $t$

$$M_{n,l}^{l-1}(t) = -\frac{2}{l!}\sum_{i=1}^{n}\sum_{j=0}^{[(l-1)/2]}\left(\frac{\pi}{n}\right)^{2j+1} p_{l,q}^{(l-2j-1)}(1)B_{l-j}\left(\frac{2i\pi}{n} - t\right).$$

Now next statement will take place.

**Theorem 9.4.7.** *Let $n, l = 2, 3, \ldots$ and $q \in [1, \infty]$. Then*

$$\inf_{M \in @_{n,l}^{l-1}} E_0(M)_q = E_0(M_{n,l}^{l-1})_q = \frac{\pi^{lq+1}N^{1/q}}{n^l}.$$

**Proof.** Let $M \in @_{n,l}^{l-1}$ is monospline of the form (9.55). Then it's obvious that

$$E_0(M)_q^q \geq \sup_{i=1} m \int_{x_i}^{x_{i+1}} \left| p_{l,q}\left(\frac{x_i + x_{i+1}}{2}; \frac{x_{i+1} - x_i}{2}; t\right)\right|^q dt =$$

$$N \sum_{i=1}^{m} h_i^{lq+1} \tag{9.58}$$

where $h_i = (x_{i+1} + x_i)/2$.

It is clear that for all $i$ we have that $h_i > 0$ and

$$\sum_{i=1}^{m} h_i = \pi. \tag{9.59}$$

Then minimizing, with help of Lagrange method, the function

$$\Phi(h_1, h_2, \ldots, h_n) = \sum_{i=1}^{m} h_i^{lq+1}$$

under conditions $h_i > 0$, $i = 1, 2, \ldots, m$ and (9.58) we obtain that the unique minimum of this function arrives at $h_1 = h_2 = \ldots h_n = \pi/n$ and the minimum value of $\Phi$ is equal to $n(\pi/m)^{lq+1}$. From this and from (9.57) we have

$$E_0(M)_q^q \geq nN\left(\frac{\pi}{m}\right)^{lq+1} \geq \frac{N\pi^{lq+1}}{n^{lq}} = E_0(M_{n,l}^{l-1})_q^q.$$

Theorem is proved.

Note that for even $l$ the monospline $M_{n,l}^{l-1}$ belongs to the set $@_{n,l}^{l-2}$ being more narrow than $@_{n,l}^{l-1}$. So the next statement will take place

**Theorem 9.4.8.** *For all* $n = 2, 3, \ldots, l = 2, 4, \ldots$ *and* $q \in [1, \infty]$

$$\inf_{M \in @_{n,l}^{l-2}} E_0(M)_q = E_0(M_{n,l}^{l-1})_q = \frac{\pi^{lq+1} N^{1/q}}{n^l}.$$

¿From the theorems 9.4.7 and 9.4.8 it follows that

**Theorem 9.4.9.** *The quadrature formula*

$$\int_0^{2\pi} f(t)dt = \frac{2}{l!} \sum_{i=1}^{n} \sum_{j=0}^{[(l-1)/2]} \left(\frac{\pi}{n}\right)^{2j+1} p_{l,q}^{(l-2j-1)}(1) f^{(2j)}\left(\frac{2i\pi}{n}\right)$$

*is the best for the class* $W_p^l$, $(p \in [1, \infty])$ *among the formulas of the type* $Q_{n,l}^{\rho}$ *for* $\rho = l - 1$, $(l = 1, 2, \ldots)$ *and for* $\rho = l - 2$, $l = 2, 4, \ldots$.

*While*

$$R(W_p^l, Q_{n,l}^{\rho}) = \frac{\pi^{l+1/p'} N^{1/p'}}{n^l}$$

*where as before* $1/p + 1/p' = 1$.

# 9.5 Commemtaries

A great number of publications are devoted to the investigation of extremal properties of monosplines and connected with its problem of optimization of quadrature (see, for example review of N.P.Kornechuk in "Supplement" to monograph S.M.Nikolsky [8], review of A.A.Dzensykbaev [3] and articles of V.P.Motornyi [2], B.D.Bojanov [4-6], [8]). Here we presented only references connected with the results,given in maim text of chapter.

Sec. 8.1. The theorem 8.1.2 is proved by V.P.Motornyi [2] (for even $m$) and A.A.Ligun [4] (for add $m$). In connection with other results of this section see the articles of V.F.Babenko and T.A.Grankina [1] and V.F.Babenklo [7].

Sec. 8.2. The theorem 8.2.1 generalized the well knoen results of V.P.Motornyi [2] and A.A.Dzensykbaev [1,3]. The wording and proof of this theorem given in monograph are presented by V.F.Babenko [4,5,19]. The theorem 8.2.4 was proved by V.F.Babenko [11,12]. The results on minimization of generalized monosplines norms one can find in articles of M.A.Chahkiev [1] and Nguen Thy Thieu Hoa [1,2].

Sec. 8.3. Inequalities of the type (9.40) for $L_p$-norms of monosplines of variable defect was proved by A.A.Ligun [6,11] for $p = 1$ and $p = \infty$ and by A.A.Dzensykbaev [4] for $1 < p < \infty$. The inequality (9.40) was proved by V.F.Babenko [10] with applications of idea of A.A.Ligun [11].

## 9.6 Supplementary results

**1.** Among all quadrature formulas of the form

$$\int_0^{2\pi} f(t)dt \asymp \sum_{k=1}^{n} \sum_{i=0}^{\rho} p_{k,i} f^{(i)}(x_k) \tag{9.60}$$

where

$$0 \le x_1 < x_2 < \ldots < x_n < 2\pi$$

and $1 \le 2[(\rho+1)]/2 < m$, the formula having equidistant nodes $x_k^* = 2\pi(k-1)/n$, $k = 1, 2, \ldots, n$ and equal coefficients $p_{k,i}^* = p_{1,i}^*$, $k = 1, 2, \ldots, n$ is the best for class $W_p^m$, $(p \in [1, \infty])$. Moreover

$$p_{1,i}^* = p_{2,i}^* = \ldots = p_{n,i}^* = 0$$

if $i$ is odd, and

$$p_{1,i}^* = p_{2,i}^* = \ldots = p_{n,i}^* > 0$$

if $i$ is even.

For $p \in (1, \infty]$ the best quadrature formula is unique with the exactness to translate of argument.

Boijanov B.D. [4-6].

**2.** Let $\omega(t)$ is convex from below modulus of continuity and $\rho = 0, 1$ and $m = 1, 3, \ldots$. The quadrature formula

$$\int_0^{2\pi} f(t)dt \asymp \frac{2\pi}{n} \sum_{k=1}^{n} f\left(\frac{2k\pi}{n}\right) \tag{9.61}$$

is the best for the class $W^r H^\omega$ among the formulas of the type (9.59). Moreover the error of the best formula on the class $W^r H^\omega$ is equal to $2\pi \| f_{n,m}(\omega) \|_\infty$.

Motornyi V.P. [1,2].

**3.** Let $\omega(t)$ is convex from below modulus of continuity and $m = 1, 3, 5, \ldots$. The quadrature formula (9.59) is the best for class $W^r H^\omega$ among all quadrature formulas of the type (9.49).

Ligun A.A. [11].

**4.** The qudrature formula (9.59) is the best for class $\tilde{W}_p^1$ among all quadrature formulas of the type

$$\int_0^{2\pi} f(t)dt \asymp \sum_{k=1}^{n} a_k f(x_k),$$

where

$$x_1 < x_2 < \ldots < x_n < x_1 + 2\pi, \quad a_1, a_2, \ldots, a_n \in R.$$

Babenko V.F. [20].

**LITERATURA**

**Aziz A.**

1.Inequalities for polynomials with a prescribed zero.-Can. J. Math.-1982.- 34, N 3.-P. 737-740.

2.Inegualities for the derivative of a polynomial.- Proc. Amer.Math.Soc.-1983.-89, N 2.-P. 259-266.

3.Inegualities for Polynomials with a prescribed zero.-J.Approxim.Theory.-1984.-41, N 1.-P. 15-20.

4.On the location of critical points of polynomials.- J.Austral Math. Soc.-1984.-36, N 1.-P. 4-11.

5.Growth of polynomials whose zeros are within or outside a circle.-Bull.Austral.Math.Soc.-1987.-35, N 2.-P.247-256.

**Aziz A., Mohammad Q.G.**

1.Simple proof of a theorem of Erdos and Lax.-Proc. Amer.Math. Soc.-1980.-80.- P. 119-122.

**Ahlberg J.H., Nilson E.N., Walsh J.L.**

1. The theory of splines and their application.- Moscow:Mir, 1972.-316 P.

**Amir D., Ziegler Z.**

1. Polynomials of extremal $L_p$-norm on the $L_\infty$-unit sphere// J. Approxim. Theory.-1976.- 18, N1.- P. 86-98.

**Arestov V.V.**

1.On best approximation of differential operator// Math. Zametki. -1967.-1, N2.-P. 149-154.(Russian)

2. On integral inequalities for trigonometric polynomials and its derivatives// Izv.Akad.Nauk USSR Ser.Math.-1982.- 45, N1.-P. 3-22. (Russian)

**Ahiezer N.I.**

1. Uber ienige Funktionen die in gegebenen am wengstem von Null alweichen// Izv.Kazan.phys.-Math.soc.-1928.- 3.-P. 1-69.

2. On best approximation of analytic functions// Dokl. AN USSR.-1938.- 18, N4/5.-P. 241-244.(Russian)

3. The general theory of P.L. Chebyshev's polynomials Science legacy of P.L.Chebyshev.-Moscow: .-1945.- P. 5-42. (Russian)

4. On integral functions with finite power deviating least from xero// Math.collection.- 1952. -31(73), N2.- P. 415-438. (Russian)

5. On weighted approximation of continuous functions by polynomials on all real axis// Uspehi Math.Nauk.-1956.- 11, N4.-P. 3-43. (Russian)

6. Lectures on approximation theory.- Moscow: Nauka, 1965.-408 P. (Russian)

**Ahiezer N.I., Babenko K.I.**

1.On weighted approximation by polynomials on all real axis// Dokl. AN USSR.- 1947.- 57, N4.- P. 315-318. (Russian)

**Babenko A.G.**

1. On one extremal problem for polynomials// Math. Zametki.-1984.- 35, N3.- P. 349-356. (Russian)

2. On one extremal problem for polynomials having fixed mean value// Approximation of functions by polynomials and splines.- Sverdlovsk USC AN USSR.- 1985.-P. 15-22. (Russian)

**Babenko V.F.**

1. Nonsymmetric approximations in the spaces of summable functions// Ukrain. Math. J.- 1982.- 34, N4.- P. 409-416. (Russian)

2.On optimization by nodes of quadrature formulas with equal coefficients on the classes of conjugate functions// Studies on modern problems of summation and approximation of functions and their applications.- Dnepropetrovsk: Dnepr.Univ.-1982.- P. 3-5. (Russian)

3. Nonsymmetric extremal problems of approximation theory// Dokl. AN USSR.- 1983.- 269, N3.-P. 521-524. (Russian)

4.Inequalities for rearrangements of differentiable periodic functions, problems of approximation and integrating// Dokl. USSR.- 272, N5.-P. 1038-1041. (Russian)

5. Inequalities for rearrangements of differentiable periodic functions and their applications//Intern. conf. on approximation theory of functions. Kiev: Inst. of Math. AN Ukrain.- 1983.- P. 12. (Russian)

6. On the widths of some classes of convolutions// Ukrain. Math.J. .-1983.- 35, N5.-P. 603-607. (Russian)

7. On one problem of optimization of the approximate integration// Studies on modern problems of summation and approximation of functions and their applications.- Dnepropetrovsk: Dnepr.Univ.-1984. -P. 3-13. (Russian)

8. Duality theorems for some problems of approximation theory// The modern problems of real and complex analysis.- Kiev: Inst. Math. AN Ukrain. -1986.- P. 7-10. (Russian)

9. Interpolation by splines of minimal defect and the best quadrature formulas with fixed nodes// Studies on modern problems of summation and approximation of functions and their applications.- Dnepropetrovsk: Dnepr.Univ.-1985.- P. 3-11. (Russian)

10. The exact inequalities for rearrangements of periodic monosplines and best quadrature formulas// Studies on modern problems of summation and approximation of functions and their applications.- Dnepropetrovsk: Dnepr.Univ.-1986.- P. 4-11. (Russian)

11. The best quadrature formulas for some classes of convolutions// Studies on theoretical and applied problems of mathematics.- Kiev, Inst. of Math. AN Ukrain.- 1986.- P. 7-10. (Russian)

12. Extremal problems of approximation theory and inequalities for rearrangements// Dokl. AN USSR.- 1986.- 290,N5.- P. 1033-1036. (Russian)

13. Exact inequalities for norms of conjugate functions and their applications// Ukrain. Math.J.- 1987.- 39,N2.- P.139-144. (Russian)

14. Inequalities for rearrangements of differentiable periodic functions and their applications//Theory of approximation of functions: Proc. Intern. conf. on theory of approximation of functions, Kiev,1983.- Moscow: Nauka, 1987.- P. 26-30. (Russian)

15. The comparison theorems and inequalities of Bernstain's type// The theory of approximation and adjacent questions of analysis and topology.- Kiev: Inst. of Math. AN Ukrain., 1987.- P. 4-8. (Russian)

16. On existence of perfect splines and monosplines having given zeros// Studies on modern problems of summation and approximation of functions and their applications.- Dnepropetrovsk: Dnepr.Univ.-1987.- P. 4-9. (Russian)

17. Nonsymmetric approximations and inequalities for rearrangements in extremal problems of approximation theory // Theory of functions and adjacent problems of analysis: Proc. of conf. on theory of functions, devoted to 80-th jubilee of S.M.Nikolski, Dnepropetrovsk,1985.- Trudy Math. Inst. V.A.Steklov AN USSR.-1987.- 180. -P. 33-35. (Russian)

18. The approximation for classes of convolutions // Siberian Math.J.- 1987.-28,N 5.-P.6-21. (Russian)

19. Approximations, widths and optimal quadrature formulae for classes of periodic functions with rearrangement invariant sets of derivatives// Anal. Math.- 1987.- 13, N4.- P. 15-28.

20. Approximations and best quadrature formulas for some classes of conjugate functions// Questions of optimal approximation of functions and summation of series.- Dnepropetrovsk: Dnepr.Univ.- 1988.- P. 5-11. (Russian)

**Babenko V.F., Grankina T.A.**

1. On the best quadrature formula on classes of convolutions with $O(M, \Delta)$-kernels// Studies on modern problems of summation and approximation of functions and their applications.- Dnepropetrovsk: Dnepr.Univ.- 1982.- P. 6-13. (Russian)

**Babenko V.F. Kafanov V.A.**

1. On constant-sign polynomials deviating least from zero in the space $L_p$// Math. Zametki.- 1985.- 37.- N2.- P. 176-185. (Russian)

**Babenko V.F., Pichugov S.A.**

1. Remark to Kolmogorov's inequality// Studies on modern problems of summation and approximation of functions and their applications.- Dnepropetrovsk: Dnepr.Univ.-1980.- P. 14-17. (Russian)

2. Exact inequality for derivative of trigonometric polynomial, having only real zeros// Math. Zametki.- 1986.- 39, N3.-P. 330-336. (Russian)

3. On inequalities for derivatives of polynomials with real zeros// Ukrain. Math. J.- 38, N4.-P. 411-416. (Russian)

**Babenko V.F., Severina O.I.**

1. On inequalities of Turan's type for trigonometric polynomials// Theory of approximation of functions and summation of series.- Dnepropetrovsk: Dnepr.Univ.- 1989.-P. 10-16. (Russian)

**Babenko K.I.**

1. Lectures on theory of approximations.- Moscow: IPM AN USSR,1970.- 60 P. (Russian)

**Bakan A.G.**

1. Ardesh's hypothesis on extremal property of Chebyshev's polynomials // Theory of approximation of functions.-Moscow: Nauka, 1987.-P. 35-36. (Russian)

**Bari N.K.**

1. Generalizations of Bernstain's and Markov's inequalities// Izv. AN USSR, Ser. Math.- 1954.- 18, N2.-P. 159-176. (Russian)

**Bernstein S.N.**

1. Extremal properties of polynomials. Moscow;Len.: GTTI.-1937.-250 P. (Russian)

2.Collection of proceed:At 4 T.-Moscow: Pub.AN USSR.-T.1.-1952.-580 P.; T.2.-1954.-626 P. (Russian)

**Boas R.**

1.Quelques generalisations d'une theoreme de S.Bernstein sur la derivee d'un poynime trigonometriqe.- Comptes Rendus.- 1948.- 227.- P. 618-619.

2.Entiere Functions.-New York: 1954.-276 c.

3.Interference phenomena for entiere functions.-Ill. J.Math.- 1957 .- 1.- P. 94-97.

5.Inequalities for polynomials with a prescribed zero.-Studies in mathematical analysis and related topics.-Stanford, Calif.: Stanford Univ. Press.-1962.- P. 42-47.

**Bohr H.**

1.Un theoreme generale sur l'integration d'un polynome trigonometrique.-Acad. Sci.- 1935.- 200.- P. 1276-1277.

**Borsuk K.**

1.Drei Satze uber die $n$-dimensional euklidiche sphare.-Fund.Math.- 1933.-20.-P. 177-193.

**Bojanov B.D.**

1. Best methods of interpolation for some classes of differentiable functions// Math. Zametki.- 1975.- 17, N4.-P. 511-524. (Russian)

2. The best recovery of differentiable periodic functions on their Fourier's coefficients// Bolgarsko Math. spis: Serdica.- 1976.- 2.- P. 300-304.

3. A note on the optimal approximation of smooth periodic functions// Dokl. Bolg. AN.-1977.- 30, N6.-P. 809-819.

4. Characterization and existence of optimal quadrature formulas for certain classes of differentiable functions// Dokl. AN USSR.-1977.-232.-N6.-P.1233-1236. (Russian)

5. Existence of optimal quadrature formulas with given multiplicities of nodes// Math.col.-1978.-105(147),N3.-P.342-370.

6. Uniqueness of optimal quadrature formulas// Dokl. AN USSR.-1979.- 248,N2.-P.272-274. (Russian)

7. A generalization of Chebyshev Polynomials// J.Approxim. Theory.-1979.- 26,N4.-P.293-300.

8. Uniqueness of the Optimal Nodes of quadrature formulae// Math. Comput.- 1981.-36,N 154.-P.525-546.

9. An Extention of the Marcov Inequality// J.Approxim. Thery.-1982.-35,N2.- P.181-190.

10. On extremal problem for polynomials// Constructive theory of functions.- Proc. Intern. conf., Varna,1-5 june 1981.- Sofia,1983.-P.229-233.

**Bojanic R.,De Vore R.**

1.On polynomials of best one-sided approximation.- Enseign.Math.-1966.-12.-P. 139-164.

**Brauder A.**

1. On Bernsteins inequality and the norm of Hermitian operatos// Amer. Math. Mon.-1971.-78.-P.821-833.

**Brudnyi Y.A., Ganzburg M.I.**

1. On one extremal problem for polynomials of $n$ variables// Izv. AN USSR, Ser. Math.-1973.-37,N2.-P.344-355. (Russian)

2. On certain exact inequality for polynomials of many variables// Math. progr.and adjacent problems. Theory of functions and functional analysis.-Moscow:1976.-P.118-123. (Russian)

**Brudnyi Y.A., Gorin E.A.**

1. Isoperimetric representations and differential inequalities.- Yaroslavl, Yaroslav. Univ.- 1981.- 92 P. (Russian)

**De Bruijn N.G.**

1.Inegualities concerning polynomials in the complex domain.-Indag.Math.- 1947.-9.-P. 591-598.

**Buslaev A.P., Tihomirov V.M.**

1. Spectrums of some nonlinear equations and widhts of Sobolev's classes// Constructive function theory: Proc. Intern. conf. on constructive function theory. Varna,1984.- Sofia: Bolg. AN.-1984.-P.14-17. (Russian)

**Barrar R., Loeb H.**

1.On a nonlinear characterization problem for monosplines.- J.Approxim.Theor 1976.-18, N 3.-P. 220-240.

**Callahan F.P.**

1. An extremal problem for polynomials// Proc. Amer. Math. Soc.- 1959.- 10,N5.-P.754-755.

**Chan T.N, Malik M.A.**

1. On Erdos-Lax theorem // Proc. Indian. Acad. Sci.- 1983.-92, N3.- P.191.

**Chahkiev M.A.**

1. Linear differential operators with real spectrum and optimal quadrature formulas // Izv. AN USSR. Ser. Math.- 1984.- 48, N5.- P. 1078. (Russian)

**Chebyshev P.L.**

1. Theory of mechamisms which are known as parallelogramms // Coll. of proceed.- T.2.- Pub. AN USSR.- 1947.- P. 23-51. (Russian)

2. On least values problems, connected with approximate representation of functions // Total coll. of proceed.- T.2.- Pub. AN USSR.- 1947.- P. 151-235. (Russian)

**Chernyh N.I.**

1. On some extremal problems for polynomials // Pub.Math.Inst.V.A.Steklov AN USSR.- 1965.- 78.- P.48-89. (Russian)

**Van der Corput J.G., Schaake G.**

1. Ungeichungen fur polynome und trigonometriche polynome// Composito Math.-1935.-2.-P.321-361.

**Van der Corput J.G., Visser C.**

1. Inequalities concerning polynomials and trigonometric polynomials// Nederl. Akad. Wetencsch., Proc.-1946.- 49.- P. 383-392.

**Duffin R.J., Schaeffer A.C.**

1.Some inequalities concerning functions of exponential type// Bull. Amer. Math. Soc.-1937.-43.-P.554-556.

2.Some properties of functions of exponential type//Ibid.-1938.-44.- P.236-240.

**Dzjadyk V.K.**

1. On the best approximation on a class of periodic functions, having restricted $s$-th derivative $(o < s < 1)$// Izv. AN USSR, Ser. Math.- 1953.-17,N2.-P.135-162. (Russian)

2. On the best approximation on classes of periodic functions, defined by integrals of linear combinations of absolute monotonous kernels// Math. Zametki.-1974.-16,N5.-P.691-703. (Russian)

3.Introduction in theory of uniform approximation of functions by polynomials.-Moscow:Nauka, 1977.-510 P. (Russian)

4. On a problem of Chebyshev and Markov// Anal. Math.-1977.-3,N3.-P.171-175. (Russian)

**Dzjadyk V.K., Dubovik V.A.**

1. To inequalities of A.N.Kolmogorov on correlations between upper bounds of derivatives of real functions, defined on all axis,ll// Ukrain. Math. J.-1975.-27,N3.-P.291-299. (Russian)

**DeVore R.A.**

1. A property of Chebyshev polynomials// J.Approxim. Theory.-1974.-12.- P.418-419.

**Dolzenko E. P.**

1. Estimations of derivatives of polynomial fractions// Izv. AN USSR, Ser. Math.-1963.-27,N1.-P.9-28. (Russian)

**Donaldson J.D., Rahman Q.I.**

1.inequalities for polynomials with a prescribed zero// Pacif. J.Math.-1972.-41.-P.375-378.

**Doronin V.G., Ligun A.A.**

1. On best one-sided approximation of certain class of functions by another// Math. Zametki.-1973.-14,N5.-P.633-638. (Russian)

2. The best one-sided approximation of some classes differentiable periodic functions// Dokl. AN USSR.-1976.-230,N1.-P.19-21.

3. Inequalities for upper bounds of functionals and best one-sided approximation some classes of periodic functions// Constructive theory of functions,77,Proc. of Int. Conf. of constr. th. of function, Blagoevgrad, 30 may-6 june 1977.- Sofia: Publ. Bolg. AN.-1980.-P.57-64. (Russian)

**Jensykbaev A.A.**

1. The best quadrature furmula for some classes of periodic functions// Izv. AN USSR, Ser. Math.-1977.-41,N5.-P.1110-1124. (Russian)

2. On best quadrature formulas for some classes of periodic functions// Dokl. AN USSR.-1977.-236,N3.-P.531-534. (Russian)

3. Monosplines of minimal norm and the best quadrature formulas// Uspehi Math. nauk.-1981.-36,N4.-P.107-159. (Russian)

4. On optimal monosplines of minimal defect//Izv. AN USSR, Ser. Math.- 1982.-46,N6.-P.1175-1198. (Russian)

**Jensykbaev A.A., Motornyi V.P.**

1. Analog of main theorem of algebra for periodic monosplines and their applications// Studies on modern problems of summation and approximation of functions and their applications.- Dnepropetrovsk: Dnepr.Univ.- 1979.-P.135-137. (Russian)

**Erdelei T.**

1. Pointvise estimates for the derivatives of a polynomial with real zeros // Acta Math. Hung.- 1987.- 49, N1.- P. 219-235.

2. Markov type estimates for derivatives of polynomials of special type // Ibid.- 1988.- N3-4.- P. 421-436.

**Erdelei T., Sabadosh J.**

1. Bernstein type inequalities for a class of polynomials // Ibid.- 1989.- 53, N1.

2. On trigonometric polynomials with positive coefficients // Stud. Sci. Math. Hung.- 1989.- 24, N1.- P. 71-91.

**Erdos P.**

1. On extremal properties of the derivatives of polynomials // Ann. of Math. (2).- 1940.- 41.- P. 310-313.

2. Extremal properties of polynomials // Ibid.- 1940.- 4.- P. 210-213.

3. Note on some elementary properties of polynomials // Bull. Amer. Math. Soc.- 1940.- P. 954-958.

4. Some remarks on polynomials // Bull. Amer. Math. Soc.- 1947.- 53.- P. 1169.

**Erdos P., Varma A.K.**

1. An extremum problem concerning algebraic polynomials // Acta Math. Hung.- 1986.- 47, N1-2.- P. 137-143.

**Erdos P., Grunwald T.**

1. On polynomials with only real roots // Ann. Math.- 1939.- 40.- P. 537-548.

**Erdos P., Turan P.**

1. on the role of the Lebesque functions in the theory of the Lagrange interpolation // Acta Math. Ac. Sci. Hung.- 1955.- N1-2.- P. 47-60.

/bf Erod J.

1. On the lower bound of the maximum of certain polynomials // Mat. Fiz. Lapok.- 1939.- 46.- P. 38-43.

**Favard J.**

1. Sur l'approximation des fonctions perioliques par polynomes trigonometriq // S.R.Acad.Sci.- 1956.- 203.- P. 1122-1124.

2. Application de la formule summatore d'Fuler a la demonstration de quelques properties extremales des fonctin periodiques // Math. Tidskript.- 1936.- 4.- P. 81-94.

3. Sur les meilleurs procedes d'approximation de certaines classes de fontions par des polinomes trigonometriques // Bull. Sci. Math.- 1937.- 61.- P. 209-224, 243.

**Fainshmidt V.L.**

1. Certain extremal problems for regular monotonous polynomials // Studies on modern problems of constructive function theory.- Moscow - 1961.- P. 96-98.

**Fan Ky, Lorentz G.G.**

1. An integral inequality // Amer. Math. Mon.- 1954.- 61.- P. 656-631.

**Fejer L.**

1. Two inequalities concerning trigonometric polynomials // J. London Math. Soc.- 1939.- 14.- P. 44-46.

**Fejer L.**

1. Uber trigonometrische Polynome Gesammlte Arbeiten, Band 1 // Acad. Kiado.- Budapest.- 1070.- P. 842-872.

**Fekete M.**

1. Uber die Verteilung der Warzelen bei gewissen algebraischen Gleichunger mit ganzzahligen Koefizieten // Math. Z.- 1923.- 17.- P. 228-249.

**Frappier C.**

1. On the inequalities of Bernstein-Markoff for an interval // J. Anal. Math.- 1983/1984.- 43.- P. 12-25.

2. Inequalities for entire functions of exponential type // Canad. Math. Bull.- 1984.- 27(4).- P. 463-471.

3. Some inequalities for trigonometric polynomials // J. Austral. Math. Soc. Ser. A.- 1985.- 39, N2.- P. 216-226.

**Frappier C., Rahman Q.I.**

1. On an inequality of S.Bernstein // Can. J. Math.- 1982.- 34, N4.- P. 932-944.

**Frappier C., Rahman Q.I., Ruschewey St.**

1. New inequalities for polynomials // Trans. Amer. Math. Soc.- 1985.- 288.

**Freud G.**

1. On two polynomial inequalities.1 // Acta Math. Sci. Hung.- 1971.- 22.- P. 109-117; ll.- 1972.- 23.- P. 137-145.

2. On Markov-Bernstein-type Inequalities and Their Applications // J. Appr. Th.- 1977.- 19, N1.- P. 22-37.

**Gabushin B.N.**

1. Inequalities for norm of functions and their derivatives in metrics $L_p$// Math. Zametki.-1967.-1,N3.-P.291-298. (Russian)

2. Exact constants in inequalities between norms of derivatives of functions// Math. Zametki.-1968.-4,N2.-P.221-232. (Russian)

**Ganzburg M.I.**

1. On exact estimation of derivative of algebraic polynomial at point// Izv.high shcool. Math.-1978.-N10.-P.29-36.

2. Jackson and Bernstein theorem in $R_N$// Uspehi Math. nauk -1979.-34,N1.-P.225-226.

3. On severaldimensional inequalities of Bernstein type// Ukrain. Math. J.- 1982.-34, N6.-P.749-753.

**Gaier D**

1. Lectures on approximation theory in complex domain.- Moscow: Mir.- 1986.-216P. (Russian)

**Geronimus M.L.**

1. Sur qualques proprietes extremales de polynomes, dont les coefficients primiers sont donnes//Zap. Chark. Math. s.-1936.-12.-P.49-60.

**Geronimus Y.L.**

1. On some extremal problems// Izv. AN USSR. Ser. Math.-1937.-N2.-P.202-225.

**Gearhart W.B.**

1. Some Chebyshev approximations by polynomias in two variables// Journ. Approx. Theory.-1973.-8,N3.-P.195-209.

**Giroux A., Rahman Q.I.**

1. Inequalities for polynomials with prescribed zeros// Trans. Amer. Math. Soc.-1974.-193.-P.67-98.

**Giroux A., Rahman Q.I., Schmeisser G.**

1. On Bernstein's inequality// Can.J.Math.-1979.-31,N2.-P.347-353.

**Govil N.K.**

1. On the derivative of a polynomial// Proc. Amer. Math. Soc.-1973.-41.- P.543-546.

2. On the maximum modulus of polynomials//J. Math. Anal. and Appl.- 1985.-112,N1.-P.253-258.

**Govil N.K., Jain V.K., Labelle F.**

1. Inequalities for polynomials satisfying $p(z) = z^n p(1/z)$// Proc. Math. Amer. Soc.-1976.-57.-P.238-242.

**Govil N.K., Rahman Q.I., Schmeisser G.**

1. On the derivative of a polynomial// Ill.J.Math.-1979.-23.-P.319.

**Golitschek M., Lorentz G.G.**

1. Trigonometric polynomials with many real zeros//Constr. Appr.-1985.- 1,N1.-P.1-13.

**Goldstein M., McDonald J.N.**

1. On extremal problem for non-negative trigonometric polynomials// J. London Math. Soc.-1984.-29,N1.-P.81-88.

**Goldstein E.G**

1. The duality theory in mathematics programming.- Moscow:Nauka,1972.- 352 P. (Russian)

**Gonchar A.A.**

1. Reverse theorems on the approximation by rational functions// Izv. AN USSR.Ser.Math.-1961.-25.-P.347-356. (Russian) 2. The estimates of growth for rational functions and their applications// Math.coll.-1967.-72,N 3.-P.489-503.

**Goncharov V.L.**

1.Theory of interpolation and approximation of functions// Moscow: Gostechizdat.-1954.-328 P. (Russian)

**Gorin E.A.**

1. Isogeometric representations and differential inequalities// Materials of conf. on algebraic and functional analysis methods for investigation of operators.- Tartu,1978.-P.9-12. (Russian)

2. Bernstein's inequalities from operators point of view// Herald of Hark. Univ. Hark: -1980.-N205.-P.77-105. (Russian)

**Goodman J., Micchelly Ch.**

1. Limits of spline function with periodic knots// J. London Math. Soc.- 1983.-28,N1.-P.103-112.

**Gusew V.A.**

1. Functionals of derivatives of algebraic polynomial and Markov's theorem// Studies on modern problems of constructive function theory.- Moscow: Fizmatgiz.-1961.-P.129-134. (Russian)

**Hall R.R.**

1. An extremal problems for polynomials // Bull. London Math. Soc.- 1985.- 17, N5.- P. 463-468.

**Hardy G.H., Littlewood I.E., Polya G.**

1. Some simple inequalities satisfied by complex functions // Mes. Math.- 1929.- 58.

2. Inequalities.- Moscow: Pub.Foreign Liter., 1948.- 456 P.

**Hvedelidze B.V.**

1. A method of Cauchy's type integrals in discontinuous problems of golomorph functions of one complex variable// Sums of science and technique. Modern problems of mathematic.- Moscow: VINITI, 1975.- P. 5-162. (Russian)

**Hormander L.**

1. A new proof and generalization of inequality of Bohr // Math. Scand.- 1954.- 2.- P. 33-45.

2. Some inequalities for functions of exponential type // Ibid.- 1955.- 3.- P. 21.

**Hall R.R.**

1. An extremal problem for polynomials // Bull. London Math. Soc.- 1985.- 17, N5.- P. 463-468.

**Ibragimov I.I.**

1. Some inequalities for algebraic polynomials// Studies on modern problems of constructive function theory.- Moscow: Fizmatgiz.- 1961.-P.139-143. (Russian)

2. Some extremal problems on the class of integral functions of finite power// Studies on modern problems of constructive function theory.- Baku: Pub. AN Azer.SSR.- 1965.-P.212-219. (Russian)

**Ivanov V.A., Lizorkin P.I.**

1. Estimations in integral norms of derivatives for harmonic polynomials and spheric polynomials// Dokl. AN USSR.-1986.-286,N1.-P.23-27. (Russian)

**Ivanov V.I.**

1. Some inequalities for trigonometric polynomials and their derivatives in various metrics// Math. Zametki.-1975.-18,N4.-P.489-498. (Russian)

2.Straight and reverse theorems of approximation theory in $L_p$-metric for $0 < p < 1$// Math. Zametki.-1974.-16,N5.-P.641-658. (Russian)

3. Some extremal properties of polynomials and reverse inequalities of approximation theory// Trudy Math. Inst. V.A.Steklov AN USSR.-1980.-145.-P. 79-110. (Russian)

**Ivanov V.I., Ydin**

1. On trigonometric system in $L_p$, $0 < p < 1$// Math. Zametki.-1980.-28,N6.- P.859-868. (Russian)

**Ioffe A.D., Tihomirov B.M.**

1. Theory of extremal problems.-Moscow:Nauka.-1974.- 479 P. (Russian)

**Kamzolov A.I.**

1. On interpolation Riesz formula and Bernstein's inequality for functions on uniform spaces// Math. Zametki.-1974.-15,N6.-P.967-978. (Russian)

**Karlin S., Stadden W.**

1. Chebyshev's systems and their applications in analysis and statistics.- Moscow: Nauka,1976.-568 P. (Russian)

**Karlin S.**

1. Total positivity 1.- Stanford: Calif. Univ. Press.-1968.-576 P.

**Kashin B.S.**

1. On some properties of the space of trigonometric polynomials with uniform norm// Trudy Math.Inst. V.A.Steklov AN USSR.-1980.-145.-P. 111-116. (Russian)

**Kellogg O.D.**

1.On bounded polynomials in several variables//Math. Z.-1927.-27.-P.55-64.

**Kemperman L.**

1. Markov type inequalities for the derivatives of a polynomial// Aspects Math. and Appl.-1986.-P.455-476.

**Kemperman L., Lorentz G.G.**

1. Bounds for polynomials with applications// Indagat. Math.-1979.-88.- P.13-26.

**Kovtunetz V.V.**

1. $K$-"grass-snakes" and properties of the best approximation polynomials// Approximation theory of functions.-Moscow: Nauka,1987.-P.206-208. (Russian)

**Kolmogorov A.N.**

1. Uber die besste Annaherung von Functionen einer gegebenen Function lassen// Ann. Math.-1936.-Bd.37.-S.107-110.

2. On inequalities between upper bounds of successive derivatives of function on infinite interval// Stud.supply of Mosc.Univ.-Mathematica. -1939.-30.-P.3-16. (Russian)

3. Selected proceed. Mathematics and mechanics.-Moscow: Nauka, 1985.- 470 P. (Russian)

**Kolmogorov A.N., Fomin C.V.**

1. The elements of function theory and functional analysis. -Moscow: Nauka,1981.- 544 P. (Russian)

**Konyagin S.V.**

1. The estimates of derivatives from polynomials// Dokl. AN USSR.-1978.-243,N5.- P.1116-1118. (Russian)

2. On V.A.Markov's inequality for polynomials in $L$-metric// Trudy Math. Inst. V.A.Steklov AN USSR.-1980.-145.- P.117-125. (Russian)

**Korkin A.I., Zolotarev E.I.**

1. Sur un certain minimum// E.I.Zolotarev. Full coll. of proceed; Issue 1.- P.138-154. (Russian)

**Korneichuk N.P.**

1. On the best uniform approximation on some classes of continuous functions// Dokl. AN USSR.-1961.-140.-P. 748-751. (Russian)

2. On the best approximation of continuous functions// Izv. AN USSR.-Ser. Math.-1963.-27.-P.29-44. (Russian)

3. Exact value of the best approximations and widths of some classes of functions// Dokl. AN USSR.-1963.-150.-P. 1218-1220. (Russian)

4. Upper bounds of the best approximation on classes of differentiable functions in metrics $C$ and $L$// Dokl. AN USSR.- 1970.-190.-P. 269-271. (Russian)

5. Extremal values of functionals and best approximation on classes of periodic functions// Izv. AN USSR, Ser. Math.-1971.-35,N1.-P.93-124. (Russian)

6. Inequalities for differentiable periodic functions and best approximation of one class of functions by another// Izv. AN USSR.- 1972.-36,N2.-P.423-434. (Russian)

7. On the best uniform approximation of periodic functions by subspaces of finite dimension// Dokl. AN USSR.-1973.- 213,N3.- P.525-528. (Russian)

8. On correlation between moduluses of continuity for functions and their rearrangements// Dokl. AN Ukrain. - 1973.- N9.-P. 794-796. (Russian)

9. On extremal subspaces and approximation of periodic functions by splines of minimal defect// Anal. Math.- 1975.-1,N2.-P. 91-101. (Russian)

10.The best approximation by splines on the classes of periodic functions in metric $L$ // Math.Zametki.-1976.-20,N 5.-P.655-664. (Russian)

11. Extremal problems of approximation theory.-Moscow: Nauka, 1976.-320 P. (Russian)

12. Extremal properties of splines// Approximation theory of functions.-Moscow: Nauka.-1977.- P.237-248. (Russian)

13. Exact error bound of approximation by interpolating splines in $L$-metric on the classes $W_P^r$, $(1 < p < \infty)$ of periodic function// Anal. Math.-1977.-3,N3.-P.109-117.

14. Exact inequalities for best approxomation by splines// Dokl. AN USSR.- 1978.-242,N2.-P.280-283. (Rassian)

15. On best approximation on segment of classes of functions with restricted $r$-th derivative by finite-dimensional subspaces// Ukrain. Math.J.-1979.-31,N1.-P. 23-31. (Rassian)

16. Inequalities for best approximation by splines of differentiable periodic functions // Ukrain. Math.J.-1979.-31,N4.-P.380-388. (Rassian)

17. Widths in $L_p$ of the classes of continuous and differentiable functions and optimal recovery of functions and their derivatives// Dokl. AN USSR.-1979.-244,N6.-P.1317-1321. (Rassian)

18. Approximation by splines in integral metric// Consructive theory of functions, 77.- Proc. Intern. conf. of constr. func. th.,Blagoevgrad, 30 may-6 june 1977.- Sofia.- 1980.- P. 99-104. (Rassian)

19. Widths in $L_p$ of the classes of continuous and differentiable functions and optimal methods of coding and recovery of functions and their derivatives// Izv. AN USSR, Ser. Math.-1981.-45,N2.-P. 266-290. (Rassian)

20. On existence and uniqueness of interpolating splines// Geometric theory of functions and tipology.-Kiev: Math. Inst. Ukrain.-1981.-P.42-55.

(Rassian)

21. On comparison of rearrangements and estimation of errors for interpolating by splines// Dokl. AN Ukrain., Ser. A.-1983.-N4.-P. 19-21. (Rassian)

22. Splines in approximation theory.- Moscow: Nauka, 1984.- 452 P. (Rassian)

23. S.M.Nikolski and progress in investigations by approximation theory of functions in USSR// Uspehi Math. Nauk.-1985.-40,N5.-P. 71-133. (Rassian)

24. Exact constants in approximation theory.-Moscow: Nauka, 1987.- 423 P. (Rassian)

25. On exact estimations for derivative of error for spline-approximation// Ukrain. Math. J.-1991.-43,N2.-P. 206-211. (Rassian)

**Korneichuk N.P., Ligun A.A.**

1. On approximation of a class by another class and extremal subspaces in $L_1$// Anal. Math.-1981.-7,N2.- P.107-119.

**Korneichuk N.P., Ligun A.A., Doronin V.G.**

1. Approximation with restrictions.- Kiev: Naukova dumka, 1982.- 250 P. (Rassian)

**Korovkin P.P.**

1. On growth of polynomials on the set// Dokl. AN USSR.-1948.-61,N5.
2. Linear operators and approximation theory.-Moscow:Fizmatgiz.-1959. (Rassian) - 211 P.

**Kofanov V.A., Kyznetzova E.I.**

1. On Zolotarev's problem for constant sign polynomials in integral metric// Problems of optimal approximation of functions and summation of serieses.-Dnepropetrovsk: Dnepr.Univ.-1988.-P. 37-41. (Rassian)

**Kristiansen G.K.**

1. Proof of an inequality for trigonometric polynomials// Proc. Amer. Math. Soc.-1974.-44,N1.-P. 49-57.

2. Proof of a polynomial conjecture// Ibid.-1974.-44,N1.-P.58-60.

3. Some inequalities for algebraic and trigonometric polynomials// J. London Math. Soc. (2).- 1979.-20.- P. 300-314.

4. Some inequalities for algebraic and trigonometric polynomials, ll// Ibid.-1983.-28,N1.-P.83-92.

**Krejn M.G., Nudelman A.A.**

1. Markov's moments problem and extremal problems.- Moscow: Nauka, 1973.- 432 P. (Rassian)

**Krylov V.I.**

1. Approximate calculation of integrals.-Moscow: Fizmatgiz.-1959.- 327 P. (Rassian)

**Kusis A.K.**

1. Introduction in theory of spaces $H^p$.-Moscow: Mir.-1984.-364 P.

**Kushpel A.K.**

1. $SK$-splines and exact estimations of widths for functional classes in the space $C_{2\pi}$.- Kiev,1985.-47 P.

2. Exact estimations of widths for classes of convolution // Izv. AN USSR, Ser. Math.-1988.-52,N6.- P. 1305-1322.

**Lax P.**

1. Proof of a conjecture of P.Erdos on the derivative of a polynomial// Bull. Amer. Math. Soc.-1944.-50.-P. 509-513.

**Lachance M., Suff E.B., Varga R.S.**

1. Inequalities for polynomials with a prescribed zeros// Math. Z.-1959.- 168.- P.105-116.

**Lebed G.K.**

1. Inequalities for polynomials and their derivatives// Dokl. AN USSR.-1957.- 117,N4.- P. 570-572. (Rassian)

2. Some inequalities for trigonometric and algebraic polynomials and their derivatives// Trudy Mat. Inst. V.A.Steklov AN USSR.- 1975.-134.- P. 142-160. (Rassian)

**Levitan B.M.**

1. On one generalization of Bernstein and Bohr inequalities// Dokl. AN USSR.- 1937.-15,N4.-P. 169-172. (Rassian)

**Ligun A.A.**

1. Exact inequalities for upper bounds of seminorms on classes of periodic functions// Math. Zametki.-1973.-13,N5.-P.647-654. (Rassian)

2. On exact constants of approximation of differentiable periodic functions// Math. Zametki.-14,N1.- P. 21-30. (Rassian)

3. Inequalities for upper bound of functionals // Anal. Math.-1976.-2.- P. 11-40. (Rassian)

4. Exact inequalities for spline-functions and best quadrature formulas for some classes of functions// Math. Zametki.- 1976.- 19,N6.- P.913-926. (Rassian)

5. On one inequality for spline-functions of minimal defect// Math. Zametki.- 1978.- 24.-N4.- P.547-552. (Rassian)

6. On best quadrature formulas for some classes of periodic functions// Math. Zametki.- 1978.- 24,N5.- P. 661-669. (Rassian)

7. Optimal method for approximate calculation of functionals on classes $W^rL_\infty$// Anal. Math.-1979.-5,N4.- P. 269-286.

8. On widths of some classes of differentiable periodic functions // Math. Zametki.- 1980.- 27,N1.- P. 61-75. (Rassian)

9. Extremal properties of splines// Constractive function theory 81.- Proc. Intern. conf. on constractive function theory, may-june,1981.- Sofia: Pub. Bolg AN.- 1983.- P. 99-105. (Rassian)

10. On inequalities between norms of derivatives of periodic functions// Math. Zametki.-1983.- 33,N3.- P. 385-391. (Rassian)

11. Exact inequalities for perfect splines and their applications// Izv. High school. Mathematics- 1984.- N5.- P. 32-38. (Rassian)

12. On exact constants in Jackson type inequalities// Dokl. AN USSR.-1985.-283,N1.- P.33-37. (Rassian)

**Litvinetz P.D., Filshtinsky V.A.**

1. On one theorem of comparison and its applications// Approximation theory of functions and its applications.- Kiev: Ukrain. Math. Inst., 1984.- P. 97-107. (Rassian)

**Lorentz G.G.**

1. Degree of approximation by polynomials with positive coefficients// Math. Ann.-1963.-151.-P. 239-251.

2. Approximation of function.- New York.- 1966.- 188 P.

3. Problems for incomplete polynomials// Appr. Th. lll,1980.- P. 41-73.

4. Covergence Theorems for polynomials with many zeros// Math. Z.-1984.- 186.- P. 117-123.

**Lukas**

1. Verscharfung der ersten Mittelwertsatzes der Integralrechnung fur Polynome// Math. Zeitschrift.-1918.-2.-P. 229-305.

**Lubinsky D.S.**

1.A weighted polynimial inequality// Proc. Amer. Math. Soc.-1984.-92,N2.- P. 263-267.

**Magaril-Iljaev G.G.**

1. Generalization of Sobolev's classes and inequalities of Bernstein's -Nikolski type// Dokl. AN USSR.-264,N5.-P. 1066-1069. (Rassian)

**Macintyre A.J., Fuchs W.**

1. Inequalities for the logarithmic derivatives of polynomial// J. London Math. Soc.-1940.-15,N1.- P. 162-168.

**Makovoz Y.I.**

1. A widths of Sobolev's classes of functions and splines deviating least from zero// Math. Zametki.- 1979.- 26,N5.- P. 805-812. (Rassian)

**Malik M.A.**

1. On the derivative of a polynomial// J. London Math. Soc. (2).-1969.- 1.- P. 57-60.

2. On the coefficients of selfinverse polynomials// Bull. Union Math. Ital.- 1982.- 2,N3.- P. 369-372.

3. An integral mean estimate for polynomial// Proc. Amer. Math. Soc.- 1984.- 91,2.- P. 281-284.

**Malozemova L.K., Malozemov V.N.**

1. Algebraic polynomials deviating least from zero together with derivative// Herald of Leningr. Univ.-1986.-N3.-P.29-34. (Rassian)

**Mamedhanov D.I.**

1. Some inequalities for algebraic polynomials and rational fractions// Studies on modern problems of constructive function theory.- Baku: Pub. AN AzerSSR.-1965.-P.255-260. (Rassian)

**Marden M.**

1. Geometry of polynomials// Amer. Math. Soc. Math. Surveys.- 1966.- 3.

**Markov A.A.**

1. On one Mendeleew problem// Izv. AN.- SPb.-1989.-62.-P. 1-24. (Rassian)

2. Selected proceed on theory of continuous fractions and theory of functons deviating least from zero.- Moscow; Leningrad: Gostechizdat.- 1948.- 410 P. (Rassian)

3. Lectures on functions deviating least from zero// Selected proceed.- Moscow; Leningrad: GNTTL.-1948.- P.244-291. (Rassian)

**Markov V.A.**

1. On functions deviating least from zero on fixed segment// SPb.- 1892.- Sanct-Peterburg. (Rassian)

2. Uber polynome die einen gegebenen intervale moglichst wening von null abweichen// Math. Ann.-1916.- 77.- P. 213-258.

**Marshall A., Olkin I.**

1. Inequality: theory of majority and its applications.- Moscow: Mir.- 1983.- 543 P. (Rassian)

**Mahler K.**

1. On the zeros of a derivative of the polynomials// Proc. Roy. Soc. London.- 1971.- 264,N 1317.- P. 145-154.

**Mairhuber J.C., Schonberg I.J., Williamson R.E.**

1. On variation diminishing transormations on the circle// Rend Circ. mat. Palermo.- 1959.-8, N2.- P. 241-270.

**Milovanovich G.**

1. An extremal problem for polynomials with nonnegative coefficients// Proc. Amer. Math. Soc.- 1985.- 94, N3.- P. 423-426.

**Milovanovich I.**

1. An extremal problem for polynomials// Numer Meth. and Appr. Th., Novi Sad, Srpt. 4-6, 1085.- Novi Sad: D. Herceg 1985.

2. An inequality concerning algebraic polynomials// Coll. Math. Soc. J. Bolyai.- 1985.- 49.- P. 623-627.

3. An extremal problem for polynomials with prescribed zero// Proc. Amer. Math. Soc.- 1986.- 97, N1.- P. 105-106.

**Mirsky L.**

1. An inequality of Bernstein-Markov type for polynomials// SIAM J. Math. Anal.- 1983.- 14, N1.- P. 1004-1008.

**Micchelly C.A.**

1. Best $L^1$ Approximation by Weak Chebyshev Systems and the uniqueness of interpolating Perfect Splines// J. Appr. Th.- 1977.- 19,N1.- P. 1-14.

**Micchelly C.A., Pinkus A.**

1. On $n$-widths in $L^\infty$// Trans. Amer. Math. Soc.- 1977.- 234.- P. 139-174.

2. Total Positivity and the exact $n$-widths of certain sets in $L_1$// Pacif. J. Math.-1977.- 71,N2.- P. 499-515.

**Motornyi V.P.**

1. On the best quadrature formula of the type $\Sigma_{k=1}^n p_k f(x_k)$ for certain classes of periodic differentiable functions// Dokl. AN USSR.- 1973.- 211, N5.- P. 1060-1062. (Rassian)

2. On the best quadrature formula of the type $\Sigma_{k=1}^n p_k f(x_k)$ for certain classes of periodic differentiable functions// Izv. AN USSR. Ser. Math.- 1974.- 38,N3.- P. 583-614. (Rassian)

**Motornyi V.P., Ruban V.I.**

1. The widths of certain classes of differentiable periodic functions in space $L_1$// Math. Zametki.- 1975.- 17, N4.- P. 531-543. (Rassian)

**Mhaskar H.N.**

1. Extremal problems for polynomials with exponential weights// Trans. Amer. Math. Soc.- 1984.- 285, N1.- P. 203-234.

**Mchaskar H.N., Saff E.B.**

1. Where does the sup norm of a weighted polynomial live ? (A generalization of incomplete polynomials)// Cons. Appl.- 1985.- 1,N1.- P. 71-91.

**Nady B.**

1. Uber gewisse Extremalfragen bei transformeiten trigonometrichen Entwicklungen// Ber. Acad. d. Wiss., Leipzig.- 1938.- 90.- P. 103-134.

**Nadjmiddinov D., Subbotin Y.N.**

1. Markov's inequalities for polynomials on triangle // Math. Zametki.- 1989.- 46, N2.- P. 76-82. (Rassian)

**Natanson I.P.**

1. Constructive function theory.- Moscow: Gostechizdat, 1949.- 686 P. (Rassian)

**Nguen Thi Thieu Hoa**

1. The best quadrature formulas and recovery methods of functions, defined by kernels, with no increase of oscillation // Math. coll.- 1986.- 310,N1.

2. The Rolle's theorem for differential operators and certain extremal problems of approximation theory// Dokl. AN USSR.- 1987.- 295, N6.- P. 1313.

**Nikolski S.M.**

1. Approximation of functions by trigonometric polynomials in mean// Izv. AN USSR. Ser. Math.- 1946.-10.- P. 207-256. (Rassian)

2. On best approximation of functions which $s$-th derivative has discontinuity of the first kind// Dokl. AN USSR.- 1946.-10.- P. 207-256. (Rassian)

3. The generalization of certain inequality of S.N.Bernstein// Dokl. AN USSR.- 1948.- 60, N9.- P. ?? (Rassian)

4. At problem on estimations of approximations by quadrature formulas// Uspehi Math. Nauk.- 1950.- 5,N2.- P. 165-177. (Rassian)

5. Inequalities for integral functions of several variables// Trudy Mat. Inst. V.A.Steklov AN USSR.- 1951.- 38.- P. 244-278. (Rassian)

6. Quadrature formulas// Izv. AN USSR. Ser. Math.- 1952.- 16,N2.- P. 181. (Rassian)

7. Approximation of functions of several variables and theorems of enclosure.-Moscow: Nauka.- 1969.- 480 P. (Rassian)

**Nikolski S.M., Lizorkin P.I.**

1. Estimations for derivatives of harmonic polynomials and spheric polynomials in $L_p$// Acta. Sci. Math.- 1985.- 48, N1.- P. 401-416. (Rassian)

2. Inequalities for harmonic, spheric and algebraic polynomials// Dokl. AN USSR.- 1986.- 289, N3.- P. 541-545. (Rassian)

3. Inequalities of Bernstein's type for spheric polynomials// Dokl. AN USSR.- 1986.- 288, N1.- P. 50-53. (Rassian)

**Novikov S.I.**

1. The widths of one class of periodic functions defined by differential operator// Math. Zametki.- 1987.- 42, N2.- P. 194-206. (Rassian)

**Newman D.J.**

1. Polynomials with large partial sums// J. Approxim. Theory.- 1978.- 23.

2. Derivative bounds for Muntz polynomials// Ibid.- 1976.- 18.- P. 360-362.

**Newman D.J., Rivlin T.J.**

1. Approximation of monomials by lower degree polynomials// Aequat. Math.- 1976.- 14.- P. 452-455.

**Oskolkov K.I.**

1. On optimality of quadrature formula with equidistant nodes on the classes of periodic functions// Dokl. AN USSR.- 1979.- 249, N1.- P. 49- 52. (Rassian)

2. On optimal Quadrature Formula on Certain Classes of Periodic Functions// Appl. Math. Optim.- 1982.- 8.- P. 245-263.

3. On exponential polynomials of the least $L$-norm// Constructive function theory.- Proc. Intern. conf., Varna, 1-5 june 1981.- Sofia, 1983.- P. 464. (Rassian)

O'Hara P.J., Rodrigues R.S.

1. Some properties of a self-inverse polynomials// Proc. Amer. Math. Soc.- 1974.- 44.- P. 331-335.

**Pashkocski S.**

1. Calculating applications of Chebyshev's polynomials and serieses.- Moscow: Nauka.- 1983.- 384 P. (Rassian)

**Pinkus A.**

1. On $n$-widths of periodic functions// J. Anal. Math.- 1979.- 35.- P. 209-235.

2. Widths in Approximation Theory.- Berlin; Heidelberg; New York; Tokyo: Springer Verlag, 1985.- 291 P.

3. $N$-widths of Sobolev spaces in $L_p$// Const. Appr.- 1985.- 1, N1.- P. 15-62.

**Pichorides S.K.**

1. On the best values of constants in the theorems of M.Riesz, Zygmund and Kolmogorov// Studia Math. 1972.- 44, N2.- P. 165-179.

2. Une remarque sur les polynomes trigonometriquch reels dont tou tes les racines sont reclles// C.r. Acad. Sci. A. 1978.- 286.- P. 17-19.

**Polia G., Szego G.**

1. Problems and theorems from analysis.- Moscow: Gostechizdat.- 1957.- V.- 396 P., V.2.- 432 P. (Rassian)

**Privalov A.A.**

1. Analog of Markov's inequality. Application to interpolation and Fourier's serieses// Trudy Mat. Inst. V.A.Steklov AN USSR.- 1983.- 1964.- P. 142. (Rassian)

**Pierre R.**

1. On explicit decomposition for positive polynomials with application to extremal ptoblems// Can. J. Math.- 1984.- 36, N6.- P. 1031-1045.

2. Extremal properties of constrained Chebyshev polynomials// Ibid.- 1986.- 38, N4.- P. 904-924.

**Pierre R., Rahman Q.I.**

1. On a problem of Turan about polynomials// Proc. Amer. Math. Soc.- 1976.- 56.- P. 231-238.

**Quade W., Collatz L.**

1. Zur interpolations theorie der reelen periodichen Funktionen// Sitzungsber. der. Press Acad. der Wiss., Phys. Math.-1938.-30.-P.383-409.

**Reimer M.**

1. On Multivariate Polynomials of least Deviation from Zero on the Unit Cube// J. Approxim. Th.- 1978.- 23, N1.- P.65-69.

**Rack H.-J.**

1. A generalization of inequality of V.Markov to multivariable polynomials// Ibid.- 1982.- 35.- P. 94-97.

2.A generalization of inequality of V.Markov to multivariable polynomials, ll// J. Approxim. Th.- 1984.- 40, N2.- P. 129-133.

3. On polynomials with largest coefficients sums// Ibid.-1989.- 56, N3.- P. 348.

**Rahman Q.I.**

1. Extremal properties of the successive of polynomials and rational functions// Math. Scand.- 1964.- 15.- P. 121-130.

2. On asymmetric entire functions.ll// Math. Annal.- 1966.- 167.- P. 49-52.

3. Functions of exponential type// Trans. Amer. Math. Soc.- 1969.- 135.- P. 295.

4. On a problem of Turan about polynomials with curved majorants// Ibid.- 1972.- 163.- P. 447-455.

5. Some inequalities for polynomials// Proc. Amer. Math. Soc.- 1976.- 56.

6. Inequalities of Markoff and Bernstein// Polynomial and Spline Approxim. Proc. NATO Adv. Study Inst.- 1079.- P. 191-201.

**Rahman Q.I., Mohammad Q.G.**

1. Remark on Schwarz's lemma// Pacif. J. Math.- 1967.- 23.- P. 139-142.

**Rahman Q.I., Schmeisser G.O.**

1. Inequalities for polynomials in the unit interval// Trans. Amer. Math. Soc.- 1977.- 231.- P. 93-100.

2. $L^p$-inequalities for polynomials// J. Appr. Th.- 1988.- 53, N1.- P. 26-32.

**Rivlin T.J.**

1. The Chebyshev Polynomials.- New York: Wiley.- 1974.- 186 P.

**Riesz M.**

1. Eine e trigonometrische interpolational formal and einige Ungleichungen] fur Polynome// Jahresber Dtsch. Math. Ver.- 1914.- 23.- P. 354-368.

**Rogosinski W., Szego G.**

1. Uber die Abschnitte von Potenzreiben die in einen Kreise beschrankt bleiben// Math. Z.- 1928.- 28.- P. 73-94.

**Roulier J.A., Varga R.S.**

1. Another property of Chebyshev Polynomials// J.Approxim. Th.- 1978.- 22.

**Ruban V.I.**

1. Even widths of the classes $W^r H^\omega$ in the space $C_{2\pi}$// Math. Zametki.- 1974.- 15, N3.- P. 387-392. (Rassian)

2.. Widths of certain class of $2\pi$-periodic functions in the space $L_p$ $(1 \leq p < \infty)$// Approximation theory and its applications.- Kiev: Inst. Math. AN Ukrain.- 1974.- P. 119-128. (Rassian)

3. Widths of the sets in the spaces of periodic functions// Dokl. AN USSR.- 1980.- 225,- N1.- P. 34-35. (Rassian)

**Rusak V.N.**

1. A generalisation of Bernstein's inequality for derivative of trigonometric polynomial// Dokl. AN Belorus.- 1963.- 7, N9.- P. 580- 583. (Rassian)

2. Extremal estimations of derivatives from rational fractions// Proc. 1 Republic conf. of Byelorassian Math.- Minsk: 1965.- P. 107-112. (Rassian)

3. A generalization of Zigmund inequality for derivative of trigonomertic polynomial// Dokl. AN Byelorassia .- 1966.- 10, N5.- P. 297-300. (Rassian)

4. Estimation of derivative from rational function// Math. Zametki.- 1973.- 13, N4. (Rassian)

5. On estimations of derivatives of algebraic fractions on finite segment// Dokl. AN Byelorassia.- 1976.- 20, N1.- P. 5-7. (Rassian)

**Ruscheweyh St.**

1. Non-negative trigonometric polynomials with constrains// Z. Anal. und Anwend.- 1986.- 5, N2.- P. 169-177.

**Rymarenko B.A.**

1. On monotonous polynomials // Studies on moderm problems of constructive functions theory.- Moscow: -1961.- P. 83-91. (Rassian)

**Sard A.**

1. Best approximate integration formulas, best approximate formulas// Amer. Math. J.- 1949.- 71.- P. 80-91.

**Schevaldin V.T.**

1. Certain problems of extremal interpoilation in the mean for linear differential operators // Trudy Mat. Inst. V.A.Steklov.- 1983.- 164.- P. 203-240. (Rassian)

2. $L$-splines and widths// Math. Zametki.- 1983.- 33, N5.- P. 735-744. (Rassian)

3.Estimations from below for the widths on the classes of functions with source-formed representations // Trudy Mat. Inst. V.A.Steklov.- 1989.- 189.- P. 185-201. (Rassian)

**Schik J.T.**

1. Inequalities for derivatives of polynomials of special type // J. Appr. Th.- 1972.- 6.- P. 354-358.

**Schaeffer A.C.**

1. Inequalities of A.Markoff and S.Bernstein for polynomials and related functions // Bull. Amer. Math. Soc.- 1941.- 47.- P. 565-579.

**Schaeffer A.C., Duffin R.I.**

1. On some inequalities of S.Bernstein and A.Markoff for derivatives of polynomials // Ibid.- 1938.- 44, N4.- P. 289-297.

2. A refinement of an inequaliy of the brothers Markoff // Trans. Amer. Math. Soc.- 1941.- 50, N3.- P. 517-522.

**Schaeffer A.C., Szego D.**

1. Inequalities for harmonic polynomials in two and three dimensions // Ibid.- 1941.- 50.- P. 157-225.

**Schonberg I.J.**

1. Notes of spline functions 1. The limit of the interpolating periodic spline functions as their degree tends to infinity //Indag. Math.- 34.- P. 412-422.

**Schur I.**

1. Uber das Maximum des absoluten Betrages eines Polynoms in einen gegebenen interval // Mat. Z.- 1919.- 4.- P. 271-287.

**Sloss J.M.**

1. Chebyshev approximation to zero // Pacif. J. Math.- 1965.- 15.- P. 305-313.

**Szabados J., Varma A.K.**

1. Inequalities for trigonometric polynomials having real zeros// Appr. Th.- lll.- Academic. Press.- 1980.- P. 881-887.

**Sivin P.**

1. Inequalities for trigonometric integrals// Duke Math. J.- 1941.- 8.- P. 656-665.

**Szasz O.**

1. Elementare example probleme uber nicht negative trigonometrische polynome// S.B.Bayer. Acad. Wiss. Math. Phys. Kl.- 1927.- P. 185-196.

**Saff E.B., Sheil-Small T.**

1. Coefficient and integral mean estimates for algebraic and trigonometric polynomials with restricted zeros// J. London Math. Soc.- 1974.- (2),9.- P. 16-22.

**Sevastjanov E.A.**

1. Some estimations of derivatives from rational fractions in integral metrics// Math. Zametki.- 1973.- 13, N4.- P. 499-510. (Rassian)

**Sewell W.E.**

1. On the polynomial derivative constant for an ellipses// Amer. Math. Monthly.- 1937.- 44.- P. 577-578.

**Szego G.**

1. Bemerkungen zu einen Satz von J.H.Crace. etc.// Math. Z.- 1922.- 13.

2. Uber einen Satz des Herrn Serge Bernstein// Schr. Konigsb. gelehrt. Ges.- 1928.- 22.- P. 59-70.

3. Orthogonal polynomials.-Moscow:Fizmatgiz.- 1962.- 500 P. (Rassian)

**Smirnov V.I., Lebedev V.I.**

1. Constructive functions theory of complex variable.- Moscow; Leningrad: Nauka, 1964.- 438 P.

**Songping Zhou**

1. On the Turan inequality in $L_p$-norm// J. Hangzhou Univ. Natur. Sci. Ed.- 1984.- 11, N1.- P. 28-33.

2. Inequalities for derivatives of Lorentz polynomials// Ibid.- 1985.- 12, N2.

3. An extension of the Turan inequality in $L_p$-space for $0 < p < 1$// J. Math. Pes. Exposit.- 1986.- N2.- P. 27-30.

**Stein E.**

1. Functions of exponential type// Ann. Math.- 1957.- 65, N3.- P. 582-592.

**Stein E., Weiss G.**

1. An extension of a theorem of Marcinkiewich and some of its applications// J. Math. and Mech.- 1959.- 8, N2.- P. 263-284.

**Stechkin S.B.**

1. A generalization of some inequalities of S.N.Bernstein// Dokl. AN USSR.- 1948.- 60, N9.- P. 1511-1514. (Rassian)

2. On best approximation for the classes of functions by any polynomials// Uspehi Math. Nauk.- 1954.- 9, N1.- P. 133-134. (Rassian)

3. On best approximation of conjugate functions by trigonometric polynomials// Izv. AN USSR. Ser. Math.- 1956.- 20.- P. 197-206. (Rassian)

4. On best approximation for certain classes of periodic functions by trigonometric polynomials// Izv. AN USSR.- 1956.- 20.- P. 643-648. (Rassian)

5. Inequalities between norms of derivatives of arbitrary function// Acta Sci. Math.- 1965.- 26, N3-4.- P. 225-230. (Rassian)

6. On inequalities between upper bounds of arbirtary function on semiaxis// Math. Zametki.- 1967.- 1, N6.- P. 665-674. (Rassian)

7. The best approximation of linear operators// Math. Zametki.- 1967.- 1,N 2.- P. (Rassian)

8. On certain extremal properties of positive trigonometric polynomials// Math. Zametki.- 1970.- 7, N4.- P. 411-422. (Rassian)

**Stechkin S.B., Subbotin Y.N.**

1. Splines in calculating mathematics.- Moscow: Nauka, 1976.- 248 P. (Rassian)

**Storozhenko E.P., Krotov B.G., Osvald P.**

1. Straight and reverse theorems of Jackson's type in the space $L_p$, $0 < p < 1$ // Math. Coll.- 1985.- 98(140)//3(11).- P. 395-415. (Rassian)

**Stechkin S.B.,Taikov L.V.**

1. On minimal continuations of linear functionals// Trudy Mat.Inst. V.A.Steklova AN USSR.- 1965.-78.- P. 12-23. (Rassian)

**Subbotin Y.N.**

1. On piece-polynomial interpolation// Math. Zametki.- 1967.-1, N1.- P. (Rassian)

2. Functional interpolation in meam with the least $n$-th derivative// Trudy Mat. Inst. V.A.Steklov AN USSR.- 1967.- 88.- P. 30-60. (Rassian)

3. The best approximation for one class of functions by another class// Math. Zametki.- 1967.- 2, N5.- P. 495-504. (Rassian)

4. The widths of class $W^r L$ in $L(0, 2\pi)$ and approximation by spline-functions // Math. Zametki.- 1970.- 7, N1.- P. 43-52. (Rassian)

5. Approximation by spline-functions and estimates of widths // Trudy Mat. Inst. V.A.Steklov AN USSR.- 1971.- 109.- P. 35-60. (Rassian)

6. Tie of spline-approximations with approximation of one class by another // Math. Zametki.- 1971.- 9, N7.- P. 501-510. (Rassian)

**Subbotin Y.N., Taikov L.V.**

1. The best approximation of differential operator in the space $L_2$ // Math. Zametki.- 1971.- 9, N7.- P. 257-264. (Rassian)

**Sun Yongsheng**

1. On best approximation of periodic differential functions by trigonometric polynomials // Izv. AN USSR, Ser. Math.- 1959.- 23, N1. (Rassian)

2. Some extremal problems on a class of perfect splines// Scientia Sinica. Ser. A.- 1984.- 27, N3.- P. 253-266.

3. Some extremal problems on the class of differentiable functions // J. of Beijing Normal Univ.- 1984, N4.- P. 11-26.

**Sun Yongsheng, Huang Daren**

1. On $N$-width of generalized Bernoully kernel // Appr. Th. and its Appl.- 1985.- 1, N2.- P. 83-92.

**Taikov L.V.**

1. One circle of extremal problems for trigonometric polynomials // Uspehi Math. Nauk.- 1965.- 2, N3.- P. 205-211. (Rassian)

2. One generalization of Bernstein's inequality // Trudy Mat. Inst. AN USSR.- 1965.- 78.- P. 43-47. (Rassian)

3. On the best approximation in the mean of certain classes of analitic functions// Math. Zametki.- 1967.- 1, N2.- P. 155-162. (Rassian)

4. Approximation in the mean of certain classes of periodic functions // Trudy Mat. Inst. V.A.Steklov AN USSR.- 1967.- 88.- P. 61-70. (Rassian)

5. Inequalities of Kolmogorov's type and the best formulas of calculating differentiation // Math. Zametki.- 1968.- 4, N2.- P. 233-238. (Rassian)

**Tarig Q.M.**

1. Concerning polynomials on the unit interval // Proc. Amer. Math. Soc.- 1987.- 99, N2.- P. 293-296.

**Talbot A.**

1. The Uniform Approximation of Polynomials by Polynomials of Lower Degree // J. Appr. Th.- 1976.- 17, N3.- P. 254-279.

**Telyakovski S.A.**

1. On norms of trigonometric polynomials and approximation of functions by linear-means of their Fourier's serieses. 1 // Trudy Mat. Inst. V.A.Steklov AN USSR.- 1961.- 62.- P. 61-97; ll// Izv. AN USSR. Ser. Math.- 1963.- 24.- P. 213. (Rassian)

2. On estimates for derivatives of trigonometric polynomials of several variables // Siberian Math. J.- 1963.- 4, N6.- P. 1404-1411. (Rassian)

3. Uniform restriction of some trigonometric polynomials of several variables // Math. Zametki.- 1987.- 42, 1.- P. 33-39. (Rassian)

**Temlyakov V.N.**

1. Approximation of functions with constrained mixed derivative // Trudy Mat. Inst. V.A.Steklov AN USSR.- 1986.- 128.- P. 3-112. (Rassian)

**Timan A.F.**

1. Approximation theory of real variable functions.- Moscow: Fizmatgiz.- 1960.- 624 P. (Rassian)

**Tihomirov V.M.**

1. On $n$-dimensional widths of some functional classes // Dokl. AN USSR.- 1960.- 130, N4.- P. 734-737. (Rassian)

2. The widths of the sets in functional spaces and the best approximation theory // Uspehi Math. Nauk.- 1960.- 15, N3.- P. 81-120. (Rassian)

3. Certain problems of approximation theory // Dokl. AN USSR.- 1965.- 160.- P. 774-777. (Rassian)

4. One remark on $n$-dimensional widths of the sets in Banah's spaces // Uspehi Math. Nauk.- 1965.- 20, N1.- P. 227-230. (Rassian)

5. The best methods of approximation and interpolation of differentiable functions in the space $C[-1,1]$ // Math. Coll.- 1969.- 80, N2.- P. 290-304. (Rassian)

6. Some problems of approximation theory // Math. Zametki.- 1971.- 9, N5.- P. 593-607. (Rassian)

7. Some problems of approximation theory .- Moscow: Pub. Moscow Univ.- 1976. (Rassian)

8. Extremal problems of approximation theory and widths of smooth functions // Approximation theory of function.- Moscow: Nauka, 1977.- P. 359-365. (Rassian)

9. Approximation theory// Sums of science and technique. Ser.Modern problems of mathematics. Fundamental ways.- 14.- Moscow: VINITI.- 1987.- P. 103-260. (Rassian)

**Turan P.**

1. Uber die Ableitung von polynomen// Composito Math.- 1939.- 7.- P. 89-98.

2. On rational polynomials// Acta. Univ. Szeged. Sect. Math.- 1946.- 11.- P.106.

**Turetzki A.H.**

1. On certain extremal problems of interpolation theory// Studies on modern problems of constructive function theory.- Baku: Pub. AN AzUSSR.- 1965.- P. 220-232.

**Tyrygin I.Ya.**

1. On constant sign polynomials deviating least from zero on the system of segments in the space $L_p$ // Approximation theory and adjacent problems of analysis and topology.- Kiev Mat. Inst. AN Ukrain.- 1987.- P. 88-96. (Rassian)

2. On inequalities of Turan type in certain integral metrics // Ukrain. Math. J. - 1988.- 40, N2.- P. 256-260. (Rassian)

3. Inequalities of P.Turan type in mixed integtal metrics // Dokl. AN Ukrain.- 1988.-N 9.- P. 14-17. (Rassian)

**Tzeretelly O.D.**

1. On integration of conjugate functions // Trudy Tbilis. Math. Inst. AN Georgia.- 1973.- 43.- P. 149-168. (Rassian)

**De la Vallee-Poussin**

1. Sur les polynomes d'approximation et la representation appro chee d'unangle// Null. Ac. Belgiue.-1910.-N2.-P.808-844.

**Varma A.K.**

1. An analogue of some inequalities of P.Turan concerning algebraic polynomials satisfying certain conditions// Proc. Amer. Math. Soc.- 1976.-55.-P.305-309.

2. An analogue of some inequalities of P. Erdosh and P.Turan concerning algebraic polynomials satisfying certain conditions// Colloquia Math. Soc. J.Bolyai, 19 Forrier.

3. Some inequalities of algebraic polynomials having real zeros// Proc. Amer. Math. Soc.-1979.-75.-P.243-250.

4. Derivatives of polynomials with positive coefficients// Ibid.-1983.- P.107-112.

**Velikin V.L.**

1. Optimal interpolation of periodic differentiable functions with constreined derivative// Math. Zametki.-1977.-22,N5.-P.663-670. (Rassian)

2. On lemitting correlation between approximations of periodic functions by splines and trigonometric polynomials// Dokl. AN USSR.-1984.- 258, N3.-P.525-529. (Rassian)

**Visser G., Van der Corput D.**

1. A Markov inequality in several dimantiong// J. Approxim. Theory.- 1974.- ??

**Voronovskaya E. V.**

1. Extremal polynomials of some simplest functionals// Dokl. AN USSR.- 1954.-99,N2.-P.193-196. (Rassian)

2. Certain problems of Chebyshev's school in modern functional-analytic methods// Trudy Third All-Union Math. cong.- Moscow: AN USSR.- 1957.-3.- P.177-187. (Rassian)

3. Functional of first derivative and making more precise of A.A.Markov theorem// Izv. AN USSR. Ser. Math.-1959.-23,N6.-P.951-962. (Rassian)

4. Method of functionals in classic problems of approxomation of function// Studies on moderm problems of constructive functions theory.- Moscow: Fizmatgiz.-1961.-P.107-116. (Rassian)

5. Method of functionals and its applications.- Leningrad: Pub. LEIS.- 1963.-181P. (Rassian)

6. Odd polynomials of least deviation// Dokl. AN USSR.-1964.-159,N4.- P.715-718.

**Voronovskaya E.V., Zinger M.Y.**

1. Estimations for polynomials in complex plane// Dokl. AN USSR.- 1962.-143.- P.1022-1025. (Rassian)

**Western P.W.**

1. Inequalities of the Markoff and Bernstein type for integral norms// Duke Math.J.-1948.-15,N3.-P.839-870.

**Widenski V.S.**

1. On inequalities as to derivatives of polynomial// Dokl. AN USSR.- 1949.- 67,N5.-P.777-780. (Rassian)

2. On estimation of derivatives of polynomial// Dokl. AN USSR.-1950.- 73,N2.- P.258-259. (Rassian)

3. On estimations of derivatives of polynomial// Izv. AN USSR. Ser. Math.- 1931.-16,N5.-P.401-420. (Rassian)

4. Generalization of Markov's inequalities// Dokl. AN USSR.-1958.- 120,N3.-P. 447-4499. (Rassian)

5. Generalization of Markov's theorem on estimation of derivative for polynomial// Dokl. AN USSR.-1959.-125,N1.-P.15-18. (Rassian)

6. Extremal estimations of derivative of trigonometric polynomials on segment, less than period// Dokl. AN USSR.-1960.-130,N1.-P.13-16. (Rassian)

7. Remark to Markov theorem on two polynomials which zeros are multiplying// Izv. AN Arm.SSR. Ser. Math.-1961.-15,N2.-P.15-24. (Rassian)

8. Extremal estimations of derivatives of trigonometric polynomials on segment// Studies on modern problems of constructive functions theory.- Moscow: Fizmatgiz.-1961.-P.98-107. (Rassian)

**Wilhelmsen D.R.**

1. A Markov inequality in several dimantiong// J.Approxim. Theory.- 1974.- 11.-P.216-220.

**Whitley R.**

1. Bernstein's asymptotic best bound for k-th derivative of polynomial// J. Math. Anal. and Appl.-1985.-105,N2.-P.502-513.

**Wolsh G.**

1. Interpolation and approximation by rational functions in complex domain.- Moscow: Nauka, 1961.- 508 P.

**Zalik R.A.**

1. Inequalities for Weigthed polynomials // J. Approxim. Theory.-1983.-37.- P.137-146

2. Some weighted polynomial inequalities//Ibid.-1984.-41,N1.-P.39-50.

**Zygmund A.**

1. A remark on conjugate series// Proc. London. Math. Soc.(2).-1932.-34.-P.392-400.

2. Trigonometric serieses.-Moscow: Mir,1965.- T.1-2.

**Zinger M.Y.**

1. A generalization of Sheffer-Duffin problem on finite functionals// Dokl. AN USSR.-1967.-172,N1.-P.14-17.

2. Estimation of modulus derivatives from algebraic polynomials on complex plane// Izv. High Schools. Mathematics.-1969.-N1(80).-P.9-11.

3. On estimation of derivatives of algebraic polynomial// Siberian Math.J.- 1969.-10,N2.-P.318-328.

4. Extremal estimations for derivatives of trigonometric polynomial on segment less than period// Izv. High Schools. Mathematics.-1969.-N5(85). -P.9-17.

5. The elements of differential theory for Chebyshev approximations.-Moscow:Nauka,1975.- 172 P.

**Zolotarev E.I.**

1. Applications of elliptic functions to problems on functions deviating least from zero// Full coll. of works. AN USSR.-1912.-P.1-59. (Rassian)

# Subject Index

# SUBJECT INDEX